Hybrid ADCs, Smart Sensors for the IoT, and Sub-1V & Advanced Node Analog Circuit Design

Pieter Harpe • Kofi A. A. Makinwa
Andrea Baschirotto
Editors

Hybrid ADCs, Smart Sensors for the IoT, and Sub-1V & Advanced Node Analog Circuit Design

Advances in Analog Circuit Design 2017

Springer

Editors
Pieter Harpe
Department of Electrical Engineering
Eindhoven University of Technology
Eindhoven, Noord-Brabant, The Netherlands

Kofi A. A. Makinwa
Department of Microelectronics
Delft University of Technology
Delft, Zuid-Holland, The Netherlands

Andrea Baschirotto
Department of Physics G. Occhialini
University of Milan
Milano, Italy

ISBN 978-3-319-61284-3 ISBN 978-3-319-61285-0 (eBook)
DOI 10.1007/978-3-319-61285-0

Library of Congress Control Number: 2017949890

Printed on acid-free paper

This Springer imprint is published by Springer Nature
The registered company is Springer International Publishing AG
The registered company address is: Gewerbestrasse 11, 6330 Cham, Switzerland

Preface

This book is part of the Analog Circuit Design series and contains contributions of all 18 speakers of the 26th workshop on Advances in Analog Circuit Design (AACD). The local organizers were Kathleen Philips, Yao-Hong Liu, and Steffie van de Vorstenbosch from Holst Centre/imec, Eindhoven, the Netherlands. The sponsors of the workshop this year were as follows: NXP Semiconductors (platinum sponsor), Dialog Semiconductor (gold sponsor), and silver sponsors Analog Devices, AnSem, Catena, Huawei, ICsense, ItoM, and Philips. The workshop was held in Eindhoven, the Netherlands, from March 28 to 30, 2017.

The book comprises three parts, covering advanced analog and mixed-signal circuit design topics that are considered highly important by the circuit design community:

- Hybrid data converters
- Smart sensors for the IoT
- Sub-1V and advanced-node analog circuit design

Each part is set up with six papers from experts in the field.

The aim of the AACD workshop is to bring together a group of expert designers to discuss new developments and future options. Each workshop is followed by the publication of a book by Springer in their successful series of Analog Circuit Design. This book is the 26th in this series. The book series can be seen as a reference for all people involved in analog and mixed-signal design. The full list of the previous books and topics in the series is given next.

We are confident that this book, like its predecessors, proves to be a valuable contribution to our analog and mixed-signal circuit design community.

Milano, Italy — Andrea Baschirotto
Delft, The Netherlands — Kofi A. A. Makinwa
Eindhoven, The Netherlands — Pieter Harpe

The Topics Covered Before in This Series

2016	Villach (Austria)	Continuous-time ΣΔ Modulators for Transceivers
		Automotive Electronics
		Power Management
2015	Neuchâtel (Switzerland)	Efficient Sensor Interfaces
		Advanced Amplifiers
		Low Power RF Systems
2014	Lisbon (Portugal)	High-Performance AD and DA Converters
		IC Design in Scaled Technologies
		Time-Domain Signal Processing
2013	Grenoble (France)	Frequency References
		Power Management for SoC
		Smart Wireless Interfaces
2012	Valkenburg (The Netherlands)	Nyquist A/D Converters
		Capacitive Sensor Interfaces
		Beyond Analog Circuit Design
2011	Leuven (Belgium)	Low-Voltage Low-Power Data Converters
		Short-Range Wireless Front-Ends
		Power Management and DC-DC
2010	Graz (Austria)	Robust Design
		Sigma Delta Converters
		RFID
2009	Lund (Sweden)	Smart Data Converters
		Filters on Chip
		Multimode Transmitters
2008	Pavia (Italy)	High-Speed Clock and Data Recovery
		High-Performance Amplifiers
		Power Management
2007	Oostende (Belgium)	Sensors, Actuators and Power Drivers for the Automotive and Industrial Environment
		Integrated PAs from Wireline to RF
		Very High Frequency Front Ends

(continued)

2006	Maastricht (The Netherlands)	High-Speed AD Converters
		Automotive Electronics: EMC issues
		Ultra Low Power Wireless
2005	Limerick (Ireland)	RF Circuits: Wide Band, Front-Ends, DACs
		Design Methodology and Verification of RF and Mixed-Signal Systems
		Low Power and Low Voltage
2004	Montreux (Swiss)	Sensor and Actuator Interface Electronics
		Integrated High-Voltage Electronics and Power Management
		Low-Power and High-Resolution ADCs
2003	Graz (Austria)	Fractional-N Synthesizers
		Design for Robustness
		Line and Bus Drivers
2002	Spa (Belgium)	Structured Mixed-Mode Design
		Multi-bit Sigma-Delta Converters
		Short-Range RF Circuits
2001	Noordwijk (The Netherlands)	Scalable Analog Circuits
		High-Speed D/A Converters
		RF Power Amplifiers
2000	Munich (Germany)	High-Speed A/D Converters
		Mixed-Signal Design
		PLLs and Synthesizers
1999	Nice (France)	XDSL and Other Communication Systems
		RF-MOST Models and Behavioural Modelling
		Integrated Filters and Oscillators
1998	Copenhagen (Denmark)	1-Volt Electronics
		Mixed-Mode Systems
		LNAs and RF Power Amps for Telecom
1997	Como (Italy)	RF A/D Converters
		Sensor and Actuator Interfaces
		Low-Noise Oscillators, PLLs and Synthesizers
1996	Lausanne (Swiss)	RF CMOS Circuit Design
		Bandpass Sigma Delta and Other Data Converters
		Translinear Circuits
1995	Villach (Austria)	Low-Noise/Power/Voltage
		Mixed-Mode with CAD Tools
		Voltage, Current and Time References
1994	Eindhoven (The Netherlands)	Low-Power Low-Voltage
		Integrated Filters
		Smart Power
1993	Leuven (Belgium)	Mixed-Mode A/D Design
		Sensor Interfaces
		Communication Circuits
1992	Scheveningen (The Netherlands)	OpAmps
		ADC
		Analog CAD

Contents

Part I
Hybrid Data Converters

The first part of this book is dedicated to recent developments in the field of hybrid data converters. While hybrid architectures, algorithms and circuits have been used for a long time, hybrid converters are much more in the picture recently, thanks to their ability to deal with technology limitations and achieve better performance in terms of speed, accuracy, efficiency and chip area. The first chapter in this part of the book gives an overview of the scope of 'hybrid', while the remaining five chapters introduce hybrid converter examples aiming for different performance directions.

The first chapter, by Kostas Doris, discusses the different dimensions of hybrid converter concepts and shows a classification of these. Combined with concrete examples, it exemplifies the broad nature of hybrid concepts in data converters.

In the second chapter, by Alessandro Venca, Nicola Ghittori, Alessandro Bosi and Claudio Nani, a time-interleaved SAR-$\Delta\Sigma$ ADC is proposed. Moreover, the DAC implementation also uses a hybrid charge redistribution/charge sharing topology. It is shown that this enables excellent performance and also reduces chip area of the integrated system.

The third chapter, by Ewout Martens, discusses several time-interleaved pipelined SAR ADCs, aiming to move the speed-precision-efficiency envelope. The various circuit implementations being used give insight into their respective limitations and advantages.

The fourth chapter, by Arindam Sanyal, Wenjuan Guo and Nan Sun, combines a SAR ADC with a $\Delta\Sigma$ modulator and a VCO-based converter, resulting in noise shaping and higher precision than achievable by a typical SAR ADC.

In the fifth chapter, Yun-Shiang Shu, Liang-Ting Kuo and Tien-Yu Lo describe a hybrid architecture for a reconfigurable SAR ADC, including SAR, $\Delta\Sigma$ and subranging and flash converter techniques. Noise and mismatch shaping techniques enable a state-of-the-art efficiency, up to 101dB SNDR and flexibility in speed and precision.

The sixth chapter, by Burak Gönen, Fabio Sebastiano, Robert van Veldhoven, and Kofi Makinwa, presents a dynamic zoom ADC, which is a combination of a SAR and a $\Delta\Sigma$ ADC. This enables very high resolution while maintaining excellent efficiency in energy and area.

Chapter 1
Hybrid Data Converters

Kostas Doris

1.1 Introduction

Novel circuit and architectural data conversion amalgams employing combinations of conventional converter architectures that exploit the advantages of nm CMOS technologies and mitigate their shortcomings continue to push the envelope in performance and power efficiency across all speed resolution domains.

In [1] the author highlighted the transformation of conventional converters to hybrids in its beginning phase. Today, nearly a decade later, with the digital assisted hype beyond it's peak point, hybrid converters are in full deployment and have even claimed dedicated sessions in recent year's ISSCC. Figure 1.1 shows this evolution of CMOS data converters in the last two decades using the Schreier's Figure of Merit. It is apparent that the performance advances are heavily correlated with advanced CMOS nodes. These advances however do not come from the mere adoption of technologies with smaller nodes but by the continuous advance in understanding how to use them optimally encapsulated in today's hybrid data converter architectures.

The objective of this introductory paper is to illustrate the broad and multidimensional nature of hybrid converters, which goes beyond the "combination of known architectures" as it is often referred. It will be argued that hybrid data converters reflect the continuation of a matching driven design policy [3]. According to this the analog to digital conversion from the wired or wireless input interface (e.g. antenna) to digital bits is conditioned optimally to the properties of CMOS technology through combinations of techniques across multiple signal domains

K. Doris (✉)
NXP Semiconductors, Eindhoven, The Netherlands
e-mail: kostas.doris@nxp.com

P. Harpe et al. (eds.), *Hybrid ADCs, Smart Sensors for the IoT, and Sub-1V & Advanced Node Analog Circuit Design*, DOI 10.1007/978-3-319-61285-0_1

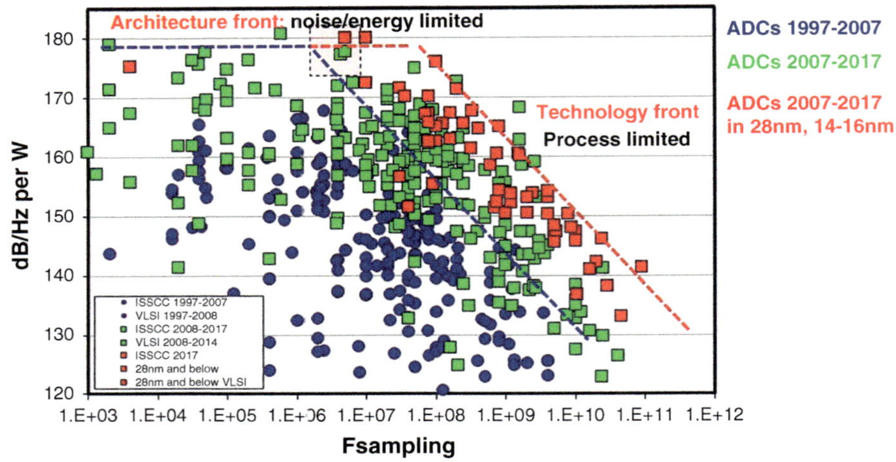

Fig. 1.1 Schreier FoM with data from [2]

and hardware abstraction layers using modulation, redundancy, scheduling, on-chip information and other concepts. Dependent on the speed resolution domain, hybrid architectures take a different shape that matches to thermal noise or process technology limitations dominating in the corresponding domain.

1.2 The Role of the Analog to Digital Conversion Interface

Figure 1.2 shows the receiver part of a communication system represented with a direct conversion architecture without loss of generality.

Physical signals serving as carrier waves propagate information across the propagation medium (the channel) prior to being received at the antenna interface. These information carrier signals are analog and are optimally matched to the physics of the propagation medium for the purpose of carrying information. They are minimally restricted in the amplitude and time domains, which means they have limited redundancy and as a consequence they are sensitive to noise, interference, distortions, etc. Abstract bits on the right side of the figure are maximally restricted in amplitude and time domains. This restriction introduces redundancy which translates to robustness against CMOS technology. In this way they become optimally matched to CMOS technology for the purpose of computation of information.

This leaves us with two signal types and media optimally matched to each other for purposes of propagation and computation of information and one analog mixed-signal domain in the middle, where signals are not optimally matched to its CMOS implementation carrier.

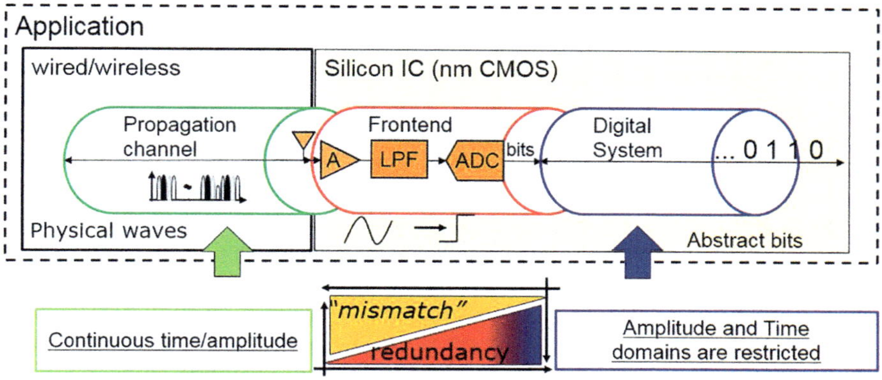

Fig. 1.2 Data converter interface as a matching network to the technology

The conversion of information from one medium to the other is done at the analog to digital interface spanning all the way from the antenna to the bit slicer, the exact point at which conversion from analog to digital fully takes place. We are neglecting for simplicity for now the conversion from the electromagnetic domain to the electrical at the antenna without loss of generality. Please note the distinction between the overall conversion from analog to digital and the data converter block, which is a boundary designers put in practice.

The primary role of this interface is to translate information from one matched domain to the other with minimum loss, from analog (no restrictions in time and amplitude) to digital (amplitude restricted to 0 and 1, sequences replacing time). As the analog signal reaches the boundary of the antenna interface it encounters a large discontinuity. Inversely, as signals go out of the bit slicer, they reach the level of best matching to CMOS technology. By restricting the signals in a stepwise manner across signal domains (continuous/discrete amplitude and time), we introduce redundancy in the signal and enable abstraction from hardware between device, circuit, architecture, algorithm and signal layers.

Abstraction and redundancy in the (mixed) signal allow a gradual functional decoupling across the signal chain. This is a typical property of digital systems but not of analog, which opens up the door for a plethora of transformations (modulation, scheduling, hardware redundancy, averaging, etc.) The hybrid combination of these techniques across the hardware abstraction hierarchy is what provides the optimal match of the conversion function (not just the converter block) to the evolving CMOS technology. This hybrid process goes stepwise along the signal path. It starts at the front-end circuits (e.g. LNA and PA) dealing with the largest discontinuity where especially at higher frequencies (e.g. mmwave) "matching" to technology means physical design approaches (e.g. resonating all nodes to overcome signal losses) and reaches full physical abstraction at the digital calibrations of the ADC.

As it can be seen, the conversion from analog to digital is not restricted to the data converter block and it is hybrid by definition, as already indicated by its dual name.

1.3 Abstraction Hierarchy and Concepts in Hybrid Converters

It is a common assumption in literature that a hybrid converter uses a combination of "single" conventional architectures such as pipelined ADCs, SAR, flash and delta-sigma in the same converter, each in the amplitude and time range where it exhibits maximum power efficiency [4].

In this context, pipelining of two single bit per cycle SARs [5, 6] is considered as a hybrid converter, and a conventional pipelined ADC using an MDAC combined with a parallel flash sub-ADC stage is not. However, since pipelining deals with the scheduling of the bit operations in time and not with the amplitude domain conversion algorithm, the opposite is actually true. A pipelined single-bit SAR ADC is not a hybrid because it uses only bit serial conversion, and the conventional pipeline ADC is a hybrid because it uses a bit-parallel (flash) algorithm embedded hierarchically into a bit serial conversion. The SAR is also hybrid when it employs multi bits per cycle.

This example illustrates that conventional architectures do not necessarily offer an orthogonal space with which to define what single and what hybrid is and this may lead to misconceptions. Instead, we can use a classification such as the one shown in Fig. 1.3 to analyse hybrid converters in more detail. At the vertical axis of the table at the right side of the figure, we can see the physical abstraction from hardware, whereas at the horizontal axis we list representative concepts.

Notice that a concept often requires (or enables) the use of another concept across this two-dimensional space. For example, in the hybrid ADC architecture shown on the left side of the figure, the sampling operation introduces redundancy in the time domain. Within one clock cycle, the time dimension is redundant. This allows time multiplexing of subsequent iterative conversion operations (sub-ADC, SAR, DAC) based on bit-parallel or serial method (e.g. flash/SAR for the ADC)

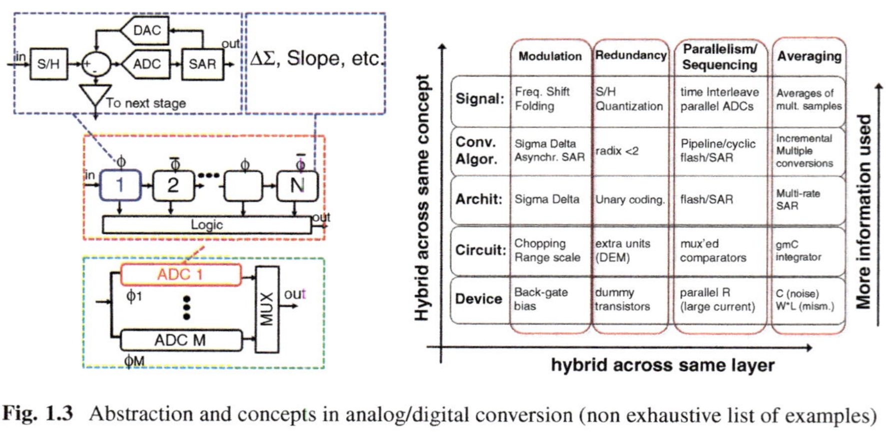

Fig. 1.3 Abstraction and concepts in analog/digital conversion (non exhaustive list of examples)

that generates an approximating quantized signal iteratively. On a higher abstraction layer, the bit conversion is parallelized in time domain through pipelining, while for the remaining bits, redundancy in time through the hold operation allows us to alter the conversion algorithm to another conversion algorithm, e.g. digital slope or even an incremental. On a higher level, parallelism can be added on sample to sample with interleaving, and even further hierarchy could be added with power splitters to isolate the ADCs electrically from the common input.

In the following, we can assume that a hybrid converter employs various independent concepts across the same horizontal or vertical layer. Some examples from literature are listed next:

- Multiple conversion algorithms in ADCs:
  - Multi-bit successive approximation [7, 8] (parallel-bit combined with serial-bit).
  - Successive approximation can be combined with slope [9, 10] and incremental converters [4] and even with oversampled converters and DEM [11–15].
- Redundancy across hardware abstraction layers in ADCs
  - Multiple ADCs for redundancy at application level or split ADC architectures used as means to deal with radiation tolerances [16].
  - Additional ADC units [17, 18]. This form of redundancy combined with multiplexing and abstraction enables frequency shifting techniques such as DEM and chopping that deal with offset and other interleave artefacts. Here, separation of the conversion function from sampling decouples the conversion function from the specific hardware making it such that the hardware can be interchanged freely. Redundancy in the number of ADCs enables the possibility to perform frequency translation fixed (or quasi-static) offset or gain and timing error signals to broadband ones resembling "noise". Similarly, by alternating the signal path, voltage offsets can be translated to higher frequencies (chopping). This makes them signal tones, which can then filtered. Inverse transformation recovers the signal back.
  - Reduced radix or over-ranging techniques increase the amplitude search range against the allowable signal range (see [5, 19] for digital and [20–23] for analog implementations) to increase tolerance to errors from offset, noise and incomplete settling of DAC MSB transitions. For as long there is sufficient range, remaining errors in conversion cycles do not count. For settling errors, this allows trading exponential settling time with linear conversion cycle time with profound benefits at high resolutions where long settling times are required.
- Asynchronous conversion in SAR [24–26] to maximize serial conversion efficiency embedded in fully synchronous overall conversion.
- Parallel processing with hardware hierarchy to address sampling rate limitations at GHz rates

  - Bit pipelining in SAR [5, 27] or pipelined ADCs [28]. Not only this offers the benefit of parallelism for speed but it also enables range scaling transformation in the amplitude domain with interstage amplification, which reduces the impact of noise in LSB conversions as well.
  - Sample to sample parallelism (interleaving).
  - Hierarchical approaches in sampling, interleaving, demultiplexing and pipelining [18, 23, 29–35].
  - Bit multiplexing with separate comparators [36, 37] or interleaved comparators [32].

- DACs with current steering flash and delta-sigma for MSB and LSBs [38], hybrid encoding with binary thermometer segmentation and combinations of flash and time interleave [39] and even functional co-integration across the receive/transmit path, e.g. mixing DACs [40].
- Hybrid circuit block implementations
  - DACs based on capacitors [24, 41, 42], resistors[5, 8, 20] or current [23, 43], current steering combined with resistors (e.g. R–2R) [44].
  - Class AB amplifiers reconfigured in and out of positive feedback [45] during operation, cascaded class A [11, 23] or class AB dynamic amplifications [27, 28, 46], charge re-use [25, 46], etc.
  - Charge sharing combined with charge redistribution DACs [4].
  - LDO's based on series-parallel concepts [47].

1.4 Reconfiguration and On-Chip Intelligence

Another key ingredient of hybrid converters is reconfiguration triggered by state or events. As the conversion process evolves from bit to bit, the data converter transforms itself through hardware reconfiguration from one to another circuit or architecture that is more optimal for the subsequent bit conversions.

Typical examples at circuit block level include comparator noise programming [25, 48, 49] in order to reduce power dissipation or reduce noise. Architecture reconfiguration has been reported in [4] changing from a 9b SAR to a 3b incremental delta-sigma to improve power efficiency and noise performance achieving 12b ADC conversion. In [48, 50] the ADCs were made reconfigurable to achieve multiple resolution levels with the same converter (here, not during conversion).

On-chip digital assistance and more generally the handling of information with on-chip intelligence are also hybrid in nature. A design policy exploiting digital assistance to address non-idealities of the converter in favour of faster operation has been the intense focus of research in multiple data converter papers the last years.

The hybrid nature of information is exploited in data converters. This can be depicted conceptually in Fig. 1.4. Information can be present in the input or output signal or in sub-signals. It is used a priori the design phase (e.g. technology spread properties, reliability and temperature behaviour) or obtained a posteriori with on-

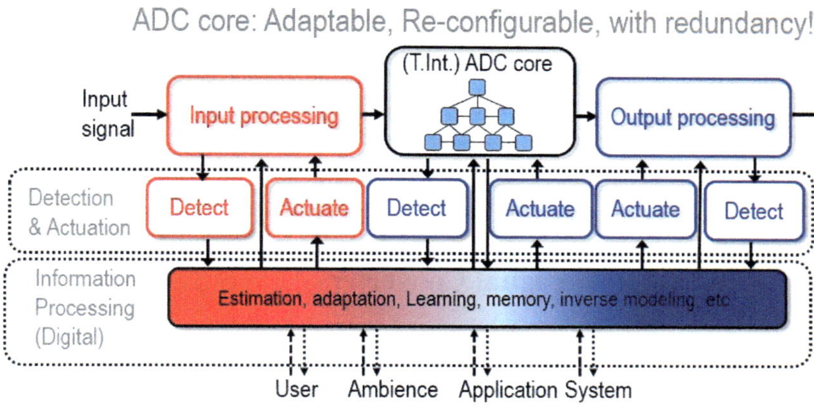

Fig. 1.4 Hybrid on-chip intelligence in data converters

chip detection (e.g. on-chip monitors) and covers user, application, ambience and technology domains. It is exploited in digital or analog form, continuous time or discrete. Converters today use a multitude of information types and apply hybrid calibration techniques to deal with errors such as time interleave artefacts, transfer function errors, programming the chip for given temperature profiles, etc., for example, time interleave error detection in digital domain and correction in the analog domain.

Such hybrid converters need to be made adaptive (reconfigurable architecture, adaptable parameters, test signals, etc.) to exploit the combination of information types available. They also need hardware redundancy. For example in a DAC, redundancy can be put in the switched hardware units [51, 52], while reconfiguration is done at the decoder. The dual ADC reported in [53] analyses the input data and chooses during operation amongst two ADCs on sample per sample basis reconfiguring the gain applied to the signal to reduce the impact of clipping in multi-carrier signals.

1.5 Functional and Technology Co-integration in Hybrid Converters

So far we have been talking mainly about hybrid data converters looking in the converter block. Another hybrid dimensions occurs along the signal path in a receiver and transmitter. Most notable examples here include the functional co-integration between the upconverter mixer and the DAC [52] for wireless systems, the power amplifier and DAC in wireline systems driving directly 50 Ω cable load, the combination of equalization and conversion in wireline systems, etc.

In the same line of thinking, references [18, 23, 43] integrate the buffer driving the large load of the ADC into the SAR loop eliminating its nonlinear behaviour. When combined with demultiplexing, this further enables to reduce the impact of wire interconnect load of the interleave array [23].

Finally, the combination of different technologies can lead also to hybrid data converters, e.g. optical and electronic ADCs.

1.6 Hybrid Analog to Digital Converters Across Speed and Resolution

The type of hybrid techniques associates strongly with the problem that needs to be addressed at the corresponding resolution speed regime.

At very high sampling rates, combinations of scheduling operations with time interleave error correction techniques dominate hybrid ADCs. Achieving tens of GS/s sampling rate and 10–20 GHz input bandwidth with low clock jitter noise overshadows the low-resolution noise requirements of the unit converters. The main issue is how to deal with the large interconnect capacitance stemming from the large converter interleaved array that is required to get to the required sampling rate (mainly clock and signal interconnect, not sampling capacitors). A large array enables high sampling rate, but limits input bandwidth and clocking performance and translates to high-power dissipation. Bandwidth also brings in additional constrains in the packaging due to input signal losses.

Hybrid concepts help to reduce the size of the interleave array, to divide and reduce the interconnect capacitance that is present at the input or at clock nodes. We observe in both signal and clocking paths hierarchical interleaving forms, multiplexing, resampling, power splitting, etc. [18, 23, 30–33]. Typically flash/SAR ADCs are preferred for either their speed or small area that both help to scale down the array size and complexity. This is achieved thanks to hybrid calibration techniques correcting time and amplitude dimension interleave errors with analog and digital correction techniques. The units need to be as fast as possible, and multiplexing is applied even at the comparator level, e.g. separate comparators [36] or interleaving [32]. Frequency domain multiplexing was also reported [54].

High-resolution levels, e.g. 10–16 b with sampling rates between 1 and 10 GS/s rates, require dealing with thermal noise, matching, large sampling and interconnect capacitance. Here there is a lot of emphasis for hybrid concepts that apply at the converter unit and associated use of on-chip information to calibrate it. Time interleaving is present but reduces drastically as the resolution increases: noise imposes restrictions due to the large sampling capacitors and clock buffers loading the input and clock nodes.

We observe hierarchical combinations of multi-bit MDAC sampling front ends with SAR sub-ADCs [55] but also pipelined multi-bit SARs [37] exploiting dynamic amplifiers and SARs with separate comparators per bit to become the alternatives of conventional pipelined ADCs using flash sub-ADCs [56]. Digital and analog corrections are implemented for nearly everything that can be calibrated in the converter for both pipeline and interleave artefacts such as comparator offsets, gains, signal transfer functions, bandwidth mismatches, interleave errors, track hold errors, clock injections, using DEM, LMS algorithms, dithering, etc. [57].

At speeds below 1 GHz [4, 28] with high-resolution levels, we see combinations of various concepts diverging from the conventional pipelined ADC. The main idea is partitioning the conversion cycle in MSB and LSB parts. A suitable architecture can be chosen for the LSB part that determines noise performance and power dissipation, and a faster architecture can be used to remove speed limitations in the MSB part. Interleaving is a degree of freedom coming on top. The work of [4] introduces an incremental delta-sigma modulator with reconfigurability, whereas [28] uses pipeline of SARs with interstage amplification based on dynamic amplifiers. In [10] an asynchronous digital slope converter is combined with a SAR exploiting also continuous time comparator techniques. The 14b ADC in [58] utilized multi-bit SARs with time interleaving of the single comparator being shared between all DACs. Analog to Digital converters with even lower speed but high resolutions follow and extend the same trend. A recent example can be seen in the audio domain sub-ranging delta-sigma ADC presented in [59].

These examples illustrated how multiple principles and techniques can be used to exploit most optimally CMOS technology to deal with fundamental limitations such as noise. However, the solution space is not restricted to these techniques only. As we move to even higher levels of abstraction, more degrees of freedom will become available. At the transmitter side, one can use spatial signal processing techniques at analog and digital domains. At the circuit layer, for example, power combining allows generating higher signal power compared to what a single amplifier unit can deliver. At the architectural level, spatial processing can be implemented with analog and digital beam forming either with phase rotation or time delays in analog and digital domains, respectively. The trend will continue at application level, for example in the future autonomous driving vehicles multiple sensors with different principles will be combined into one sensor architecture (ultra sound, radar, lidar, camera, etc.) to enable the car process a broad spectrum of signals from low frequencies up to optical wavelengths with high accuracy and reliability.

1.7 Conclusions

To appreciate properly the nature of hybrid converters, one needs to see them from multiple angles and abstraction layers and appreciate the whole conversion from antenna to bits.

In the past, hybrid converters were mostly all about the circuit layer of abstraction. Nowadays, hybrid data converters expand across all abstraction layers, especially with regard to scheduling, algorithmic conversions, handling of information for corrections and the transceiver signal path.

The hybrid analog to digital or digital to analog converter has a dual name to begin, analog and digital. It operates in multiple signal amplitude and time domains and utilizes combinations of conversion techniques and concepts such as redundancy, modulation, parallelism, sequencing and sub-ranging across hardware abstraction hierarchy. Through reconfigurable and adaptable hardware, it exploits multiple forms of information to deal with nm CMOS imperfections. The ideal hybrid converter breaks completely the boundaries of conventional converter blocks from antenna to bits and becomes a converter truly *matched* to the application and the properties of silicon technology and propagation channel.

References

1. Doris, K.: Time interleaved analog-to-digital converters: an algorithmic melting pot. In: 2009 IEEE International Solid-State Circuits Conference (ISSCC), Jan 2009
2. Murmann, B.: ADC performance survey. http://www.stanford.edu/~murmann/adcsurvey.html
3. van Roermund, A.: Shifting the frontiers of analog and mixed-signal electronics. Advances in Electronics, vol. 2014 (2014)
4. Venca, A., et al.: A 0.076 mm^2 12 b 26.5 mW 600 MS/s 4-way interleaved sub-ranging SAR-$\Delta\Sigma$ ADC with on-chip buffer in 28 nm CMOS. IEEE J. Solid State Circuits **51**(12), 2951–2962 (2016)
5. Louwsma, S.M., et al.: A 1.35 GS/s, 10 b, 175 mW time-interleaved AD converter in 0.13 μm CMOS. IEEE J. Solid State Circuits **43**(4), 778–786 (2008)
6. Verbruggen, B., et al.: A 2.6 mW 6 bit 2.2 GS/s fully dynamic pipeline ADC in 40 nm digital CMOS. IEEE J. Solid State Circuits **45**(45), 2080–2090 (2010)
7. Cao, Z., et al.: A 32 mW 1.25 GS/s 6b 2b/Step SAR ADC in 0.13 μm CMOS. In: IEEE International Solid-State Circuits Conference, Digest of Technical Papers, pp. 542–543 (2008)
8. Wei, H., et al.: A 0.024 mm^2 8b 400MS/s SAR ADC with 2b/Cycle and resistive DAC in 65 nm CMOS. In: IEEE International Solid-State Circuits Conference, Digest of Technical Papers, pp. 188–190 (2011)
9. Ding, M., et al.: A 5bit 1GS/s 2.7 mW 0.05 mm^2 asynchronous digital slope ADC in 90 nm CMOS for IR UWB radio. In: 2012 IEEE Radio Frequency Integrated Circuits Symposium, pp. 487–490, June 2012
10. Liu, C.C., et al.: A 12 bit 100 MS/s SAR-assisted digital-slope ADC. IEEE J. Solid State Circuits **51**(12), 2941–2950 (2016)
11. Harsener, M., et al.: A 14b 40 MS/s redundant SAR ADC with 480 MHz clock in 0.13 μm CMOS. In: IEEE International Solid-State Circuits Conference, Digest of Technical Papers, pp. 248–249 (2007)
12. Hurrel, C., et al.: An 18b 12.5 MHz ADC with 93 dB SNR. In: IEEE International Solid-State Circuits Conference, Digest of Technical Papers, pp. 378–379 (2010)
13. Fredenburg, J., Flynn, M.: A 90MS/s 11MHz bandwidth 62dB SNDR noise-shaping SAR ADC. In: IEEE International Solid-State Circuits Conference, Digest of Technical Papers, pp. 468–471 (2012)
14. Liu, C.-C., et al.: A 0.46 mW 5MHz-BW 79.7 dB SNDR noise-shaping SAR ADC with dynamic-amplifier-based FIR-IIR Filter. In: 2017 IEEE International Solid-State Circuits Conference (ISSCC), pp. 466–467, Jan 2017
15. Shu, Y.S., et al.: An oversampling SAR ADC with DAC mismatch error shaping achieving 105 dB SFDR and 101 dB SNDR over 1 kHz BW in 55 nm CMOS. IEEE J. Solid State Circuits **51**(12), 2928–2940 (2016)

16. Xu, H., et al.: A 78.5dB-SNDR radiation- and metastability-tolerant two-step split SAR ADC operating up to 75MS/s with 24.9 mW power consumption in 65 nm CMOS. In: 2017 IEEE International Solid-State Circuits Conference (ISSCC), pp. 477–477, Jan 2017
17. Ginsburg, B.P., Chandrakasan, A.P.: Highly interleaved 5-bit, 250-MSample/s, 1.2-mW ADC with redundant channels in 65-nm CMOS. IEEE J. Solid State Circuits **43**(12), 2641–2650 (2008)
18. Janssen, E., et al.: An 11b 3.6GS/s time-interleaved SAR ADC in 65 nm CMOS. In: 2013 IEEE International Solid-State Circuits Conference Digest of Technical Papers, pp. 464–465, Feb 2013
19. Kuttner, F.: A 1.2V 10b 20 MSample/s non-binary successive approximation ADC in 0.13 μm CMOS. In: IEEE International Solid-State Circuits Conference, Digest of Technical Papers, pp. 176–177 (2002)
20. Boyacigiller, Z., et al.: An error-correcting 14b/20ps CMOS A/D converter. In: IEEE International Solid-State Circuits Conference, Digest of Technical Papers, pp. 62–63 (1981)
21. Draxelmayr, D.: A self calibration technique for redundant A/D converters providing 16b accuracy. In: IEEE International Solid-State Circuits Conference, Digest of Technical Papers, pp. 204–205 (1988)
22. Liu, W., et al.: A 12b 22.5/45MS/s 3.0 mW 0.059 mm^2 CMOS SAR ADC achieving over 90 dB SFDR. In: IEEE International Solid-State Circuits Conference, Digest of Technical Papers, pp. 380–381 (2010)
23. Doris, K., et al.: A 480 mW 2.6GS/s 10b 65 nm time-interleaved ADC with 48.5 dB SNDR up to Nyquist. In: IEEE International Solid-State Circuits Conference, Digest of Technical Papers, pp. 180–182 (2011)
24. Chen, S.-W.M., Brodersen, R.W.: A 6-bit 600-MS/s 5.3-mW asynchronous ADC in 0.13-μm CMOS. IEEE J. Solid State Circuits **41**(12), 731–740 (2006)
25. Craninckx, J., Van der Plas, G.: A 65J/conversion-step 0-to-50MS/s 0-to-0.7 mW 9b charge-sharing SAR ADC in 90 nm digital CMOS. In: IEEE International Solid-State Circuits Conference, Digest of Technical Papers, pp. 246–247 (2007)
26. Harpe, P., et al.: A 30fJ/Conversion-Step 8b 0-to-10MS/s asynchronous SAR ADC in 90 nm CMOS. In: IEEE International Solid-State Circuits Conference, Digest of Technical Papers, pp. 387–389 (2010)
27. Verbruggen, B., Iriguchi, M., Craninckx, J.: A 1.7 mW 11b 250MS/s 2x interleaved fully dynamic pipelined SAR ADC in 40 nm Digital CMOS. In: IEEE International Solid-State Circuits Conference, Digest of Technical Papers, pp. 466–469 (2012)
28. van der Goes, F., et al.: A 1.5 mW 68 dB SNDR 80 Ms/s 2x interleaved pipelined SAR ADC in 28 nm CMOS. IEEE J. Solid State Circuits **49**(12), 2835–2845 (2014)
29. Gupta, S., et al.: A 1GS/s 11b time-interleaved ADC with 55-dB SNDR, 250 mW power realized by a high bandwidth scalable time-interleaved architecture. IEEE J. Solid State Circuits **41**, 2650–2657 (2006)
30. Greshishchev, Y., et al.: A 40GS/s 6b ADC in 65 nm CMOS. In: IEEE International Solid-State Circuits Conference, Digest of Technical Papers, pp. 390–391 (2010)
31. Doris, K., et al.: Interleaving of sar adcs in deep submicron CMOS technology. Advances in Analog and RF IC Design for Wireless Communication Systems, 1st edn. Elsevier. ISBN:9780123983268
32. Kull, L., et al.: 22.1 A 90GS/s 8b 667 mW 64x; interleaved SAR ADC in 32 nm digital SOI CMOS. In: 2014 IEEE International Solid-State Circuits Conference Digest of Technical Papers (ISSCC), pp. 378–379, Feb 2014
33. Duan, Y., Alon, E.: A 12.8 GS/s time-interleaved ADC with 25 GHz effective resolution bandwidth and 4.6 ENOB. IEEE J. Solid State Circuits **49**(8), 1725–1738 (2014)
34. Wu, J., et al.: A 4GS/s 13b pipelined ADC with capacitor and amplifier sharing in 16nm CMOS. In: 2016 IEEE International Solid-State Circuits Conference (ISSCC), pp. 466–467, Jan 2016
35. Wu, J., et al.: A 5.4GS/s 12b 500 mW pipeline ADC in 28 nm CMOS. In: 2013 Symposium on VLSI Circuits, pp. C92–C93, June 2013

36. Kull, L., et al.: A 10b 1.5GS/s pipelined-SAR ADC with background second-stage common-mode regulation and offset calibration in 14 nm CMOS FinFet. In: 2017 IEEE International Solid-State Circuits Conference (ISSCC), pp. 474–476, Jan 2017
37. Vaz, B., et al.: A 13b 4GS/s digital assisted dynamic 3-stage asynchronous pipelined-SAR ADC. In: 2017 IEEE International Solid-State Circuits Conference (ISSCC), pp. 476–477, Jan 2017
38. Su, S., Chen, M.S.W.: A 12b 2GS/s dual-rate hybrid DAC with pulsed timing-error pre-distortion and in-band noise cancellation achieving >74dBc SFDR up to 1 GHz in 65 nm CMOS. In: 2016 IEEE International Solid-State Circuits Conference (ISSCC), pp. 456–457, Jan 2016
39. Olieman, E., et al.: A 110 mW, 0.04 mm^2, 11gs/s 9-bit interleaved DAC in 28 nm fdsoi with 50 dB SFDR across nyquist. In: 2014 Symposium on VLSI Circuits Digest of Technical Papers, June 2014, pp. 1–2
40. Bechthum, E., et al.: A wideband RF mixing-DAC achieving IMD lt; -82 DBC up to 1.9 GHz. IEEE J. Solid State Circuits **51**(6), 1374–1384 (2016)
41. McCreary, J., Gray, P.: All-MOS charge redistribution analog-to-digital conversion techniques. IEEE J. Solid State Circuits **10**(6), 371–379 (1975)
42. Alpman, E., et al.: A 1.1V 50 mW 2.5GS/s 7b time-interleaved C-2C SAR ADC in 45 nm LP digital CMOS. In: IEEE International Solid-State Circuits Conference, Digest of Technical Papers, pp. 76–77, 77a (2009)
43. Kramer, M.J., et al.: A 14 b 35 MS/s SAR ADC achieving 75 dB SNDR and 99 dB SFDR with loop-embedded input buffer in 40 nm CMOS. IEEE J. Solid State Circuits **50**(12), 2891–2900 (2015)
44. Poulton, K., et al.: A 7.2-GSa/s, 14-bit or 12-GSa/s, 12-bit DAC in a 165-GHz fT BiCMOS process. In: 2011 Symposium on VLSI Circuits – Digest of Technical Papers, pp. 62–63, June 2011
45. Hershberg, B., et al.: Ring amplifiers for switched capacitor circuits. IEEE J. Solid-State Circuits **47**(12), 2928–2942 (2012)
46. van Elzakker, M., et al.: A 10-bit charge-redistribution ADC consuming 1.9μW at 1MS/s. IEEE Journal of Solid-State Circuits **45**(5), 1007–1015 (2010)
47. Breems, L., et al.: A 2.2 GHz continuous-time $\Delta\Sigma$ ADC with 102 dBc THD and 25 MHz bandwidth. IEEE J. Solid-State Circuits **51**(12), 2906–2916 (2016)
48. Harpe, P., et al.: A 7-to-10b 0-to-4MS/s flexible SAR ADC with 6.5-to-16fJ/conversion-step. In: IEEE International Solid-State Circuits Conference, Digest of Technical Papers, pp. 472–475 (2012)
49. Kramer, M.J., et al.: A 14-bit 30-MS/s 38-mW SAR ADC using noise filter gear shifting. IEEE Trans. Circuits Syst. Express Briefs **64**(2), 116–120 (2017)
50. Yip, M., Chandrakasan, A.: A resolution-reconfigurable 5-to-10b 0.4-to-1V power scalable SAR ADC. In: IEEE International Solid-State Circuits Conference, Digest of Technical Papers, pp. 190–191 (2011)
51. Tang, Y., et al.: A 14 bit 200 MS/s DAC with SFDR $>$ 78 dBc, im3 < -83 dBc and NSD < -163 dbm/Hz across the whole Nyquist band enabled by dynamic-mismatch mapping. IEEE J. Solid State Circuits **46**(6), 1371–1381 (2011)
52. de Vel, H.V., et al.: 11.7 a 240 mW 16b 3.2gs/s DAC in 65 nm CMOS with -80 dBc im3 up to 600 MHz. In: 2014 IEEE International Solid-State Circuits Conference Digest of Technical Papers (ISSCC), pp. 206–207, Feb 2014
53. Lin, Y.: et al.: An 11b 1GS/s ADC with parallel sampling architecture to enhance SNDR for multi-carrier signals. In: 2013 Proceedings of the ESSCIRC (ESSCIRC), pp. 121–124, Sept 2013
54. Greshishchev, Y.: CMOS ADCs for optical communications. In: Proceedings of the 20th Workshop on Advances in Analog Circuit Design (AACD), Apr 2012
55. Brandolini, M., et al.: 26.6 A 5GS/S 150 mW 10b SHA-less pipelined/SAR hybrid ADC in 28 nm CMOS. In: 2015 IEEE International Solid-State Circuits Conference – (ISSCC) Digest of Technical Papers, pp. 1–3, Feb 2015

56. Devarajan, S., et al.: A 12b 10GS/s interleaved pipeline ADC in 28 nm CMOS technology. In: 2017 IEEE International Solid-State Circuits Conference (ISSCC), pp. 288–289, Jan 2017
57. Ali, A.M.A., et al.: A 14-bit 2.5GS/s and 5GS/s RF sampling ADC with background calibration and dither. In: 2016 IEEE Symposium on VLSI Circuits (VLSI-Circuits), pp. 1–2, June 2016
58. Kapusta, R., et al.: A 14b 80 Ms/s SAR ADC with 73.6 db SNDR in 65 nm CMOS. IEEE J. Solid State Circuits **48**(12), 3059–3066 (2013)
59. Gonen, B., et al.: 15.7 A 1.65 mW 0.16 mm2 dynamic zoom-ADC with 107.5 dB DR in 20 kHz BW. In: 2016 IEEE International Solid-State Circuits Conference (ISSCC), pp. 282–283, Jan 2016

Chapter 2
Hybrid and Segmented ADC Techniques to Optimize Power Efficiency and Area: The Case of a 0.076 mm^2 600 MS/s 12b SAR-$\Delta\Sigma$ ADC

Alessandro Venca, Nicola Ghittori, Alessandro Bosi, and Claudio Nani

2.1 Introduction

Integration of low-power and area-efficient ADCs is a key differentiator in modern mixed-signal SoCs. The increasing usage of software radio architectures for the design of wireline communication receivers calls for the design of high-speed medium to high-resolution ADCs with high power efficiency [1, 2].

Traditionally, these challenges have been addressed in scaled technologies using SAR architectures, often in combination with techniques like redundancy, asynchronous operation, and time interleaving to meet the required application sampling rate. However, for high-resolution ADCs (9b + ENOB), a traditional SAR is intrinsically energy inefficient since it reuses the same low-noise comparator to perform both coarse conversions (where little accuracy is needed) and fine conversions (where thermal noise is of paramount importance) [8].

In the last few years, this limit has been addressed by combining SAR with other conversion algorithms like averaging [4], noise shaping with oversampling [5, 7], single slope [6], and pipelining. As an example, in [7], flash, SAR, and noise shaping with oversampling techniques are combined together in a single design to achieve a Schreier FoM of 180 dB in 4 KHz bandwidth. For high-resolution and medium sampling rate converters, excellent power efficiency figures have been achieved with hybrid SAR-pipeline architectures [8, 9] that employ SAR as subADCs, but whose noise performance is determined by a high-efficiency interstage amplifier as opposed to the comparator. All these works focus on combining different conversion techniques in a single design and use each of them in a resolution and

---

A. Venca • N. Ghittori • A. Bosi • C. Nani (✉)
Marvell Italia s.r.l., viale Repubblica 38, 27100 Pavia, Italy
e-mail: nanis.cn@gmail.com

P. Harpe et al. (eds.), *Hybrid ADCs, Smart Sensors for the IoT, and Sub-1V & Advanced Node Analog Circuit Design*, DOI 10.1007/978-3-319-61285-0_2

signal amplitude range where they perform at its maximum power efficiency. This approach has become so widespread that the resulting ADC architectures have been denoted as "hybrid ADCs."

At the same time, emerging wireline standards like the G.hn Gen 2 [3] that employ MIMO strategies often require the integration of a large number of analog front ends (AFE) in a single SoC. Such AFEs should therefore require little or no external components as this may have a significant impact on the complete system bill of materials (BoM). In order to tackle these application demands, the design of wide-bandwidth high-resolution ADCs with extremely low power *and* area is clearly of key importance.

Also in the field of ADC area, the adoption of the SAR architecture and the increasing usage of digital calibrations allowed data converters to reap the benefits of technology scaling. As an example, the introduction of digital DAC linearity calibrations [12] resulted in a significant reduction of matching requirements, allowing scaling down the DAC capacitance to the kT/C limit.

In noise-limited SAR ADCs, for every extra bit of resolution, the sampling and DAC capacitance grows four times, while, at the same time, the ripple on the converter reference voltage has to be reduced two times to preserve linearity. Such a trend sets very tough requirements for the reference generator of high-resolution charge-redistribution DAC (CR-DAC) SAR ADCs, especially when no external components can be used, hence requiring the integration of very large on-chip capacitors [9]. Another approach to address this issue in literature is using DAC switching schemes that optimize the current drawn from the reference [8, 9] or with DAC topologies more immune to reference ripple, like current-steering [10] or charge-sharing DACs (CS-DAC) [11].

In this paper, a 12b four-way interleaved 600 MS/s ADC with on-chip input signal and reference buffer is presented. The energy efficiency challenge is addressed with a hybrid ADC architecture that employs a SAR as a coarse ADC and an incremental Delta-Sigma as a fine ADC. A significant area reduction is achieved with a segmented charge-sharing charge-redistribution DAC architecture that relaxes significantly the accuracy requirements on the reference generator and can be scaled down to kT/C limit as a conventional CR-DAC.

The 28 nm CMOS ADC prototype delivers 58 dB SNDR at Nyquist for 26.5 mW of power with a total area of only 0.076 mm^2 and does not require any external component.

Section 2.2 discusses the thermal noise performance of incremental $\Delta\Sigma$ ADCs and compares it with the SAR architecture. The subrange SAR-$\Delta\Sigma$ ADC architecture is then introduced in Sect. 2.3. Sections 2.4 and 2.5 review the requirements for the reference generator of high-resolution charge-redistribution DACs and the scalability limits of the charge-sharing DAC architecture. The segmented charge-sharing charge-redistribution DAC architecture is then presented in Sect. 2.6. The overall ADC architecture and calibrations are presented in Sect. 2.7, while in Sect. 2.8 the circuit level implementation is presented. Section 2.9 presents the measurement results, and finally conclusions are drawn in Sect. 2.10.

2.2 Incremental $\Delta\Sigma$ Conversion Algorithm

In literature, a very power-efficient conversion technique from a thermal noise perspective is the incremental $\Delta\Sigma$ algorithm [13–15] whose basic block diagram is shown in Fig. 2.1. It consists of a T/H stage followed by a $\Delta\Sigma$ modulator whose internal state can be reset. After the sampling phase, the input signal is held constant at the input of the $\Delta\Sigma$ modulator, and it is converted for N clock cycles ($NT_{\mathrm{clk}} = T_{\Delta\Sigma}$). At the end of conversion, the state of the $\Delta\Sigma$ modulator, which is the loop filter capacitances, is reset, and the N samples output sequence is then filtered by a FIR filter and decimated by N.

A key property of continuous-time (CT) incremental $\Delta\Sigma$ modulators is their behavior with respect to thermal noise added at the modulator inputs, mainly ascribed to DAC and loop filter integrators (V_n in Fig. 2.1). Unlike the input signal, this noise source is not sampled and therefore is filtered by the CT-$\Delta\Sigma$ modulator signal transfer function and appears at the quantizer output. The digital FIR decimation filter then suppresses the out-of-band noise before the final decimation. By assuming a white Gaussian input thermal noise power spectral density S_{Vn}, the final output noise power is given by

$$P_n = \int_0^{f\mathrm{clk}/2} S_{Vn}|H_{\mathrm{FIR}}(f)|^2 \cong S_{Vn}Bw_n \ \text{ with } Bw_n = \frac{\sum_{i=1}^{N} c_i^2}{2T_{clk}} = \frac{1}{\varphi T_{\Delta\Sigma}} \tag{2.1}$$

where $H_{\mathrm{FIR}}(f)$ is the FIR frequency response and Bw_n is the equivalent analog noise bandwidth of the FIR filter. The φ parameter depends on the actual FIR filter coefficients, and it can be proven that it is maximized ($\varphi = 2$) when all the FIR coefficients are equal (dumped integrator). If we implement the loop filter using a simple open-loop integrator (first-order $\Delta\Sigma$ modulator), this result is similar to the one in [8]; however in this scheme, the integrator, being inside a $\Delta\Sigma$ loop, does not need gain calibration nor stringent linearity performance.

We can compare the incremental $\Delta\Sigma$ and SAR thermal noise performance by considering the $\Delta\Sigma$ integrator and the SAR dynamic comparator as the only noise (and power) limiting blocks (see Fig. 2.2). The latter is usually implemented as a dynamic integrator followed by a CMOS latch (see Fig. 2.2a), and its input-referred

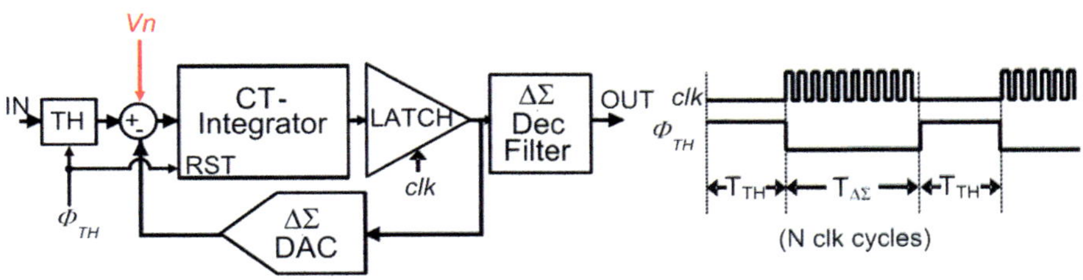

Fig. 2.1 Incremental $\Delta\Sigma$ converter block diagram and timing

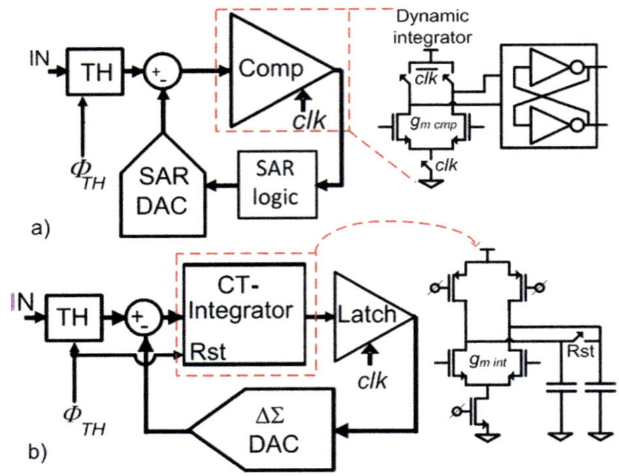

Fig. 2.2 (**a**) SAR and (**b**) first-order ΔΣ ADC block diagrams and circuit level implementations of the blocks considered as main contributors to noise power performance

thermal noise is dominated by the contribution of the input differential pair that can be written [16] as

$$P_{n\ \mathrm{SAR}} = S_{Vn}Bw_n \cong \frac{8kT\gamma}{g_{m\ \mathrm{cmp}}}\frac{1}{2T_{\mathrm{int}}} \tag{2.2}$$

where T_{int} is the duration of the dynamic integrator integration phase. If a simple open-loop integrator is used in an incremental ΔΣ ADC (assuming no linearity constraints, see Fig. 2.2b), the resulting noise power using Eq. (2.1) is

$$P_{n\ \Delta\Sigma} = S_{Vn}Bw_n \cong \frac{8kT\gamma}{g_{m\,\mathrm{int}}}\frac{1}{\varphi T_{\Delta\Sigma}} \tag{2.3}$$

Equations (2.2) and (2.3) have very similar structure and show that for both architectures the thermal noise is proportional to the inverse of the product $g_m{\cdot}T$. If we assume the same input pair overdrive (V_{ov}), the product $g_m{\cdot}T = 2I_{\mathrm{d}}/V_{\mathrm{ov}}{\cdot}T$ is proportional to the total charge drawn by the integrators.

While in an incremental ΔΣ ADC this charge ($g_{m\,\mathrm{int}}{\cdot}T_{\Delta\Sigma}$) is drawn only once during the complete ΔΣ conversion time ($T_{\Delta\Sigma}$ see Fig. 2.3b), in a SAR approximately the same charge ($g_{m\,\mathrm{cmp}}{\cdot}T_{\mathrm{int}}$) must be drawn every time the comparator is activated (see Fig. 2.3a). That means that in an N-bit SAR converter, the total charge drawn by the SAR comparator is N times the one used by an incremental ΔΣ integrator for the same level of thermal noise. In redundant SAR [17], the comparator power and noise can be optimized during conversion, but also in this case, more than one conversion cycle with the final comparator noise level is required when the quantization noise approaches the comparator thermal noise.

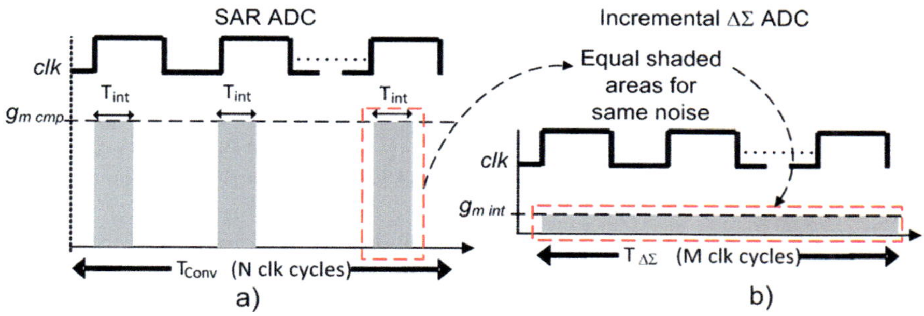

Fig. 2.3 Timing diagram with grey highlights when g_m is used to suppress thermal noise (**a**) for SAR ADC and (**b**) for a first-order $\Delta\Sigma$ ADC

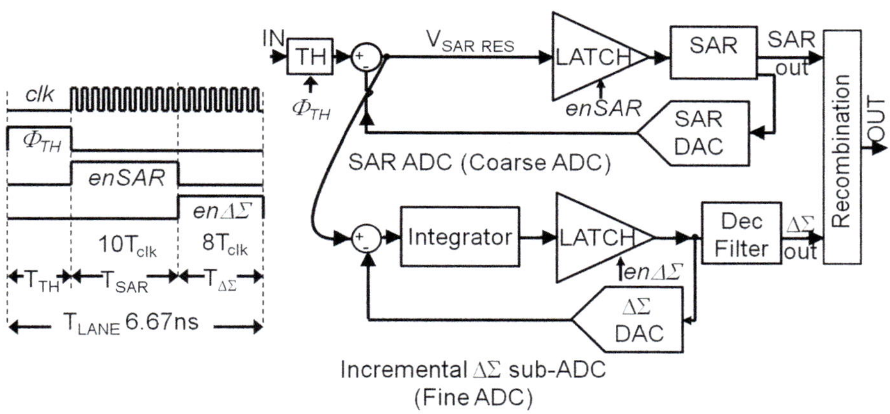

Fig. 2.4 SAR-$\Delta\Sigma$ subrange lane ADC conceptual block diagram and timing

2.3 SAR-$\Delta\Sigma$ Subrange Hybrid ADC Architecture

While CT-incremental $\Delta\Sigma$ converters are very efficient in mitigating the thermal noise components from the loop filter integrators, their key weakness is the large oversampling ratio, that is, number of clock cycles required to bring quantization noise to the target level.

A possible solution is then to build a hybrid converter by combining a mid-resolution coarse SAR ADC (where SAR algorithm provides maximum power efficiency) with CT-incremental $\Delta\Sigma$ fine modulator that deals with the thermal noise while providing only few bits of quantization. The resulting hybrid subrange ADC block diagram is shown in Fig. 2.4. By resolving a significant number of bits using the SAR algorithm, the number of clock cycles required by the $\Delta\Sigma$-subADC to get to the target final resolution can be reduced drastically.

In the presented 12b design, the coarse non-binary SAR ADC provides 9b of equivalent resolution, and the fine ADC ($\Delta\Sigma$-subADC) implemented as 1-bit first-order CT-incremental $\Delta\Sigma$ ADC resolves the remaining 3b to get to the final 12b resolution. 1b overrange is added between the coarse SAR ADC and the $\Delta\Sigma$-subADC in order to correct for the coarse SAR ADC error induced by the SAR loop thermal noise leading to a final $\Delta\Sigma$-subADC resolution of 4b after filtering and decimation.

The operation phases are shown in Fig. 2.4. After the sampling and the coarse SAR conversion phase (ten clock cycles), the SAR residual error is converted by the fine $\Delta\Sigma$-subADC in eight clock cycles. The eight $\Delta\Sigma$-subADC comparator decisions are filtered using an 8-taps low-pass FIR filter and decimated by eight before being recombined with the SAR output to get a 12b final code.

Once the $\Delta\Sigma$-subADC and coarse SAR output are recombined, the only residual noise is the sampling kT/C noise and the $\Delta\Sigma$-subADC noise, i.e., $\Delta\Sigma$ quantization noise, latch thermal noise, and loop filter thermal noise. All these components are low-pass filtered by the $\Delta\Sigma$-subADC FIR filter with the equivalent bandwidth of the FIR (approx $1/\varphi T_{\Delta\Sigma}$ where $T_{\Delta\Sigma}$ is the $\Delta\Sigma$-subADC conversion time) as shown in Sect. 2.2. Since both quantization and latch thermal noise are first-order shaped, they are strongly suppressed by the FIR leaving as dominant terms the sampling (kT/C) noise and the input-referred noise of the $\Delta\Sigma$ integrator (with integration time $T_{\Delta\Sigma}$). Artifacts due to first-order shaping are dithered by the thermal noise added by the latch.

The $\Delta\Sigma$-subADC is implemented with minimum hardware overhead by merging the $\Delta\Sigma$-DAC with the SAR-DAC and reusing the comparator latch. Reconfiguring the SAR comparator static preamp stage (G_{mint}) into the $\Delta\Sigma$ integrator only requires stopping the reset of its load capacitor (C_{int}) after the last SAR cycle (Fig. 2.5).

A consequence of this choice is that the transconductance of the resetting integrator in the SAR phase and of the loop integrator in the $\Delta\Sigma$ phase are the same. The resetting integrator is designed to have a 3σ thermal noise to be less than the coarse LSB (LSB_{SAR}) to guarantee the SAR noise (quantization + thermal) to be within the input range of the fine $\Delta\Sigma$ ADC. The integrator thermal noise power at the $\Delta\Sigma$ output can then be derived from (Eq. 2.2) and (Eq. 2.3) as

$$P_n = P_{n\,\text{SAR}} \cdot \frac{2T_{\text{int}}}{\varphi T_{\Delta\Sigma}} \tag{2.4}$$

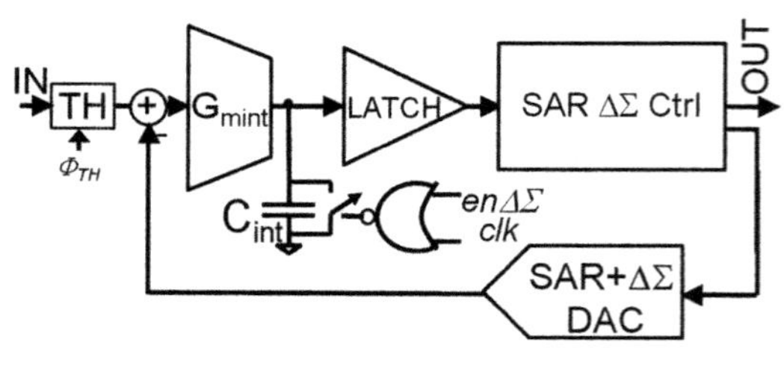

Fig. 2.5 SAR-$\Delta\Sigma$ subrange lane ADC minimum hardware implementation

where the suppression of the integrator thermal noise can be easily appreciated since $T_{\Delta\Sigma} = N{\cdot}T_{CLK}$ where N is the number of $\Delta\Sigma$ clock cycles, whereas T_{int} is a fraction of T_{CLK}.

Since a first-order $\Delta\Sigma$ suppresses the quantization noise with the inverse of the oversampling ratio (OSR $\cong T_{\Delta\Sigma}/T_{CLK}$) at the power of 3, it can be calculated that a first-order modulator is enough to suppress the quantization noise in $N = 8$ $\Delta\Sigma$ cycles to roughly the same level as the thermal noise.

The FIR filter frequency response has been designed in order to maximize the final ADC ENOB performance in the presence of all noise sources (including latch thermal noise) as well as quantization noise. The resulting filter shows a φ factor of approximately 1.8. Given the limited number of $\Delta\Sigma$-subADC conversion cycles, the digital FIR filter can be implemented in a simple direct form as weighted sum of the eight $\Delta\Sigma$-subADC comparator decisions.

In the presented implementation, all timing signals required for the conversion are generated internally using a self-timed loop approach [18]. Jitter on the internal clock during the $\Delta\Sigma$ phase is minimized by allocating a fixed time to comparator decisions.

2.4 Charge-Redistribution SAR-DAC

In the previous section, we have seen that it is possible to improve ADC power efficiency by applying a combination of conversion algorithms each of them used in a resolution range where it performs at its peak energy efficiency. A similar approach can be also applied to minimize the converter area, that is, by implementing blocks with a combination of different architectures, it is possible to achieve significant area saving and/or remove the need for external components. In the specific case of SAR-$\Delta\Sigma$ ADC, we are focusing on the DAC and its reference generator blocks which are often the largest contributor to total converter area.

The D/A converter of SAR ADCs is commonly a capacitive-based topology due to the inherent high matching of metal capacitors in deep-submicron technology. Among them, the charge-redistribution DAC (CR-DAC) (Fig. 2.6) [19] is extensively used mainly for its small C-array area. As the DAC switches are on the bottom plate (the side opposite to the comparator), the CR-DAC is inherently insensitive to switch parasitics (C'_n, C'_{n-1}, ..., C'_1), since these undesired capacitances are always connected to the reference generator. This property allows scaling down the DAC capacitance eventually to the kT/C noise limit, and it represents the main reason for the small C-array area achieved by the CR-DAC topology.

A key limitation of CR-DAC is that it requires a very accurate reference voltage not to impact the linearity performance. In a CR-DAC, in fact, the reference voltages are connected through the DAC switches and capacitors to the comparator inputs during the SAR phase of the ADC. The transfer function from the reference voltage node to the input of the comparator is not linearly code dependent, ranging from a unity transfer function when the DAC has all the capacitors connected between

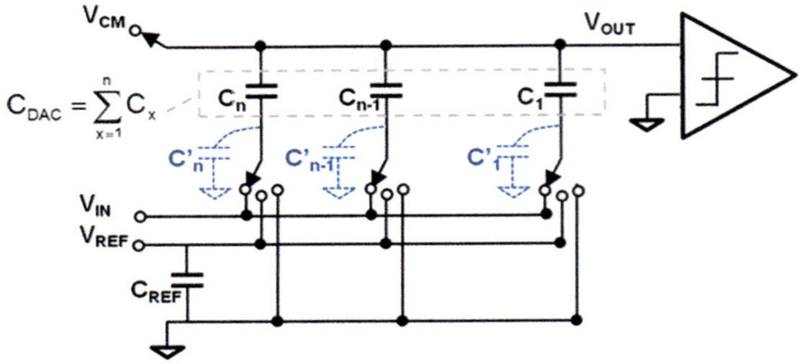

Fig. 2.6 Single-ended charge-redistribution DAC (CR-DAC) in track mode

the comparator input node and the reference voltage to virtually infinite attenuation when the DAC has all the capacitors connected between the comparator input node and the ground voltage. Therefore, ripples on the reference voltage greater than 1LSB during the SAR phase can cause errors in the comparator decisions that if not corrected may produce distortion in the converted signal. Moreover the net charge absorbed by the reference generator during one conversion, due to the switching of the DAC capacitors, is not linearly code dependent as well, as described in [19]. The net switching charge per conversion is absorbed by the finite impedance of the reference generator and produces a low-frequency code-dependent voltage ripple with a major spectral component at the ADC input signal second harmonic (in differential structures). This ripple, when mixed with the input signal fundamental, produces a third harmonic at the ADC output.

The two nonlinear mechanisms discussed above translate into very tough requirements for the reference generator of high-resolution charge-redistribution SAR ADCs. To mitigate the effects of the nonlinear reference-to-comparator transfer function, a large decoupling capacitance is usually required to reduce the high-frequency ripple, while low output peak impedance of the reference buffer is needed to attenuate the effects of the low-frequency nonlinear current drawn from the references by the switching DAC activity.

In a SAR ADC using binary-weighted CR-DAC, the reference voltage is required to settle within 1LSB accuracy at every conversion cycle. Non-binary-weighted CR-DAC implementations tolerate larger reference errors in the redundant part of the DAC, but still they require 1LSB settling accuracy during the last few conversion cycles where errors cannot be recovered by redundancy. The requirements on the reference voltage accuracy in a generic non-binary CR-DAC can be evaluated from Fig. 2.7a where the reference voltage ripple is sketched in steady state (ripple recovered within T_S). In this qualitative example, the reference generator has a bandwidth much smaller than the ADC conversion frequency f_S, and the decoupling capacitance is equal to C_{REF}. In this condition, by approximating the ripple peak amplitude generated by the first few MSB transitions as $\varepsilon_{\mathrm{MAX}} = \alpha C_{\mathrm{DAC}}/C_{\mathrm{REF}}$ and assuming a linear reference voltage recovery, the required C_{REF} can be estimated as

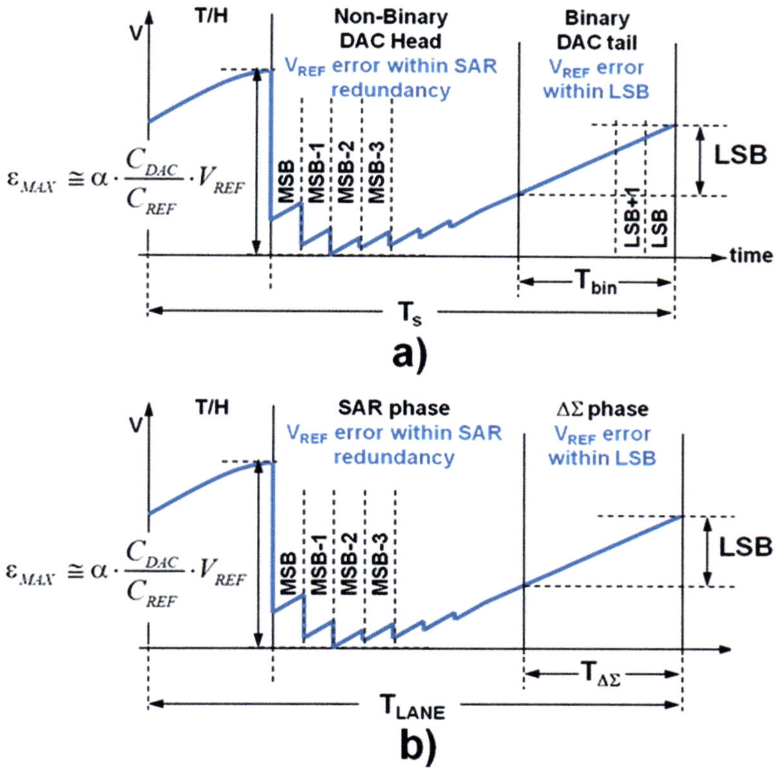

Fig. 2.7 Qualitative reference voltage ripple (**a**) in a non-binary CR-DAC and (**b**) in a non-binary SAR-$\Delta\Sigma$ architecture using a CR-DAC

$$C_{\text{REF}} \geq \alpha \cdot \frac{T_{\text{bin}}}{T_S} \cdot 2^{N-1} \cdot C_{\text{DAC}} \tag{2.5}$$

where C_{DAC} is the total DAC capacitance and the factor α takes into account for the switching scheme adopted in the DAC. In [20], for example, a switching scheme with α equal to 0.25 has been presented. This factor can be reduced further but often results in a code-dependent DAC output common mode [21].

In the case of a noise-limited charge-redistribution SAR ADC, it is interesting to analyze the scaling of the size of the C_{REF} as the required resolution increases, by using (Eq. 2.5). In Fig. 2.8 an example for 1 bit of additional resolution requirement is shown. Here the C_{DAC} has been multiplied by a factor 4 to meet the kT/C noise requirement (Fig. 2.8b). To recover the original ripple, the C_{REF} must be increased by a factor 4 as well (Fig. 2.8c). However since the new LSB is half the original one (N is increased by 1), to preserve the linearity specification, the ripple must be reduced by another factor 2, hence requiring C_{REF} to be multiplied by a factor 8 (Fig. 2.8d). As a consequence of that in a noise-limited charge-redistribution SAR

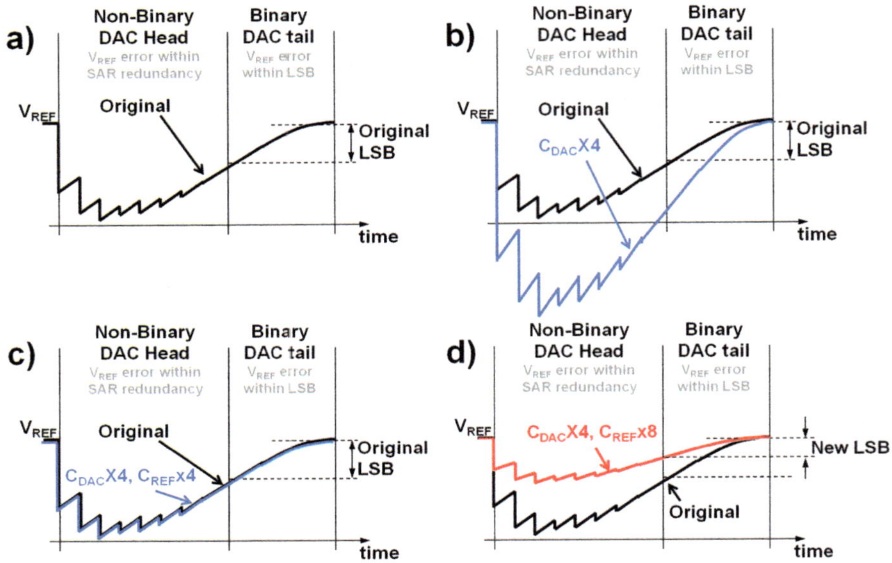

Fig. 2.8 Reference ripple and requirement on C_{REF} for one additional bit of resolution (T/H phase omitted to simplify the visualization). (**a**) Original ripple (*black*), (**b**) ripple (*blue*) with C_{DAC} increased four times to meet kT/C requirements, (**c**) ripple (*blue*) with C_{REF} increased four times to obtain the original ripple, and (**d**) ripple (*red*) with C_{REF} increased eight times to meet linearity requirements

ADC as the resolution increases, the size of the reference capacitance grows with double rate with respect to its core DAC capacitance eventually becoming the main integrated area contributor for resolutions beyond 9/10bit ENOB.

For the specific case of the 12b SAR-$\Delta\Sigma$ architecture using a CR-DAC, the reference ripple can be tolerated to a certain extent in the SAR phase, thanks to the DAC redundancy and to the $\Delta\Sigma$ overrange, but during the fine $\Delta\Sigma$ phase, the reference error must be kept within 1LSB as in a non-binary-weighted SAR ADC (Fig. 2.7b). The required C_{REF} can be then calculated using (Eq. 2.5) as

$$C_{\mathrm{REF}} \geq \alpha \cdot \frac{T_{\Delta\Sigma}}{T_{\mathrm{LANE}}} \cdot 2^{N-1} \cdot C_{\mathrm{DAC}} \cong 50\mathrm{pF} \qquad (2.6)$$

where $N = 12$, $C_{\mathrm{DAC}} = 300\mathrm{fF}$, $\alpha = 0.25$, $T_{\Delta\Sigma} = 2.22$ ns, and $T_{\mathrm{LANE}} = 6.66$ ns.

Integrating such a large capacitance may result in exceeding the area budget as it will most likely dominate the converter area [9]. This example clearly shows that while the CR-DAC is a very area-efficient topology to implement DACs with very small LSB units and/or total DAC capacitance, it becomes less area effective to implement large DACs where total capacitance is dominated by kT/C noise requirements.

2.5 Charge-Sharing SAR-DAC

An effective way to reduce the area of the reference generator is the charge-sharing DAC (CS-DAC) topology [11]. A CS-DAC is depicted in Fig. 2.9 together with the T/H capacitance on which the input signal is sampled at the end of the T/H phase. Simultaneously the reference voltage V_{REF} is sampled on C_{DAC}. After sampling, the V_{REF} is disconnected from C_{DAC} for the entire SAR conversion phase. In this way the comparator decisions are virtually insensitive to the noise/ripple on the references. Moreover the C_{DAC} is reset to zero at the end of the conversion phase by the SAR algorithm itself making the net charge absorbed by the reference generator during one conversion and the associated reference voltage ripple code independent. The combination of these two properties relaxes the noise/ripple requirements on the CS-DAC reference generator allowing a drastic reduction of the C_{REF} and easing the specifications on the reference buffer output peak impedance even for kT/C noise-dominated DACs.

The CS-DAC however is not suitable to implement DACs with very small LSB unit capacitance. Each CS-DAC element requires, in fact, a top plate switch (that is a switch on the comparator side) in order to share the charge between C_{DAC} and C_{TH} in additive/subtractive way during the conversion phase. The parasitic capacitances associated with the top plate switches ($C'x$, $C''x$) affect the DAC element effective

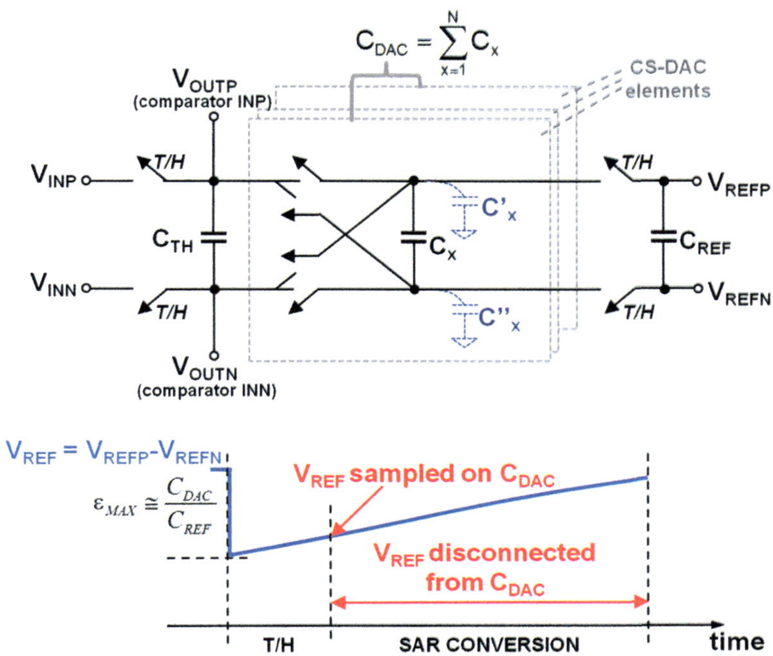

Fig. 2.9 Charge-sharing DAC (CS-DAC) and T/H

capacitance value. The scaling of the LSB capacitance in a CS-DAC is therefore dictated by the capacitance of a minimum size switch instead of kT/C noise. In deep-submicron technologies, this usually leads to a significant core area penalty with respect to a CR-DAC.

2.6 Segmented Charge-Sharing Charge-Redistribution SAR-DAC

In order to build an area-efficient kT/C noise-dominated DAC with a very small LSB unit capacitance, we can combine the two DAC topologies described above in a segmented SAR-DAC architecture (CS-CR-DAC) using a CS-DAC for the 4-MSBs and a CR-DAC for the remaining 6-LSBs (Fig. 2.10).

The parasitic insensitive 6-LSBs DAC capacitors can be scaled down as in a conventional CR-DAC to meet the kT/C noise requirements on the complete array, while the 4-MSBs CS-DAC capacitances are large enough that the associated switch parasitics can be easily maintained proportionally small, hence impacting negligibly the DAC linearity. As the CR-DAC total capacitance (C_{CR}) is limited to the LSBs, the associated high-frequency ripple ($\varepsilon_{\mathrm{MAX}}$) and the low-frequency code-dependent ripple are reduced roughly by 2 to the power of 4 (number of MSBs). Moreover the transfer function from the reference voltage nodes to the comparator input benefits from the large attenuation (β) provided by the 4-MSBs of the C_{DAC} capacitance (C_{CS}) referred to ground. By keeping separate the reference voltages for

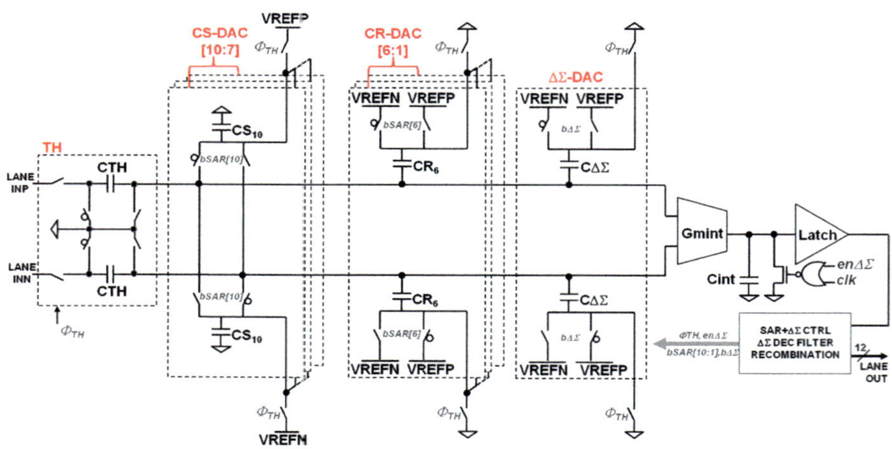

TH and DAC weights (1=600aF)											
CTH	CS[10:7]				CR[6:1]						CΔΣ
512	236	127	69	37	20	11	6	3	2	1	2

Fig. 2.10 CS-CR-DAC and ΔΣ DAC

the CR-DAC (V_{REFCR}) and for the CS-DAC (V_{REFCS}) segments, the former reference capacitance can be sized according to

$$C_{REFCR} \geq \alpha \cdot \beta \cdot \frac{T_{\Delta\Sigma}}{T_{LANE}} \cdot 2^{N-1} \cdot C_{CR} \cong 1.6\text{pF} \qquad \text{with } \beta = \frac{C_{CR}}{C_{CS}} \tag{2.7}$$

where $N = 12$, $C_{CR} = 25.8\text{fF}$, $\alpha = 1$, $\beta = 0.09$, $T_{\Delta\Sigma} = 2.22$ ns, and $T_{LANE} = 6.66$ ns. In our implementation, we have chosen a decoupling capacitance C_{REFCS} equal to 3.4 pF limiting the drop on the sampled reference value to 10%. The presented SAR ADC (with a core capacitance of 300fF single ended) requires therefore a single-ended reference capacitor of only 5 pF ($C_{REFCS} + C_{REFCR}$), a quite large area saving when compared to the 50 pF initially required with a charge-redistribution-only solution.

A drawback of the CS-DAC is an input signal attenuation of 6 dB due to the sharing of the signal charge between sampling capacitor C_{TH} and C_{CS} capacitor. This increases the ADC input-referred thermal noise from the comparator whose contribution to total noise is, however, reduced with respect to a conventional SAR ADC with the same power budget, by the SAR-$\Delta\Sigma$ ADC subrange architecture. Additionally, from a kT/C perspective, the CS-CR-DAC has roughly 3 dB more noise for the same DAC capacitance than a conventional CR-DAC since the reference voltages are sampled on the C_{CS} at the end of the tracking phase. This kT/C additional noise contributor is denoted in Fig. 2.16 as "Reference".

2.7 ADC Architecture and Calibrations

The overall 600MS/s ADC architecture and the calibration scheme are shown in Fig. 2.11. The ADC uses four interleaved lanes driven with a 25% duty-cycle clock generated from a 600 MHz PLL input clock. The V_{REFCR} and V_{REFCS} buffers are shared among the four time-interleaved lanes for a total ADC reference capacitance of 20 pF. Since they have relaxed specifications on the peak output impedance, each of them consumes only 100 μA from a 1.5 V power supply. The ADC uses both foreground and background calibrations. At start-up the offset and gain mismatches among lanes are calibrated as well as SAR-DAC mismatches. The latch offset calibration is performed both in foreground and in background. This is required to prevent $\Delta\Sigma$-subADC saturation.

2.8 Circuit Level Implementation

The static integrator (Fig. 2.12) driving the dynamic latch uses a complementary input transconductor and a folding stage to increase the output impedance.

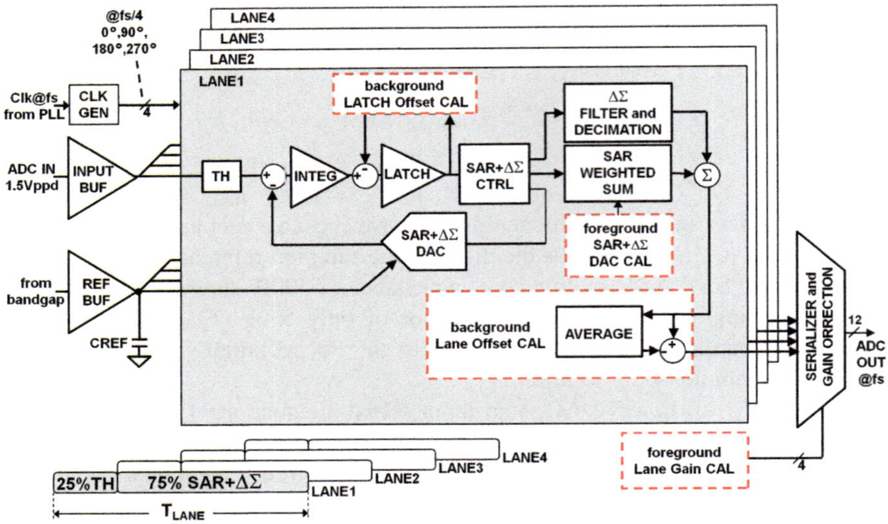

Fig. 2.11 Four-way interleaved SAR-ΔΣ ADC block diagram and calibrations

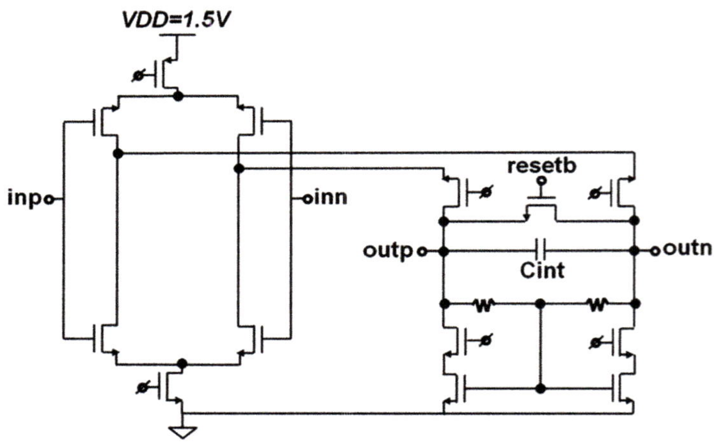

Fig. 2.12 Integrator with reset

The input unity gain signal buffer receives a 1.5 V_{ppd} differential signal and drives the four interleaved ADC lanes T/H stages. The 25% duty-cycle T/H sampling clock has been chosen such that always one lane is connected to the input signal buffer at a given time for best power efficiency. The T/H stage (see Fig. 2.10) consists of a couple of sampling capacitors ($C_{TH} = 300$ fF) in bottom plate sampling configuration. Due to charge-sharing operation, the C_{TH} are reset to a common mode voltage at the end of the conversion phase before being switched to the buffer output for tracking. The buffer consists of an OTA in differential inverting configuration

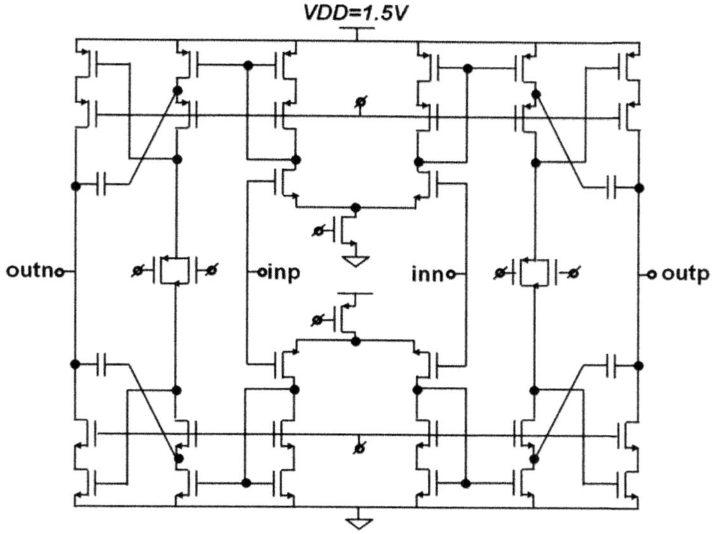

Fig. 2.13 Input buffer OTA

with a resistive ($R = 700\ \Omega$) feedback network. The OTA (Fig. 2.13) is a two-stage single-slope topology with 2 GHz unity gain bandwidth and a class AB cascode-compensated output stage.

2.9 Measurements

The presented ADC is fabricated in 28 nm CMOS process and consumes 17.5 mW from a dual 1.2 V/1.5 V supply. 1.5 V is used only for the reference generator and the integrator in order to improve the power supply rejection ratio. During operation, the power consumption of the digital calibration is less than 0.2 mW. The on-chip buffer, which drives the ADC full scale of 1.5 V_{ppd}, has an SNR of 65 dB at Nyquist and consumes 9 mW from 1.5 V. As the buffer cannot be bypassed, all the measurements include buffer power consumption, distortions, and noise. The total area including input and reference buffer as well as digital calibration is 0.076 mm^2, while the ADC core is 0.04 mm^2 (see Fig. 2.14).

With a 2 MHz input, the SNR and SNDR before calibration with $\Delta\Sigma$ off are 42 dB and 44 dB, respectively (Fig. 2.15a). After activating the fine $\Delta\Sigma$-subADC and performing calibration, the SNR and SNDR performance increases to 61.13 dB and 60.7 dB, respectively (Fig. 2.15b). The simulated breakdown of the different contributors to the final SNR is shown in Fig. 2.16.

The ADC is integrated in a power-line communication (PLC) SoC, and it is clocked by an on-chip PLL with 3 ps-rms jitter which cannot be bypassed. PLL

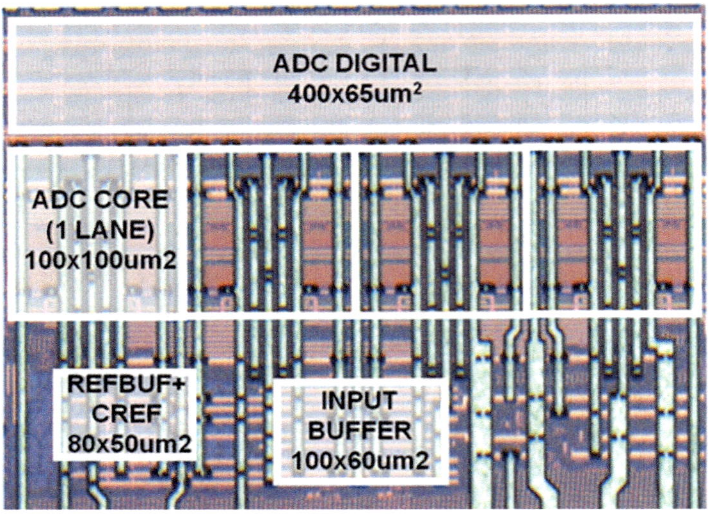

Fig. 2.14 Chip micrograph

jitter is measured between 300 kHz and 20 MHz frequency offset using an auxiliary clock output and a phase noise meter. Below 300 kHz, phase noise is considered as part of the input signal power, while above 20 MHz, the measurement is dominated by white noise of the auxiliary path buffer chain. This jitter figure is enough for the target application as the ADC input signal has a peak-to-average power ratio (PAPR) which is much larger compared to a sine wave. A spectrum with a 265 MHz input signal and PLL jitter included is shown in Fig. 2.17. The effect of the PLL jitter is visible in the skirts around the input tone. The SNR performance as a function of the signal amplitude for a 2 MHz and a 265 MHz input frequency (see Fig. 2.18) shows the typical jitter-induced roll-off of the SNR when the input signal approaches full scale at high frequency. We can further verify this jitter measurement result by evaluating the ADC performance as a function of the input signal frequency (Fig. 2.19). The SNDR roll-off with the input signal frequency matches very well with a 3 ps-rms sampling time jitter-induced SNR depicted in Fig. 2.19 as red curve. Given the good agreement between the jitter measurement techniques, that is, the phase noise measurement and the SNDR roll-off with input signal amplitude and frequency, we have extrapolated the ADC SNR performance from a small signal SNR measurement. The resulting SNDR performance has then been calculated by adding to the extrapolated SNR all the tones in the original spectrum. The results of this operation are depicted in Fig. 2.19 as dashed lines. After de-embedding the PLL jitter, the final SNDR performance at high frequency is 58 dB. The extrapolated SNDR degradation with the input signal frequency can be mainly ascribed to sampling time interleaving errors which are not calibrated in this design and to an increase in THD.

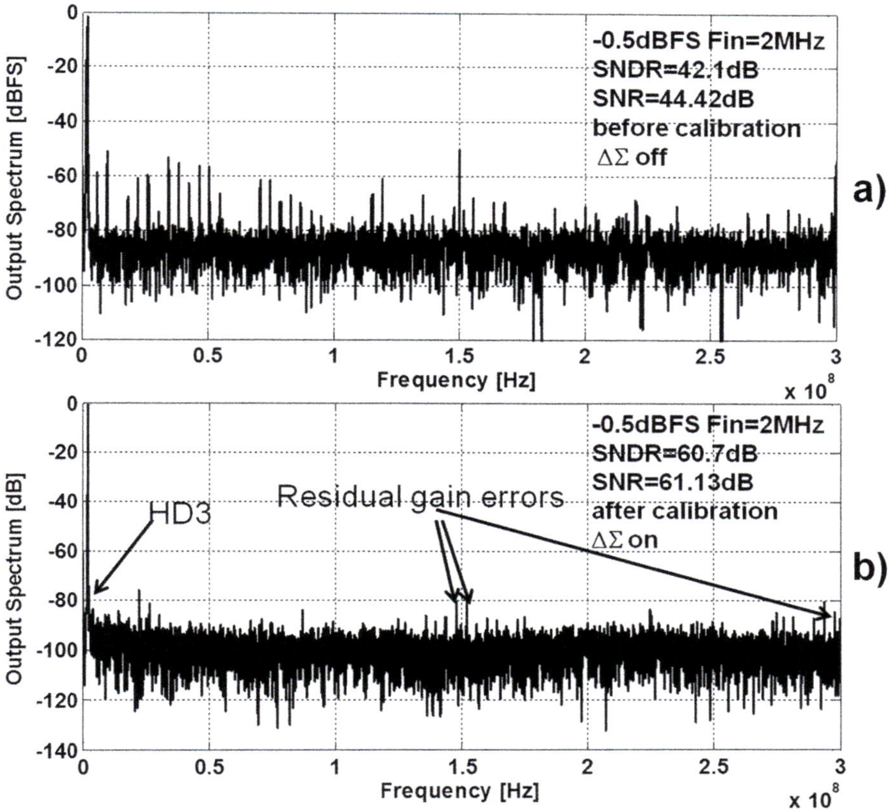

Fig. 2.15 ADC output spectrum at fin = 2 MHz. (**a**) Before calibration with $\Delta\Sigma$ off and (**b**) after calibration with $\Delta\Sigma$ on

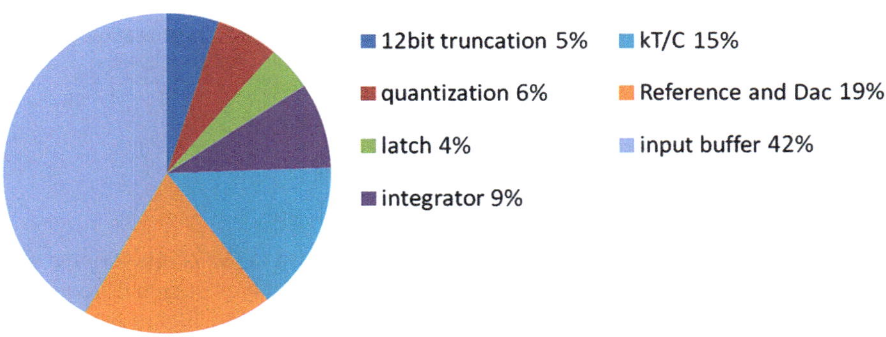

Fig. 2.16 Noise contributors of the SAR-$\Delta\Sigma$ ADC including input buffer at low frequency. SNR = 61.13 dB

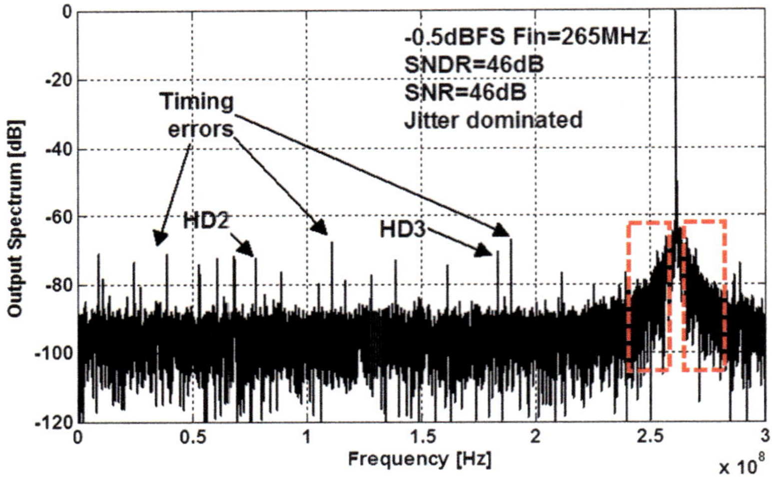

Fig. 2.17 ADC output spectrum at fin = 265 MHz (PLL jitter dominated)

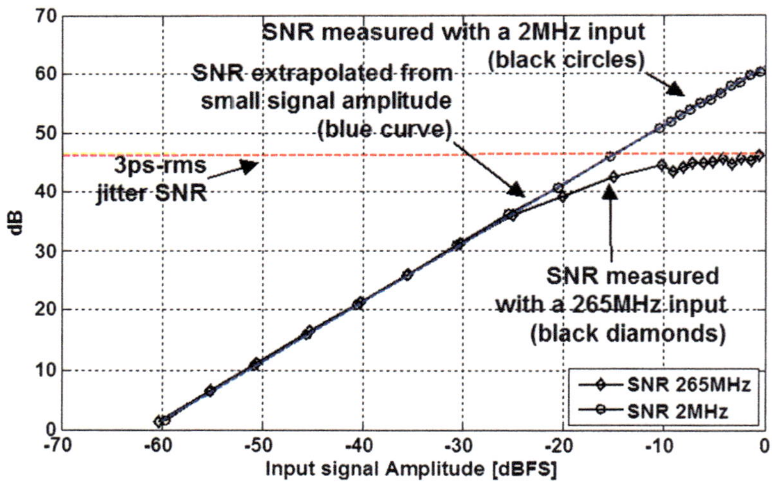

Fig. 2.18 ADC performance at 600MS/s vs. input amplitude at fin = 265 MHz

Figure 2.20 shows the Schreier FoM plot as a function of the Nyquist frequency for all converters published between 1997 and 2015 with Nyquist signal frequency and sampling speed larger than 10 MHz and SNDR at Nyquist higher than 57 dB [25]. The presented work delivers 58 dB of SNDR with 26.5 mW of total power including input signal buffer, reference generator, and biasing without requiring any additional BoM. The resulting HF Schreier FoM is 158.5 dB. Table in Fig. 2.21 shows a comparison between this design and other state-of-the-art ADCs with on-chip input buffers in the same speed and SNDR range. This work gives significant improvement in both FoMs and a drastic reduction in area.

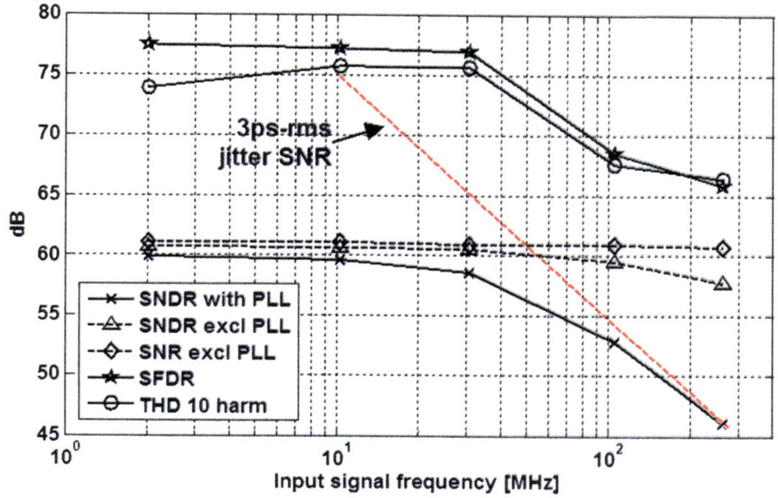

Fig. 2.19 600MS/s ADC performance vs. input frequency with and without PLL jitter

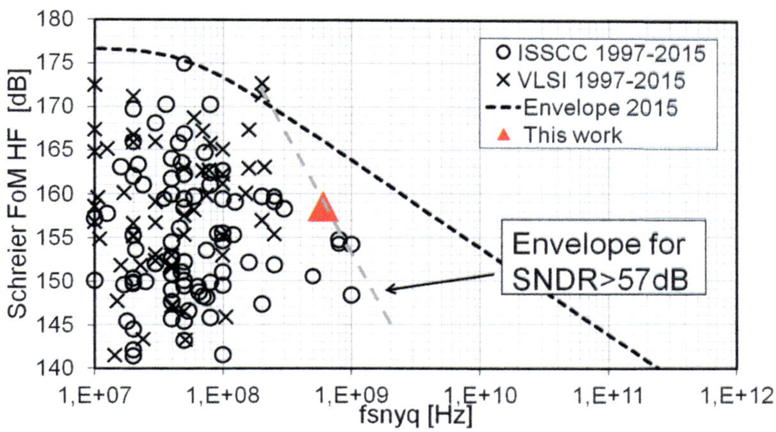

Fig. 2.20 Nyquist Schreier FoM vs. Nyquist sampling frequency for ADCs published at ISSCC and VLSI between 1997 and 2015 with SNDR at Nyquist larger than 57 dB [25]

2.10 Conclusions

In this paper we have shown how hybrid ADC techniques and segmented DAC architectures can be used to improve ADC power efficiency and area. The presented SAR-$\Delta\Sigma$ hybrid ADC architecture achieves better power efficiency compared to a conventional SAR, while the usage of the segmented charge-sharing charge-redistribution DAC reduces drastically the total area compared to a conventional charge-redistribution DAC. The 12b 600MS/s CMOS prototype ADC implementing

		[22] J. Mulder ISSCC 2015	[23] J. Mulder ISSCC 2011	[24] El Chammas ISSCC 2015	This work
Technology		28nm CMOS	40nm CMOS	180nm BiCMOS	28nm CMOS
Architecture		TI Pipeline	TI Pipeline	Pipeline	TI SAR-ΔΣ
Resolution	bit	13	12	14	12
Fs	MS/s	800	800	500	600
Supply	V	1-1.8	1-2.5	1.8-3.3	1.2-1.5
Max input swing	Vppdiff	NA	NA	1.25	1.5
SNDR LF	dB	59.19	59	65	60.7
SNDR HF	dB	57	59	64	58
Power	mW	76.4	105	550	26.5
Walden FoM HF	fJ/c.s.	162.4	180.2	849.3	68.0
Schreier FoM HF	dB	154.3	154.8	150.6	158.5
Area tot	mm^2	0.23	0.88	2.5	0.076

Fig. 2.21 Performance summary and comparison with state-of-the-art ADCs (SNDR HF > 57 dB, Fs > 300MS/s) including on-chip buffering

these techniques delivers 58 dB SNDR up to Nyquist for only 26.5 mW of total power including input signal buffer, reference generator, biasing, and digital calibration. The total area of this design in 28 nm CMOS is only 0.076 mm^2, and it does not require any external component.

References

1. Janssen, E., et al.: A direct sampling multi-channel receiver for DOCSIS 3.0 in 65nm CMOS. In: VLSI Circuits (VLSIC), 2011 Symposium on, Honolulu, pp. 292–293 (2011)
2. Wu, J., et al.: 27.6 A 4GS/s 13b pipelined ADC with capacitor and amplifier sharing in 16nm CMOS. In: 2016 IEEE International Solid-State Circuits Conference (ISSCC), San Francisco, pp. 466–467 (2016)
3. Oksman, V., Galli, S.: G.hn: the new ITU-T home networking standard. IEEE Commun. Mag. **47**(10), 138–145 (2009)
4. Morie, T., et al.: A 71dB-SNDR 50MS/s 4.2mW CMOS SAR ADC by SNR enhancement techniques utilizing noise. Solid-State Circuits Conference Digest of Technical Papers (ISSCC), 2013 IEEE International, San Francisco, pp. 272–273 (2013)
5. Fredenburg, J., Flynn, M.: A 90MS/s 11MHz Bandwidth 62dB SNDR Noise-Shaping SAR ADC. ISSCC Dig. Tech. Papers, pp. 468–469, Feb. (2012)
6. Liu, C.C.: 27.4 A 0.35mW 12b 100MS/s SAR-assisted digital slope ADC in 28nm CMOS. 2016 IEEE International Solid-State Circuits Conference (ISSCC), San Francisco, pp. 462–463 (2016)

7. Shu, Y.S., et al.: 27.2 an oversampling SAR ADC with DAC mismatch error shaping achieving 105dB SFDR and 101dB SNDR over 1kHz BW in 55nm CMOS. 2016 IEEE International Solid-State Circuits Conference (ISSCC), San Francisco, pp. 458–459 (2016)
8. van der Goes, F. et al.: A 1.5 mW 68 dB SNDR 80 Ms/s 2 × Interleaved Pipelined SAR ADC in 28 nm CMOS. IEEE J. Solid-State Circuits **49**(12), 2835–2845, Dec. (2014)
9. Verbruggen, B., et al.: A 70 dB SNDR 200 MS/s 2.3 mW dynamic pipelined SAR ADC in 28 nm digital CMOS. VLSI circuits digest of technical papers, 2014 Symposium on, Honolulu, pp. 1–2 (2014)
10. Doris, K., et al.: A 480mW 2.6GS/s 10b 65nm CMOS time-interleaved ADC with 48.5dB SNDR up to Nyquist. Solid-state circuits conference digest of technical papers (ISSCC), 2011 IEEE International, San Francisco, pp. 180–182 (2011)
11. Craninckx, J., van der Plas, G.: A 65fJ/Conversion-Step 0-to-50MS/s 0-to-0.7mW 9b Charge-Sharing SAR ADC in 90nm Digital CMOS. Solid-state circuits conference, 2007. ISSCC 2007. Digest of technical papers. IEEE International, San Francisco, pp. 246–600 (2007)
12. Liu, W., et al.: A 600MS/s 30mW 0.13μm CMOS ADC Array Achieving over 60dB SFDR with Adaptive Digital Equalization. IEEE International Solid-State Circuits Conference, Digest of Technical Papers, pp. 82–83 Feb. (2009)
13. Quiquempoix, V., et al.: A low-power 22-bit incremental ADC. IEEE J. Solid State Circuits. **41**(7), 1562–1571 (July 2006)
14. Markus, J., et al.: Theory and applications of incremental $\Delta\Sigma$ converters. In: IEEE Transactions on Circuits and Systems I: Regular Papers, vol. 51, no. 4, pp. 678–690, April 2004.
15. Robert, J., et al.: A 16-bit low-voltage CMOS A/D converter. IEEE J. Solid State Circuits. **22**(2), 157–163 (Apr 1987)
16. van Elzakker, M., et al.: A 10-bit charge-redistribution ADC consuming 1.9 μW at 1 MS/s. IEEE J. Solid State Circuits. **45**(5), 1007–1015 (May 2010)
17. Harpe, P., et al.: A 7-to-10b 0-to-4MS/s flexible SAR ADC with 6.5-to-16fJ/conversion-step. 2012 IEEE international solid-state circuits conference, San Francisco, pp. 472–474 (2012)
18. Chen, S.W.M., et al.: A 6-bit 600-MS/s 5.3-mW asynchronous ADC in 0.13-μm CMOS. IEEE J. Solid State Circuits. **41**(12), 2669–2680 (Dec. 2006)
19. Ginsburg, B.P., Chandrakasan, A.P.: An energy-efficient charge recycling approach for a SAR converter with capacitive DAC. IEEE International Symposium on, Circuits and Systems, 2005. ISCAS 2005, vol. 1, pp. 184–187 (2005)
20. Hariprasath, V., et al.: Merged capacitor switching based SAR ADC with highest switching energy-efficiency. Electron. Lett. **46**(9), 620–621, April 29 (2010)
21. Liu, C.C., et al.: A 10-bit 50-MS/s SAR ADC with a monotonic capacitor switching procedure. IEEE J. Solid State Circuits. **45**(4), 731–740 (April 2010)
22. Mulder, J., et al.: An 800MS/S 10b/13b receiver for 10GBASE-T Ethernet in 28nm CMOS. ISSCC Digest of Technical Papers, pp. 462–463, Feb. 2015
23. Mulder, J., et al.: An 800MS/s dual-residue pipeline ADC in 40nm CMOS. ISSCC Digest of Technical Papers, pp. 184–185, Feb. 2011
24. El-Chammas, M., et al.: A 90dB-SFDR 14b 500MS/S BiCMOS switched- current pipelined ADC. ISSCC Digest of Technical Papers, pp. 286–287, Feb. 2015
25. Murmann, B.: ADC performance survey 1997–2015. [Online]. Available: http://web.stanford.edu/~murmann/adcsurvey.html

Chapter 3
Interleaved Pipelined SAR ADCs: Combined Power for Efficient Accurate High-Speed Conversion

Ewout Martens

3.1 Introduction

High-speed data communication requires analog-to-digital converters (ADCs) with high conversion speeds and a medium to high number of bits. The higher the accuracy and/or sampling speed, the more the analog part of the receiver chain can be simplified [1]. This results in digital-intensive receivers which offer more flexibility and more gain of technology scaling and consume less power. The main challenge then becomes to realize high accuracy with high sampling rate without excessively increasing the power budget, especially in mobile applications. Different ADC architectures usually focus on one of these aspects of the design space. Hence, hybrid structures which combine the best aspects of the different topologies have gained lots of interest in recent years.

Figure 3.1 shows the ENOB versus bandwidth that has been demonstrated in the last 20 years for some ADC architectures [2]. To compare the efficiency, the Walden figure of merit is also shown which is defined as follows [3]:

$$FoM_{\text{Walden}} = \frac{\text{Power}}{2^{\text{ENOB}+1}\text{Bandwidth}}.$$

Flash converters offer the highest conversion speeds [4], but they require a lot of overhead in terms of both area and power and suffer from high input loads. The number of comparators increases exponentially with the required number of bits, and techniques like interpolating [5], folding [6], and asynchronous binary search algorithms [7] are used to reduce the required number of comparisons. However,

E. Martens (✉)
imec, Kapeldreef 75, Leuven, Belgium
e-mail: Ewout.Martens@imec.be

P. Harpe et al. (eds.), *Hybrid ADCs, Smart Sensors for the IoT, and Sub-1V & Advanced Node Analog Circuit Design*, DOI 10.1007/978-3-319-61285-0_3

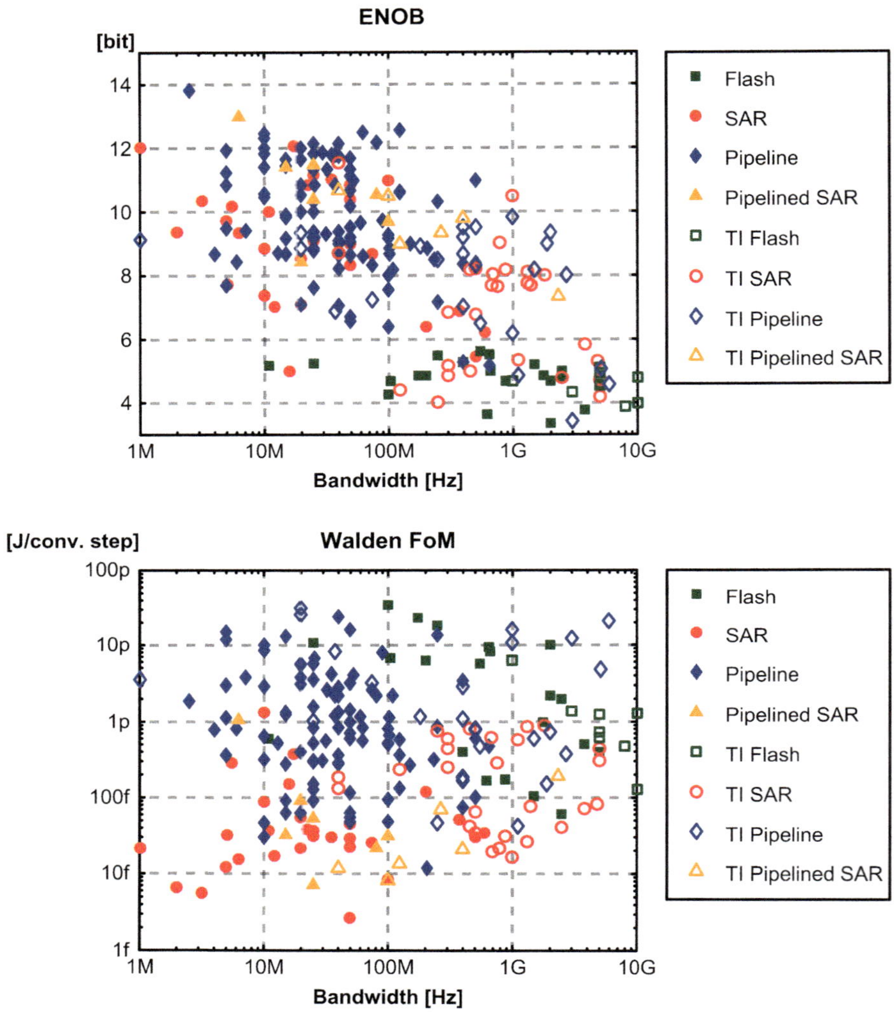

Fig. 3.1 Flash, SAR and pipelined ADCs, and combinations of them published at ISSCC and VLSI between 1997 and 2016 [2]

for a high number of bits, less noise can be tolerated in the comparators resulting in a rapid increase of the energy per comparator and limiting the practical number of bits.

Analog-to-digital architectures based on successive approximation (SAR) [8] are the most energy-efficient choice, since they require only a small number of operations, and all of them can be made fully dynamic avoiding any static current. For N-bit accuracy, only N comparisons are required, and N-1 residues need to be generated, for example, by switching a capacitive digital-to-analog converter (DAC). No amplification is needed. However, since all operations happen subsequently, only moderate conversion speeds can be achieved. Multiple bits per cycle can be resolved to reduce the required number of cycles [9], but they tend to become very complicated. Also, without any amplification, the noise of the comparators becomes a bottleneck to achieve a high number of bits with basic SAR architectures.

Pipelined ADCs [10] are a popular choice since they cover a wide area of conversion accuracies and speeds. They can achieve more than 12 ENOB and sampling frequencies up to 1 GHz [11]. The high speed is achieved by splitting up the conversion into different stages which do only a low-bit flash conversion, generate a residue, and amplify it toward the next stage. Thanks to the amplification and redundancy between the stages, the noise of the comparators is usually not critical, and higher accuracies can be achieved. However, the presence of extra amplification operations increases the power consumption which results in worse figures-of-merit.

In a pipelined SAR architecture [12], an efficiency comparable to SAR converters is obtained by using multi-bit SAR converters in the pipelined stages. The pipelined processing allows to realize higher conversion speeds and resolutions than for SAR ADCs. The use of multi-bit stages reduces the number of required amplifications compared to conventional pipelined ADCs. The amplifiers between the stages avoid the need for ultralow-noise comparators to achieve a higher number of bits without excessive power consumption in the comparators. To increase the bandwidth further, parallel processing in the form of time interleaving [13] is adopted resulting in an interleaved pipelined SAR ADC [14].

In the remainder of this chapter, the architectural choices of interleaved pipelined SAR ADCs are first discussed. Then, the design challenges of the main building blocks are elaborated, and how they influence the ADC performance. The last part of the chapter shows some practical implementations and their measurement results.

3.2 Architecture

Figure 3.2 shows a generic diagram of an interleaved pipelined SAR ADC. The architectural choices include the number of interleaved channels, the number of pipelined stages per channel, and the number of bits resolved by the SAR converters in each stage.

3.2.1 *Number of Stages*

Each stage of the pipeline should have enough time to receive the signal from the previous stage, do N comparisons (T_{cmp}), generate N residues (T_{res}), and amplify

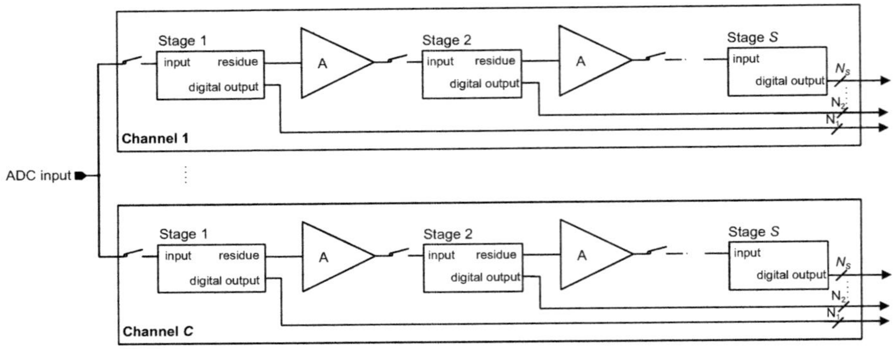

Fig. 3.2 Generic architecture of the interleaved pipelined SAR ADC

the last residue to the next stage (T_{amp}). These requirements result in the following timing constraints for the first, intermediate, and last stages:

$$T_{\mathrm{track}} + \sum_{i=1}^{N_1} \left(T_{\mathrm{cmp},1,i} + T_{\mathrm{res},1,i}\right) + T_{\mathrm{amp}} + T_{\mathrm{reset}} < CT_{\mathrm{sample}},$$

$$\sum_{i=1}^{N_k} \left(T_{\mathrm{cmp},k,i} + T_{\mathrm{res},k,i}\right) + 2T_{\mathrm{amp}} + T_{\mathrm{reset}} < CT_{\mathrm{sample}},$$

$$\sum_{i=1}^{N_S-1} \left(T_{\mathrm{cmp},S,i} + T_{\mathrm{res},S,i}\right) + T_{\mathrm{cmp},S,N_S} + T_{\mathrm{amp}} + T_{\mathrm{reset}} < CT_{\mathrm{sample}}.$$

with the track time T_{track} usually between $T_{\mathrm{sample}}/2$ and T_{sample}. Based on these constraints, it becomes clear that a pipeline of only a small number of stages is usually preferred. Indeed, the available time in the intermediate stages for comparisons and residue generation is rather limited since twice the amplification time needs to be allocated in these stages. Reducing the amplification time T_{amp} or increasing the interleaving factor C leaves more time for the SAR conversions. That gives also extra room to do extra conversions in the first and last stages. Hence, for moderate quantization levels, two stages are usually sufficient. On the other hand, multiple stages cannot be avoided to achieve a high resolution. Also, when a slightly higher channel speed is desired, reducing one or two conversions from the first and last stages and making an extra intermediate stage become more efficient than increasing the number of channels.

An example of an architecture and its timing diagram is shown in Fig. 3.3. With only two stages, a quantization level of 14 bits is achieved for channels which are designed for 200 MHz. To increase the robustness of the ADC, redundancy between the different stages is used resulting in 6 bits to be resolved in the first stage and 10 bits in the second one. Further, the last two comparisons are low-noise comparisons with a 1-bit redundancy to correct for settling and comparator errors.

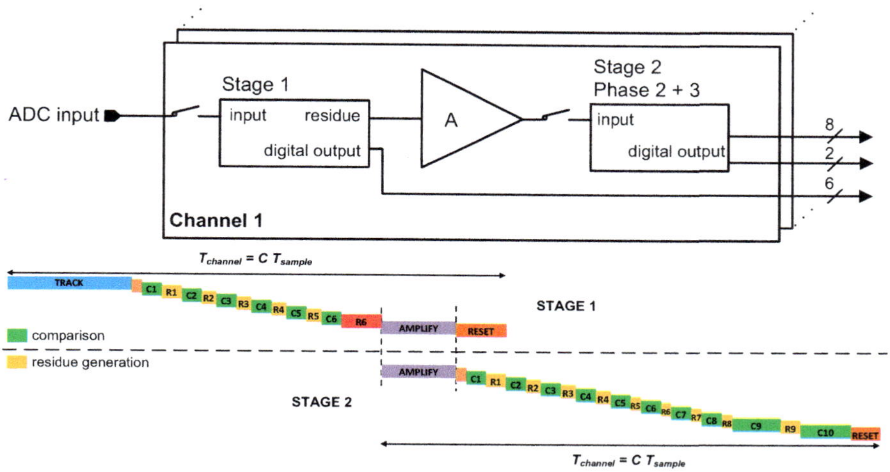

Fig. 3.3 Architecture and timing diagram of interleaved pipelined SAR with two stages and 14-bit quantization level targeting 200 MHz per channel

3.2.2 Number of Channels

Doubling the number of interleaved channels also doubles the available time per stage, so a higher number of lanes seem to be attractive. However, mismatches between the channels degrade the signal-to-noise ratio [15–17]:

- Channel offset mismatches result in spurs at multiples of f_s/C. For $C > 2$, it ends up within the Nyquist band.
- Channel gain mismatches act as amplitude modulation and cause two images of the signal to appear around multiples of f_s/C.
- Clock skew results in sample-time errors and hence phase modulation. It causes similar images as gain mismatches, but their magnitude becomes frequency dependent.
- Bandwidth mismatches between the channels also cause phase modulation and hence images around multiples of f_s/C.

Offset errors are readily corrected in digital domain [20]. The pipelined SAR ADC allows to easily tune the gain of a channel via tunable capacitors to ground in the first stage resulting in a fine analog compensation of the gain mismatch errors [19]. Skew mismatches can be reduced by using a full-rate front-end sampler which is easily combined with the top-plate sampling of the first stage [18]. Finally, bandwidth mismatches can be tuned via tunable boost voltages in the sampling switch [21].

Table 3.1 List of error sources in interleaved pipelined SAR ADCs

Block	Noise source	Distortion source
Sampler	Sampling (kT/C) noise Clock jitter	Nonlinear distortion Clock skew Bandwidth mismatch
DAC	DAC settling Drivers noise	MSB mismatch DAC settling Reference distortion
Comparator	Comparator noise Comparator metastability	Comparator offset
Residue amplifier	Amplifier noise	Amplifier offset Amplifier distortion
ADC	Quantization noise	Channel offset Channel gain mismatch

3.2.3 Error Sources

The total SNDR of an interleaved pipelined SAR ADC is degraded by several error sources listed in Table 3.1. Gain and offset errors are easily calibrated out using either foreground and/or background calibration techniques [22, 23]. Mismatch in the DAC is caused by capacitor mismatch and can also be calibrated for using a one-time effort offline calibration. Other noise sources end up in the overall noise budget.

A typical example of a noise budget is shown in Fig. 3.4 which corresponds to a 12-bit ENOB pipelined SAR ADC with 200 MHz channels as shown in Fig. 3.3. There are four major noise sources which are given an equal weight: the sampling noise, the amplifier noise, the comparator noise, and the rest including quantization noise and jitter. Consequently, the power budget usually follows a similar pattern. Next section discusses the contributions of the different blocks into more detail.

3.3 Main Building Blocks

This section focuses on the four main building blocks of the interleaved pipelined SAR ADC: the sampling stage, the DAC, the comparator, and the residue amplifier. Although it also has a major impact on the performance of high-accuracy high-speed designs, low-jitter clock buffering and distribution are not discussed here.

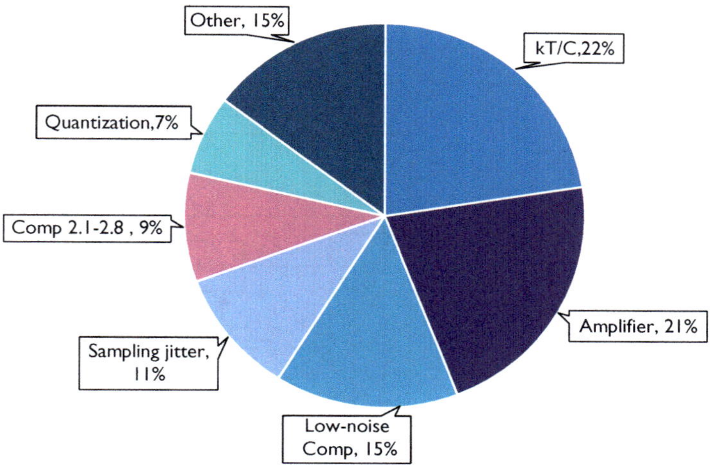

Fig. 3.4 Example of noise budget split-up over different noise sources in interleaved pipelined SAR ADC of Fig. 3.3

3.3.1 Sampling Stage

High-speed, high-accuracy sampling requires low values for the resistance in the sampling network. This poses strict requirements on the output impedance of the input buffer and the switch resistance. When the tracking phase starts, the voltage on the first-stage sampling capacitance C_{dac} should settle to the value of the input signal. Assuming a maximum error of LSB/α is tolerable, the required settling time is then

$$T_{\text{settling}} = \frac{1}{\beta f_{\text{sample}}} > RC_{\text{dac}} \left(N \log 2 + \log \alpha\right),$$

with R the total resistance of the sampling network and N the number of bits, and $1/\beta$ indicates the tracking time fraction of the sampling period. This means that the bandwidth of the sampling network scales as follows:

$$\frac{BW}{f_{\text{sample}}} > \frac{\beta}{2\pi} \left(N \log 2 + \log \alpha\right).$$

Further, assuming the kT/C noise should be smaller than LSB/γ, the maximum resistance can be computed as follows:

$$R < \frac{V_{\text{ptp}}^2}{\beta f_{\text{sample}} kT 2^{2N} \gamma^2 \left(N \log 2 + \log \alpha\right)},$$

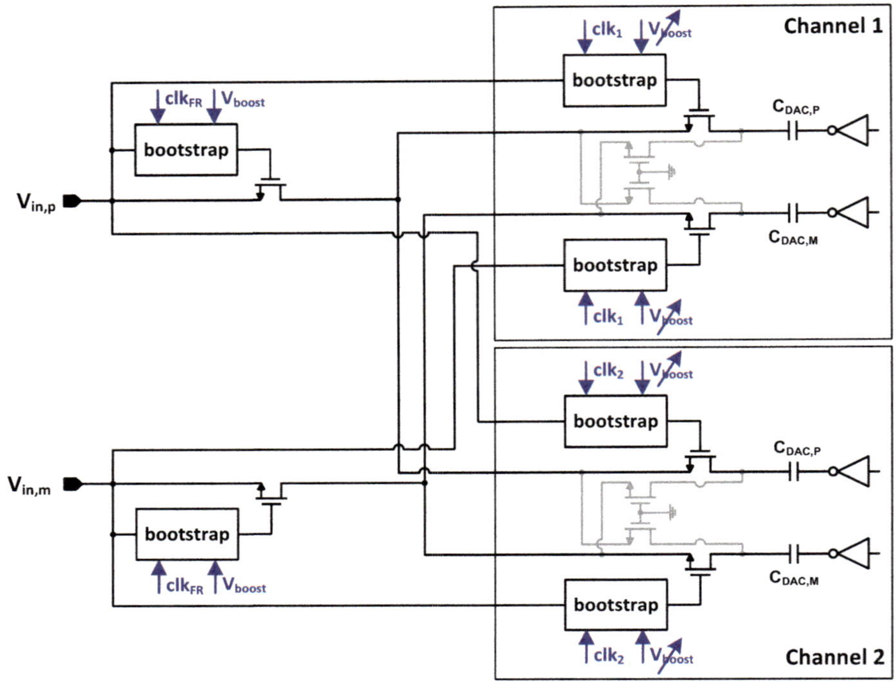

Fig. 3.5 Example of sampling network for a two-time interleaved pipelined SAR

where the input range V_{ptp} is proportional to the supply voltage. Hence, with smaller technology nodes and lower supply voltages, the requirement on the resistance continuously becomes tougher.

High accuracy also means low distortion. A smaller switch resistance reduces the signal dependency of the resistance. Since most pipelined SAR ADCs use a capacitive DAC and switch the bottom plates of the first-stage DAC, they cannot easily use bottom-plate sampling which makes them more vulnerable to non-idealities such as charge injection and clock feedthrough.

To meet these requirements, careful design of a sampling network with bootstrapped switches is needed [24]. Figure 3.5 shows an example of the sampling network of a high-speed high-accuracy two-time interleaved pipelined SAR ADC [21]. To minimize the skewing errors between the channels, a front-end sampler operating at the full clock rate clk_{FR} is added. To increase the linearity, the bootstrap circuits use the original input rather than the signal at the source of the sampling switch inside the channels. The bootstrap voltage used inside the channels can be tuned to calibrate mismatches between the bandwidths of the sampling networks of the channels.

Since a small switch resistance is required, the sampling transistors become quite large, and they consequently have large parasitic capacitances. To avoid coupling

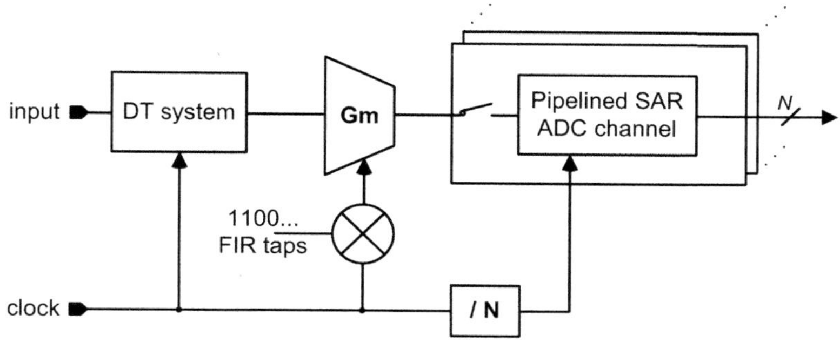

Fig. 3.6 Interleaved pipelined SAR ADC embedded in a discrete-time system with integrated filtering function

over the switches when the SAR algorithm is executed, dummy switches are added to cancel the coupling in the first order and avoid cross talk from one channel to another.

When the ADC is part of a discrete-time system, it can easily be driven by an amplifier which acts as a Gm stage that puts some charge on the DAC of the first stage, in the same way as the residue amplifier between the stages operates. This configuration offers an easy way to integrate a filtering function with the interleaved pipelined SAR ADC [25]. As shown in Fig. 3.6, a FIR function is realized when a high-rate clock adds the charge of some samples onto the DAC capacitor after which the ADC starts the conversion.

3.3.2 DAC

Most pipelined SAR ADCs use capacitive DACs to generate the residue. The total capacitance is dictated by the noise and mismatch requirements. In Fig. 3.7, a comparison is made for a 6-bit DAC in 28 nm technology between the kT/C noise and the accuracy due to mismatch (based on Monte Carlo simulations). A differential input range of 2 V_{DD} is assumed. Mismatch dominates the SNDR degradation, and for more than 10-bit accuracy, one or more MSB units of the DAC can be calibrated in an offline calibration step. The improvement is also shown in Fig. 3.7 for a different number of MSB units that are calibrated with a resolution of 200aF. A common-centroid layout is adopted to improve the matching of the DAC.

For high-accuracy designs, the switching energy of the DAC significantly contributes to the total conversion energy (corresponding to the portion of the kT/C noise in Fig. 3.4). With a different switching scheme, some energy can be saved. Several switching schemes have been developed, and four of them are compared in Fig. 3.8. These schemes don't need extra voltages and have a simple logic. The total DAC size is always the same, and the DACs are reset after the conversion.

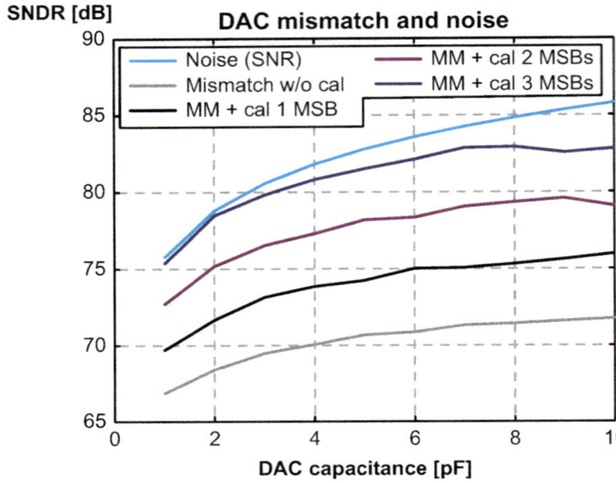

Fig. 3.7 Comparison between noise and mismatch limitation for 6-bit capacitive DAC in the first stage of a pipelined SAR ADC with ideal second stage

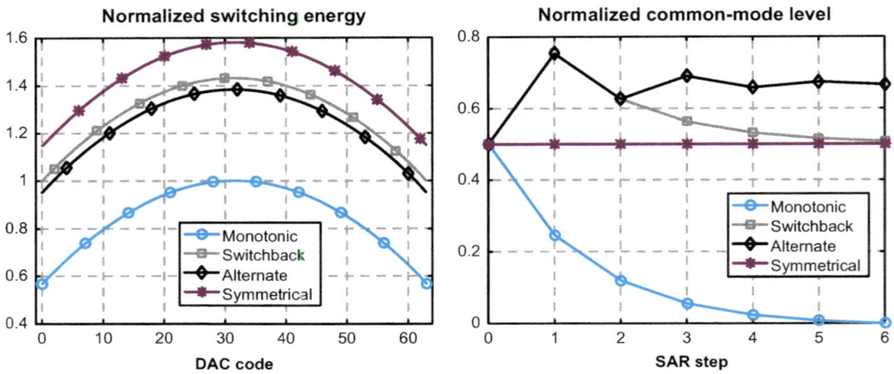

Fig. 3.8 Switching energy and common-mode level of 6-bit DAC used in the first stage of a pipelined SAR ADC with different switching schemes

In the conventional symmetrical scheme, both P and N sides are switched up or down in each step ensuring a constant common mode as shown in the right part of Fig. 3.8. The monotonic scheme [26] switches only down, either the P or N side depending on the comparator decision. It consumes the least amount of energy, but the common-mode level changes drastically making it more difficult for the comparators to operate accurately since the offset, noise, and speed of the comparator change with the common-mode input level. The switchback scheme [27] first switches up, either the P or N side, and then switches down to reduce the variation of the common mode at the cost of more switching energy. The last scheme [28] alternates between switching up and down with the best compromise between energy and common-mode variations.

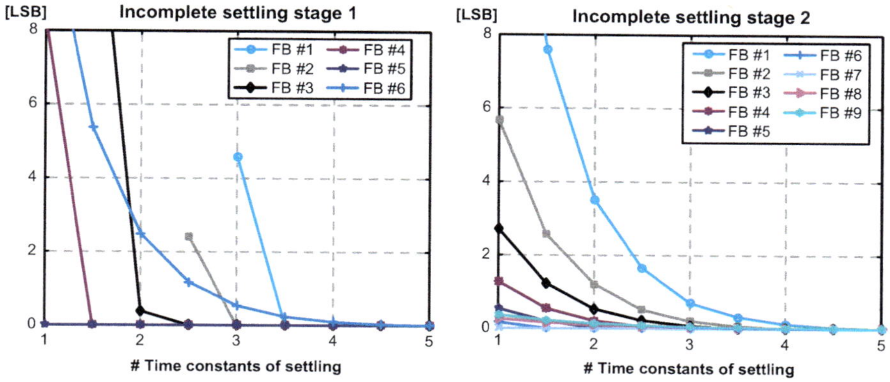

Fig. 3.9 Input-referred noise of ADC due to finite DAC settling in interleaved pipelined SAR ADC of Fig. 3.3

The bottom-plate drivers of the DAC can be simple inverters. Each time the DAC switches, it draws some signal-dependent charge from the reference, which can be just the supply voltage in a pipelined SAR ADC. This ripple on the reference voltage introduces nonlinearities. To compensate for this effect at high accuracy and high speed, a large decoupling capacitance can be added [21]. A more area-efficient approach is to use a reference buffer, which however needs a large bandwidth and hence becomes quite power hungry [29].

The finite settling time of the DAC results in errors in the top-plate voltage which results in wrong decisions by the comparators. In the pipelined SAR architecture, the redundancy between the stages makes these errors less critical. Although the effect is deterministic, the subsequent DAC feedbacks and the redundancy in the ADC make the finite settling time errors act more as random noise rather than as distortion, somewhat similar to the conversion of quantization errors to quantization noise. Figure 3.9 shows the input-referred noise created by finite DAC settling for the first and second stage of a two-stage pipelined SAR ADC. The settling time before the final residue generation in front of the amplifier is most critical. Hence, extra settling time is usually foreseen as also shown in the timing diagram of Fig. 3.3 (R6 in stage 1 and R9 in stage 2).

Another error source introduced by the DAC is noise from the drivers. This noise gets filtered on the top plate of the DAC:

$$V_{n,\text{top}}^2 = 4kT \sum_i \left| \frac{1}{1 + (R_i + j\omega C_i) \sum_{k \neq i} 1/(R_k + j\omega C_k)} \right|^2 \rho_i,$$

with C_i the units of the DAC, R_i the output resistance of the driver, and ρ_i its equivalent noise resistance. When the stage is followed by a residue amplifier like the one of Fig. 3.12, the noise is further filtered by a sinc characteristic because of the integrating functionality of this dynamic amplifier.

3.3.3 Comparators

In the comparators, a trade-off between noise, comparison time, and power consumption should be made. To avoid static power consumption, a dynamic comparator [30] is often used. When the noise of the input pair is dominant, the input-referred noise is in first order given by

$$V_{n,\mathrm{in}} \sim \sqrt{\frac{kT}{C_D \frac{g_m}{I_{DS}} V_{\mathrm{tl}}}},$$

with C_D the capacitance on the drain terminals and V_{tl} the threshold of the latch [31]. Reducing the noise by half then consumes four times more comparison energy. Further, the decision time consists of a part needed to build up V_{tl} on C_D ($C_D V_{\mathrm{tl}}/I_{DS}$) and a part due to the regeneration time of the latch:

$$t_{\mathrm{comp}} = t_{1\mathrm{mV}} - \tau \log\left(\frac{V_{\mathrm{in}}}{1\ \mathrm{mV}}\right),$$

with V_{in} the input signal and $t_{1\mathrm{mV}}$ the delay with a 1 mV input signal.

The pipelined SAR architecture helps in relaxing the specifications for noise and time so that energy can be saved. First, there is redundancy between the different stages so that errors made in the front section of the pipeline become less important. Second, the gain of the residue amplifiers relaxes the noise requirements of the comparators near the end. Also, the larger the input signals of the comparator, the lower the decision time.

For example, Fig. 3.10 shows the input-referred noise of the different comparators in the two-stage pipelined SAR ADC of Fig. 3.10. For each comparator, a Gaussian noise model is assumed that allows to compute the variation of the output value of the ADC which can be translated to the input to find the input-referred noise. As expected, the comparators of the first stage (left) can be fast and noisy without too much overall SNDR degradation. The input-referred noise of the comparators of the second stage is suppressed by the gain of the amplifier for both the high-noise (middle) and low-noise comparators (right). These low-noise comparators only resolve the last bits of the 14-bit quantization level so that the maximum error settles to about 1 LSB.

With asynchronous operation of the SAR algorithm [32], each comparator gets as much time as it needs, depending on the input signal it sees. However, with an asynchronous operation, there is always a certain probability that there is not enough time available for the complete conversion. Based on the delay model described above, this probability has been computed in Fig. 3.11 for various amplifier gains and comparator delays (with a 1 mV input signal). The gain of the amplifier relaxes the requirement on the delay of the comparator enabling higher-speed operation.

Each channel of the interleaved pipelined SAR can be driven completely asynchronously: the comparators and the amplifier give a ready signal which starts the next part of the conversion. However, due to finite probability on time-outs, a

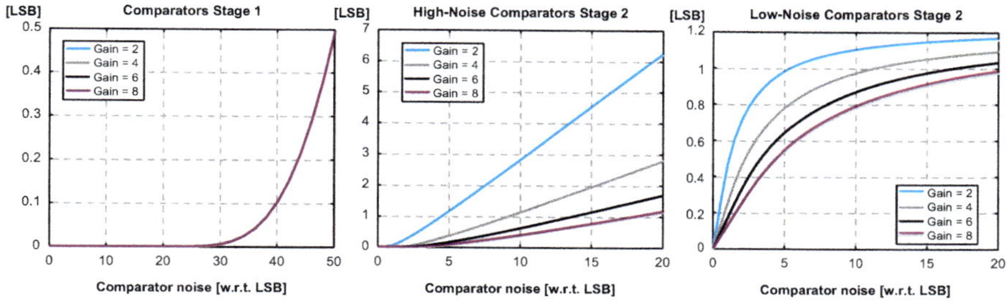

Fig. 3.10 Input-referred noise of comparators in different stages corresponding to the timing diagram of Fig. 3.3

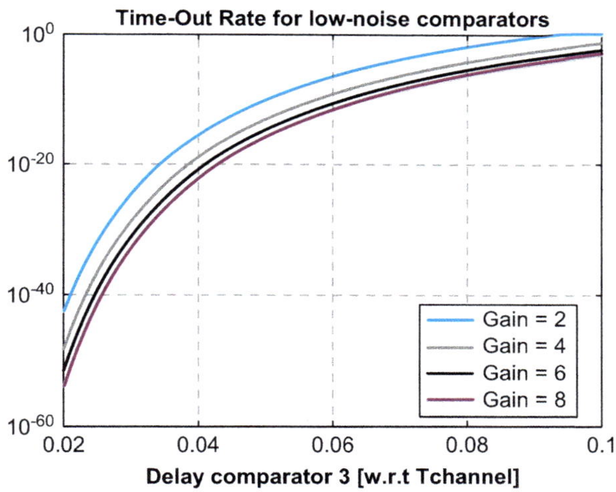

Fig. 3.11 Example of computation of probability on time-out in low-noise comparators of Fig. 3.3 for different nominal comparator delays and gain factors

stage S in the pipeline cannot be sure that stage $S + 1$ has finished and that the amplification between the two stages can start. To resolve this issue, stage S needs to set an interrupt signal to stage $S + 1$.

The offset of a comparator depends on its common-mode input signal which increases the DNL of the ADC when the common-mode voltage changes during the SAR algorithm when one of the nonsymmetrical switching schemes of Fig. 3.8 is used. Therefore, a different comparator for each decision can be used, and the offset is calibrated out at the correct common-mode level. This scheme leads to a comparator-driven controller [7]: the valid signal of the comparator can directly be used as clock for the next one. Furthermore, with one of the switching schemes of Fig. 3.8, the output of the comparators can directly drive the bottom plate of the corresponding unit in the DAC. Also, all comparators can be reset in parallel ensuring complete reset for all of them independent of the time needed for the DAC to settle.

3.3.4 Residue Amplifier

Energy-efficient amplification is a key element to realize a low-power high-accuracy, high-speed interleaved pipelined SAR ADC. Various kinds of amplifiers have been proposed for pipelined ADCs besides conventional opamp-based closed-loop amplifiers, for example, open-loop amplifiers [33], bucket-brigade devices [34], ring amplifiers [35], charge-steering opamps [36], and zero-crossing-based circuits [37]. Some of them have also been used in pipelined SAR ADCs [38].

Due to the SAR algorithm in a pipelined SAR architecture, the residue that needs to be amplified is small compared to the full-scale range and equal to $V_{\text{ptp}}/2^{N_1}$. As discussed above, a trade-off needs to be made between the amplifier and the comparators. The gain should be large enough to relax the specifications for the comparators in the subsequent phases, but a gain factor much smaller than 2^{N_1} is usually sufficient to end up with a comparator noise requirement of a few 100 μV. With a smaller gain, the outputs of the amplifier are smaller making it in general easier to realize a more linear amplifier. A smaller gain also means that less charge should be put on the DAC of the next stage saving energy and amplification time.

Various types of dynamic amplifiers have been developed [39]. Basically, the DAC capacitors at the output are first reset to a fixed voltage, and then current is put on the capacitors via a transconductor in the form of a basic differential pair. Assuming a certain g_m/I_d biasing for this differential pair, the gain then becomes

$$A = \frac{g_m T_{\text{amp}}}{C_{\text{dac}}},$$

where T_{amp} is the amplification time. This value can be determined intrinsically by observing the common-mode level of the outputs. More flexibility is offered with a tunable delay which allows to set the gain of the amplifier.

The input-referred noise of the amplifier is directly translated to the input of the ADC. In first order, it is given by

$$V_{n,\text{in}} \sim \sqrt{\frac{4kT\gamma}{g_m T_{\text{amp}}}}.$$

For a certain g_m/I_d biasing, the required energy per amplification then becomes

$$E_{\text{amp}} \sim \frac{2V_{DD}}{g_m/I_D} g_m T_{\text{amp}} = \frac{2V_{DD}}{g_m/I_D} A\, C_{\text{dac}} \sim \frac{2V_{DD}}{g_m/I_D} \frac{4kT\gamma}{V_{n,\text{in}}^2},$$

showing the trade-off between power consumption, amplification time, gain, and noise.

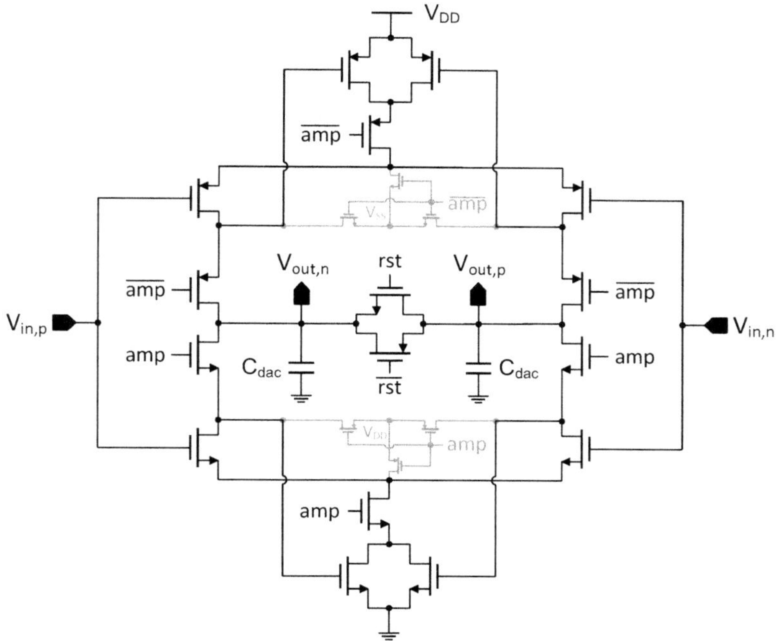

Fig. 3.12 Complementary dynamic amplifier [21]

The complementary dynamic amplifier shown in Fig. 3.12 improves the efficiency via current and charge reuse. After finishing the SAR algorithm and resetting the DAC of the next stage, the previous values are back at the top plates. The first switches of the amplifier then use charge sharing to restore the common-mode level before the next amplification starts. A local common-mode feedback loop provides some common-mode rejection. When the amplifier is not active, it loads the DACs of the previous and next stages. To avoid large nonlinear capacitance, the terminals of the input transistors are reset, and the gates of the transistors for the CM loop are connected to an internal node rather than directly to the outputs.

The SAR algorithm reduces the input of the amplifier which improves its linearity. Hence, the linearity of the amplifier is relaxed. This is shown in Fig. 3.13, where the THD of the ADC is plotted as a function of the HD2 and HD3 of the amplifier (measured with a 0.75 LSB_1 input for the amplifier).

Since the amplifier is not used during most of the time, it can be shared among different channels to save some area [40]. However, this requires some multiplexing between the channels (e.g., at the output via all switches driven by "amp" or its inverse) which also increases the cross talk between the channels.

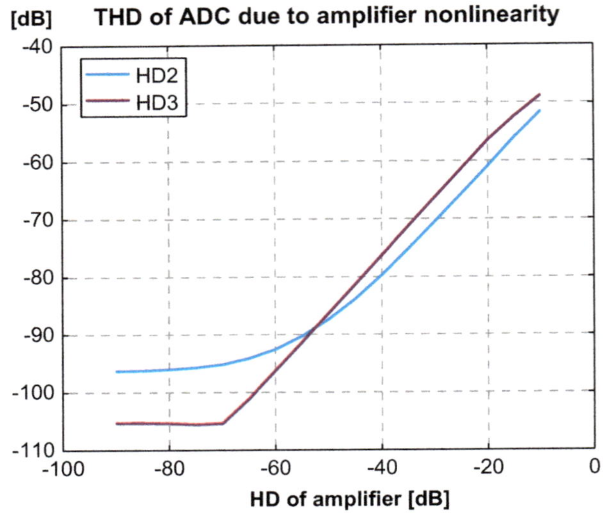

Fig. 3.13 Impact of amplifier nonlinearity on THD of ADC of Fig. 3.3

3.4 Design Examples

Table 3.2 lists the specifications and architectural details of the three interleaved pipelined SAR ADCs shown in Fig. 3.14. They all have been designed in a 1P9M 28 nm CMOS process. An architecture with two interleaved channels and two stages per channel has been chosen for all three designs. They differ in the design point:

- The first example [19] has the highest sampling speed of 410 MS/s with an 11-bit resolution. Digital calibration is put on chip to calibrate comparator offsets, amplifier gain, and MSB mismatch errors. This calibration engine takes up an area of more than the actual ADC as shown in the left part of Fig. 3.14.
- ADC 2 [25] consumes only 1 mW at 250 MS/s which is the lowest power consumption of the three designs. It is part of a larger discrete-time system with an integrated filtering function like the system of Fig. 3.6. Therefore, its input range is much smaller than for the other two ADCs.
- The third ADC [21] demonstrates the feasibility of 14-bit resolution with an interleaved pipelined SAR architecture at a fair high speed of 280 MS/s. These high-speed, high-resolution specifications result in a large decoupling capacitance which is also denoted in the right part of Fig. 3.14.

ADC 2 resolves only 3 bits in the first stage, whereas the others use a 6-bit SAR conversion. Since its input range is also small, the residue to be amplified is of the same order, and an open-loop dynamic amplifier can be used without increasing the nonlinearity too much. Nevertheless, the harmonic distortion of the amplifier is about 8 dB higher, and spurs are clearly observed in the output spectrum shown in Fig. 3.15.

Table 3.2 Overview of properties of the three interleaved pipelined SAR ADCs

		ADC 1 [19]	ADC 2 [25]	ADC 3 [21]
Technology		CMOS 28 nm	CMOS 28 nm	CMOS 28 nm
Channels		2	2	2
Sampling speed		410 MS/s	250 MS/s	280 MS/s
Resolution		11 bit	10 bit	14 bit
$ENOB_{lf}$		10 bit	7.4 bit	11.8 bit
$ENOB_{hf}$		9.5 bit	N/A	9.8 bit
SFDR		70 dB	57 dB	70 dB
Power		1.9 mW	1.0 mW	3.2 mW
Input range		1.2 V_{ptp}	200 mV_{ptp}	1.6 V_{ptp}
Channel time $T_{channel}$		4.9 ns	8 ns	7.1 ns
Stage 1	Bits	6	3	6
	DAC size	860 fF	1.6 pF	3.2 pF
	DAC type	Monotonic	Symmetrical	Switchback
	Comp. 1 noise	1.3 LSB	2.0 LSB	6.5 LSB
	Comp. 1 delay	0.02 $T_{channel}$	0.02 $T_{channel}$	0.02 $T_{channel}$
Amplifier	Type	Two-stage	Complementary	Complementary
	Gain	4	6	4
	T_{amp}	0.11 $T_{channel}$	0.1 $T_{channel}$	0.09 $T_{channel}$
	Noise	0.2 LSB	0.2 LSB	0.4 LSB
	THD	55 dB	47 dB	56 dB
	Energy	0.38 E_{conv}	0.25 E_{conv}	0.27 E_{conv}
Stage 2	Bits	5 + 2	7 + 2	8 + 2
	DAC size	500 fF	830 fF	3.1 pF
	DAC type	Monotonic	Symmetrical	Monotonic
	Comp. 2 noise	0.8 LSB	2.0 LSB	3.4 LSB
	Comp. 2 delay	0.03 $T_{channel}$	0.02 $T_{channel}$	0.02 $T_{channel}$
	Comp. 3 noise	0.6 LSB	1.3 LSB	1.5 LSB
	Comp. 3 delay	0.04 $T_{channel}$	0.02 $T_{channel}$	0.08 $T_{channel}$

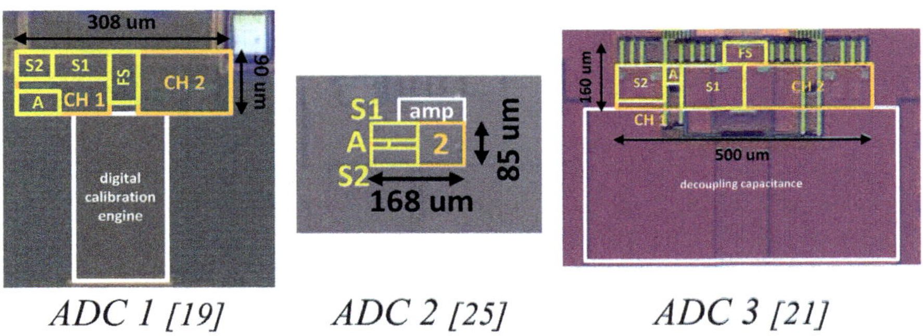

Fig. 3.14 Chip photographs of the three interleaved pipelined SAR ADCs

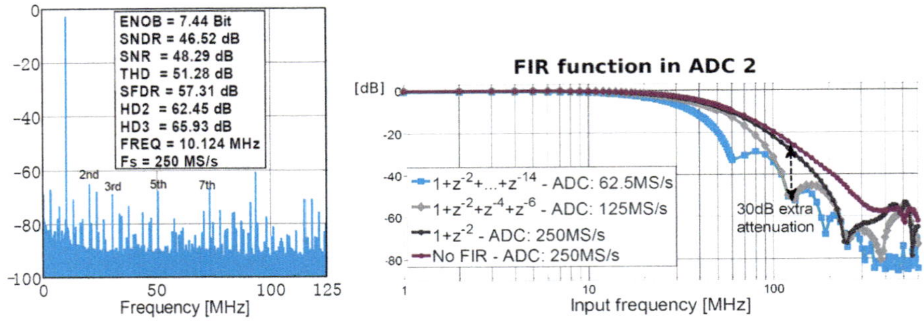

Fig. 3.15 Output spectrum and filtering function of ADC 2

The three ADCs have similar timing diagrams with 25–50% tracking time, about 10% amplification time, and the rest for SAR conversion and reset. To realize the 14-bit quantization with only two stages, the high-noise comparators of the second stage in ADC 3 are made a bit faster, while the low-noise comparators are made much slower, and a longer channel time is needed to fit ten comparisons in a stage resulting in a slower ADC. Its timing diagram is shown in Fig. 3.3.

Different switching schemes are used in the three designs. ADC 2 uses the symmetrical scheme with a constant common mode (see Fig. 3.8). This configuration allows to use only one comparator in the first stage and only two in the second stage, whereas the other ADCs use the comparator-based controller architecture with a different comparator per step. Since also the DAC is quite small and the resolution only 10 bit, ADC 2 has the most compact area. However, more control is needed, which takes up a large part of the total energy conversion.

The high resolution of ADC 3 requires a large DAC in the first stage to limit kT/C noise. The linearity is enhanced by calibration of the first three MSBs. The switching energy of this first-stage DAC takes up a larger part of the total conversion energy compared to the other two ADCs.

Fig. 3.15 shows measurement results of ADC 2 at 250 MS/s for a 10 MHz input. The spectrum shows an SNDR of almost 47 dB which is mainly limited by the SNR. The right part shows the FIR filtering profile. The FIR can create notches at specific frequencies which can be used in a receiver to filter out unwanted interferers. More notches can be added by increasing the number of filter taps.

For ADCs 1 and 3, the trade-off between bandwidth, accuracy, and power consumption is depicted in Fig. 3.16. ADC 1 achieves 9.64-bit ENOB at 410 MS/s with a low-frequency input which drops to 8.85 bit with a Nyquist input. The peak efficiency is 5.5 fJ/c.s. which remains less than 12 fJ/c.s. at 410 MS/s with a Nyquist input signal, including the power for the (slow) background calibration. For ADC 3, the peak ENOB with low-frequency input is 11.8 bit at 280 MS/s. It remains 9.8 bit with Nyquist input in part degraded by jitter from the measurement setup. With a high-frequency Walden FoM of 12.8 fJ/c.s., it realizes a higher accuracy at the cost of a slower speed but with comparable power consumption. All examples compare favorably to the state of the art shown in Fig. 3.1.

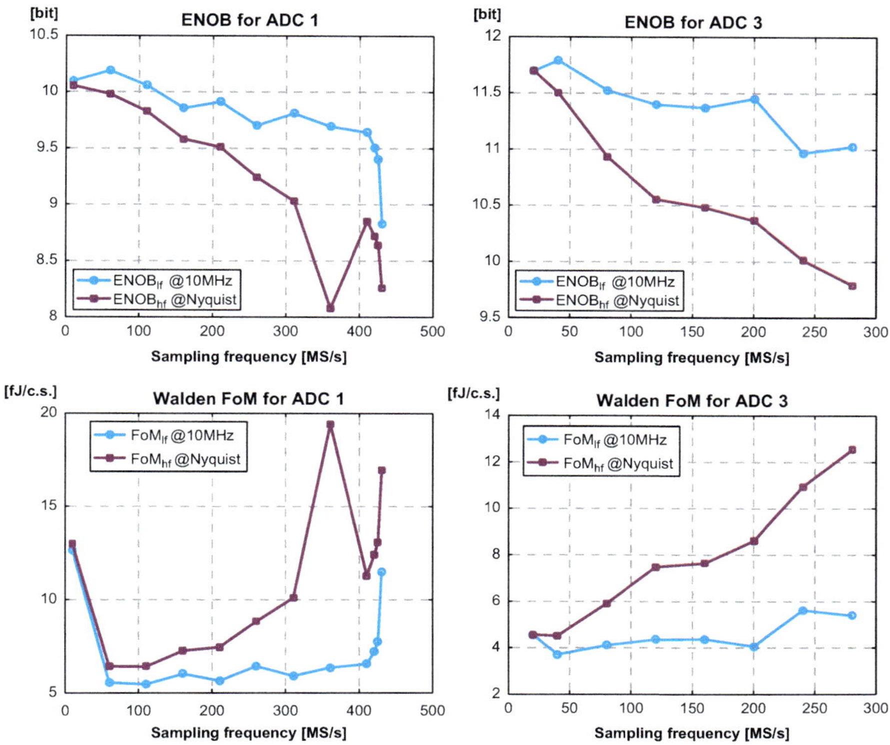

Fig. 3.16 ENOB and efficiency for ADC 1 and 3

3.5 Conclusions

Combining energy-efficient SAR converters via pipelining and interleaving results in the interleaved pipelined SAR ADC. The architecture relaxes some of the specifications of the building blocks to realize higher accuracy and larger bandwidths than basic SAR converters can offer while still being power efficient. Design examples have demonstrated sampling speeds of more than 400 MS/s, resolutions of 14 bit, and power consumption of 1 mW. This shows that this architecture is a good candidate for software-defined receivers.

Acknowledgment The author would like to thank Bob Verbruggen, Badr Malki, Masao Iriguchi, Kazuaki Deguchi, and Jan Craninckx for their contributions to this work.

References

1. Martens, E., et al.: RF-to-baseband digitization in 40 nm CMOS with RF bandpass $\Delta\Sigma$ modulator and polyphase decimation filter. IEEE J. Solid State Circuits. **47**(4), 990–1002 (2012)
2. Murmann, B.: ADC Performance survey 1997–2016. [online] Available: http://web.stanford.edu/~murmann/adcsurvey/html
3. Walden, R.H.: Analog-to-digital converter survey and analysis. IEEE J. Sel. Areas Commun. **17**(4), 539–550 (1999)
4. Spagnolo, A., Verbruggen, B., D'Amico, S., Wambacq, P.: High-speed analog-to-digital converters in downscaled CMOS. 2015 International Conference on IC Design & Technology (ICICDT), Leuven, pp. 1–4 (2015)
5. Steyaert, M., Roovers, R., Craninckx, J.: A 100 MHz 8 bit CMOS interpolating A/D converter. Custom Integrated Circuits Conference, 1993., Proceedings of the IEEE 1993, San Diego, pp. 28.1.1–28.1.4 (1993)
6. Van De Plassche, R.J., Van Der Grift, R.E.J.: A high-speed 7 bit A/D converter. IEEE J. Solid State Circuits. **14**(6), 938–943 (1979)
7. Plas, G.V.D., Verbruggen, B.: A 150 MS/s 133 μW 7 bit ADC in 90 nm digital CMOS. IEEE J. Solid State Circuits. **43**(12), 2631–2640 (2008)
8. McCreary, J.L., Gray, P.R.: All-MOS charge redistribution analog-to-digital conversion techniques. I. IEEE J. Solid State Circuits. **10**(6), 371–379 (1975)
9. Cao, Z., Yan, S., Li, Y.: A 32mW 1.25GS/s 6b 2b/step SAR ADC in 0.13μm CMOS. 2008 IEEE International Solid-State Circuits Conference – Digest of Technical Papers, San Francisco, pp. 542–634 (2008)
10. Martin, K.W.: A high-speed, high-accuracy pipelined A/D converter. IEEE 15th Asilomar Conference on Circuits, Systems, and Computers (1981)
11. Ali, A.M.A., et al.: 29.3 A 14b 1GS/s RF sampling pipelined ADC with background calibration. 2014 IEEE International Solid-State Circuits Conference Digest of Technical Papers (ISSCC), San Francisco, pp. 482–483 (2014)
12. Jinghua, L.I., Maloberti, F.: Pipeline of successive approximation converters with optimum power merit factor. 9th International Conference on Electronics, Circuits and Systems, vol. 1, pp. 17–20 (2002)
13. Black, W.C., Hodges, D.A.: Time interleaved converter arrays. IEEE J. Solid State Circuits. **15**(6), 1022–1029 (1980)
14. Verbruggen, B., Iriguchi, M., Craninckx, J.: A 1.7 mW 11b 250 MS/s 2-times interleaved fully dynamic pipelined SAR ADC in 40 nm digital CMOS. IEEE J. Solid State Circuits. **47**(12), 2880–2887 (2012)
15. Jamal, S.M., Daihong, FU, Hurst, P.J., Lewis, S.H.: A 10b 120MSample/s time-interleaved analog-to-digital converter with digital background calibration. 2002 IEEE International Solid-State Circuits Conference. Digest of Technical Papers (Cat. No.02CH37315), San Francisco, pp. 132–436 (2002)
16. Poulton, K., Neff, R., Muto, A., Liu, W., Burstein, A., Heshami, M.: A 4GSample/s 8b ADC in 0.35/spl mu/m CMOS. 2002 IEEE International Solid-State Circuits Conference. Digest of Technical Papers (Cat. No.02CH37315), San Francisco, pp. 126–434 (2002)
17. Razavi, B.: Design considerations for interleaved ADCs. IEEE J. Solid State Circuits. **48**(8), 1806–1817 (2013)
18. Gupta, S., Choi, M., Inerfield, M., Wang, J.: A 1GS/s 11b Time-interleaved ADC in 0.13/spl mu/m CMOS. 2006 IEEE International Solid State Circuits Conference – Digest of Technical Papers, San Francisco, pp. 2360–2369 (2006)
19. Verbruggen, B., et al.: A 2.1 mW 11b 410 MS/s dynamic pipelined SAR ADC with background calibration in 28nm digital CMOS. 2013 Symposium on VLSI Circuits, Kyoto, pp. C268–C269 (2013)

20. Le Dortz, N. et al.: A 1.62GS/s time-interleaved SAR ADC with digital background mismatch calibration achieving interleaving spurs below 70dBFS. 2014 IEEE International Solid-State Circuits Conference Digest of Technical Papers (ISSCC), San Francisco, pp. 386–388 (2014)
21. Verbruggen, B., Deguchi, K., Malki, B., Craninckx, J.: A 70 dB SNDR 200 MS/s 2.3 mW dynamic pipelined SAR ADC in 28 nm digital CMOS. 2014 Symposium on VLSI Circuits Digest of Technical Papers, Honolulu, pp. 1–2 (2014)
22. Murmann, B.: Digitally assisted data converter design. 2013 Proceedings of the ESSCIRC (ESSCIRC), Bucharest, pp. 24–31 (2013)
23. Verbruggen, B.: Digitally assisted analog to digital converters. 2014 Workshop on Advances in Analog Circuit Design, Lisbon.
24. Abo, A.M., Gray, P.R.: A 1.5-V, 10-bit, 14.3-MS/s CMOS pipeline analog-to-digital converter. IEEE J. Solid State Circuits. **34**(5), 599–606 (1999)
25. Malki, B., Verbruggen, B., Martens, E., Wambacq, P., Craninckx, J.: A 150 kHz–80 MHz BW discrete-time analog baseband for software-defined-radio receivers using a 5th-order IIR LPF, active FIR and a 10 bit 300 MS/s ADC in 28 nm CMOS. IEEE J. Solid State Circuits. **51**(7), 1593–1606 (2016)
26. Liu, C.C., Chang, S.J., Huang, G.Y., Lin, Y.Z.: A 10-bit 50-MS/s SAR ADC with a monotonic capacitor switching procedure. IEEE J. Solid State Circuits. **45**(4), 731–740 (2010)
27. Huang, G.Y., Chang, S.J., Liu, C.C., Lin, Y.Z.: 10-bit 30-MS/s SAR ADC using a switchback switching method. IEEE Trans. Very Large Scale Integr. VLSI Syst. **21**(3), 584–588 (2013)
28. Lin, K.T., Cheng, Y.W., Tang, K.T.: A 0.5 V 1.28-MS/s 4.68-fJ/conversion-step SAR ADC with energy-efficient DAC and trilevel switching scheme. IEEE Trans. Very Large Scale Integr. VLSI Syst. **24**(4), 1441–1449 (2016)
29. Kuppambatti, J., Kinget, P.R.: Current reference pre-charging techniques for low-power zero-crossing pipeline-SAR ADCs. IEEE J. Solid State Circuits. **49**(3), 683–694 (2014)
30. Miyahara, M., Asada, Y., Paik, D., Matsuzawa, A.: A low-noise self-calibrating dynamic comparator for high-speed ADCs. 2008 IEEE Asian Solid-State Circuits Conference, Fukuoka, pp. 269–272 (2008)
31. Nuzzo, P., De Bernardinis, F., Terreni, P., Van der Plas, G.: Noise analysis of regenerative comparators for reconfigurable ADC architectures. IEEE Trans. Circuits Syst. I, Reg. Papers. **55**(6), 1441–1454 (2008)
32. Chen, S.-W.M., Brodersen, R.W.: A 6b 600MS/s 5.3mW Asynchronous ADC in 0.13/spl mu/m CMOS. 2006 IEEE International Solid State Circuits Conference - Digest of Technical Papers, San Francisco, pp. 2350–2359 (2006)
33. Murmann, B., Boser, B.E.: A 12 b 75 MS/s pipelined ADC using open-loop residue amplification. 2003 IEEE International Solid-State Circuits Conference, 2003. Digest of Technical Papers. ISSCC., San Francisco, vol. 1, pp. 328–497 (2003)
34. Anthony, M., Kohler, E., Kurtze, J., Kushner, L., Sollner, G.: A process-scalable low-power charge-domain 13-bit pipeline ADC. 2008 IEEE Symposium on VLSI Circuits, Honolulu, pp. 222–223 (2008)
35. Hershberg, B., Weaver, S., Sobue, K., Takeuchi, S., Hamashita, K., Moon, U.K.: Ring amplifiers for switched-capacitor circuits. 2012 IEEE International Solid-State Circuits Conference, San Francisco, pp. 460–462 (2012)
36. Chiang, S.H.W., Sun, H., Razavi, B.: A 10-Bit 800-MHz 19-mW CMOS ADC. 2013 Symposium on VLSI Circuits, Kyoto, pp. C100–C101 (2013)
37. Brooks, L., Lee, H.S.: A zero-crossing-based 8-bit 200 MS/s pipelined ADC. IEEE J. Solid State Circuits. **42**(12), 2677–2687 (2007)
38. Lim, Y., Flynn, M.P.: 26.1 A 1mW 71.5dB SNDR 50MS/S 13b fully differential ring-amplifier-based SAR-assisted pipeline ADC. 2015 IEEE International Solid-State Circuits Conference – (ISSCC) Digest of Technical Papers, San Francisco, pp. 1–3 (2015)

39. Malki, B., Verbruggen, B., Wambacq, P., Deguchi, K., Iriguchi, M., Craninckx, J.: A complementary dynamic residue amplifier for a 67 dB SNDR 1.36 mW 170 MS/s pipelined SAR ADC. ESSCIRC 2014 – 40th European Solid State Circuits Conference (ESSCIRC), Venice Lido, pp. 215–218 (2014)
40. Huang, Y.C., Lee, T.C.: A 10b 100MS/s 4.5mW pipelined ADC with a time sharing technique. 2010 IEEE International Solid-State Circuits Conference – (ISSCC), San Francisco, pp. 300–301 (2010)

Chapter 4
Hybrid VCO Based 0-1 MASH and Hybrid ΔΣ SAR

Arindam Sanyal, Wenjuan Guo, and Nan Sun

4.1 Introduction

With technology scaling, traditional voltage-domain (VD) analog-to-digital converters (ADCs) face severe challenges due to reduction in transistor intrinsic gain and supply voltage. This has resulted in increased interest and research focus on ADCs that have highly digital nature. Both successive approximation register (SAR) and voltage-controlled oscillator (VCO) based ADCs have turned out to be very popular in scaled CMOS technologies.

SAR ADC has very high energy efficiency for medium resolutions. However, as the target resolution goes beyond 10-bit, its efficiency quickly diminishes due to its tight requirement on comparator noise. Moreover, the exponentially increasing capacitance of the digital-to-analog converter (DAC) array not only costs large chip area and power but also makes it difficult to drive. There have been efforts to reduce switching power in the SAR DAC by adopting bidirectional single-sided switching [1, 2]. The techniques in [3, 4] have adopted statistical estimation to reduce the effect of comparator thermal noise which reduces the comparator power consumption significantly.

Ring VCO is widely used as time-domain (TD) quantizer due to its highly digital nature and simplicity of design. By acting as an integrator in phase domain, a ring

A. Sanyal
State University of New York at Buffalo, Buffalo, NY, USA
e-mail: arindams@buffalo.edu

W. Guo
Intel Corporation, Austin, Austin, TX, USA
e-mail: wenjuan.guo@intel.com

N. Sun (✉)
The University of Texas at Austin, Austin, TX, USA
e-mail: nansun@mail.utexas.edu

P. Harpe et al. (eds.), *Hybrid ADCs, Smart Sensors for the IoT, and Sub-1V & Advanced Node Analog Circuit Design*, DOI 10.1007/978-3-319-61285-0_4

VCO ADC can also provide intrinsic first-order quantization noise shaping, and thus a ring VCO can be used as a first-order TD $\Delta\Sigma$ ADC. However, frequency tuning gain of a ring VCO is highly nonlinear and is sensitive to variations in process, voltage, and temperature (PVT), which seriously undermines accuracy and robustness of VCO-based ADCs. In addition, existing VCO-based ADCs usually have energy efficiency around 100 fJ/conversion-step [5–11]. Recent research has tried to address the PVT sensitivity of VCO-based ADCs by using a pipelined, hybrid VCO ADC with digital background nonlinearity calibration [12, 13] and embedding a first-order VCO inside a loop [14–16].

This work presents a scaling friendly and energy-efficient hybrid 0-1 MASH $\Delta\Sigma$ ADC [17]. A coarse 8-bit first-stage SAR ADC is combined with a fine VCO which acts as the second stage. The VCO is effective at quantizing small voltages in time domain. Since the VCO sees only a small SAR residue, VCO nonlinearity is greatly suppressed, and no nonlinearity correction is needed. The VCO cancels out SAR quantization noise as well as the comparator thermal noise. Thus, the quantization noise at the ADC output comes only from the VCO and is first-order shaped. The PVT variation of VCO tuning gain can still cause SAR quantization noise leakage into the output and degrade SNDR. To address this issue, a simple digital background calibration technique is developed. The proposed background calibration technique enables precise tracking of the VCO gain and results in 12-bit ADC linearity.

While the proposed SAR + VCO architecture can be used as a high-resolution $\Delta\Sigma$ ADC, there have been efforts to develop highly digital $\Delta\Sigma$ ADCs which do not require calibration. The first noise-shaping (NS) SAR ADC is published in [18], but it still needs an OTA to realize a first-order noise transfer function (NTF) zero at $z = 0.64$. It also requires a finite impulse response (FIR) DAC that introduces extra noise and increases chip area. Later, a fully passive first-order NS SAR ADC is published in [19]. It obviates the need for any OTA, but its noise-shaping performance is very limited, as its NTF zero is located at 0.5 rather than 1. Moreover, its input signal is attenuated by two times during normal conversion, leading to 6-dB penalty in SNR or quadrupled analog power for the same SNR. In addition, it requires double capacitance, increasing chip area. We propose a novel NS SAR architecture [20] that is simple, robust, and low power. It gets rid of OTA-based active integrators and realizes a NTF zero at 0.75 with a passive integrator. The passive integrator only requires one switch and two capacitors. The zero location is fully determined by the capacitor ratio, which is insensitive to PVT variations and ensures the robustness of the architecture. Compared to [18] ($z = 0.64$) and [19] ($z = 0.5$), the proposed NS SAR ADC achieves the best noise-shaping performance with a zero closest to 1. Figure 4.1 compares the NTF magnitude with zeros from [18, 19] and this work. As can be seen, this work achieves around 3 dB more in-band attenuation than [18] and 6 dB more in-band attenuation than [19]. Furthermore, the proposed architecture does not cause any signal attenuation and requires less capacitance than [19]. With minimum modification to the original SAR ADC architecture, the proposed NS SAR ADC altogether shapes the quantization noise, comparator noise, and DAC noise with a NTF of $(1 - 0.75z^{-1})$. It allows the use of a low-resolution DAC and relaxes the requirement on comparator noise, making it possible to reach high resolution and high power efficiency simultaneously.

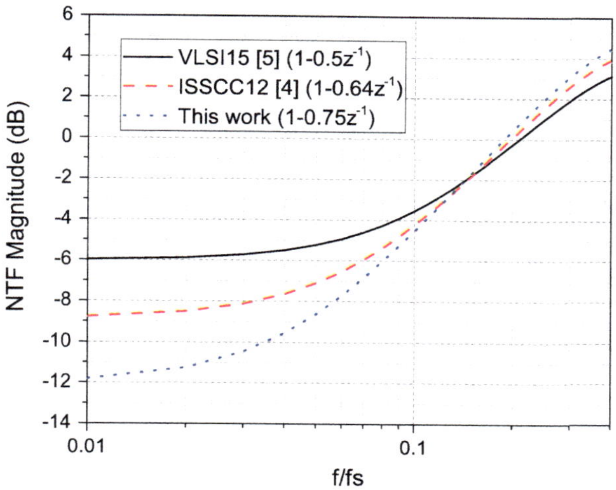

Fig. 4.1 NTF magnitude comparisons with zeros from [18, 19] and this work

The rest of the paper is organized as follows. Section 4.2 discusses the proposed SAR + VCO hybrid ADC in details along with chip measurement results. The noise-shaping hybrid ΔΣ SAR ADC is presented in Sect. 4.3 along with chip measurement results. The conclusion is brought up in Sect. 4.4.

4.2 SAR + VCO 0-1 MASH ΔΣ ADC

4.2.1 ADC Architecture

The proposed SAR + VCO ADC is shown in Fig. 4.2 along with the timing diagram. An 8-bit SAR forms the first stage of the ADC. During the phase ϕ_1, the input is sampled differentially on the bottom plates of the capacitive DAC array. The bottom-plate sampling switches are closed during ϕ_{1e} and are opened slightly before the input sampling switches are opened. The sampled input is quantized by the SAR during the phase ϕ_2. After quantization, the first-stage residue is fed to a pseudo-differential dual VCO during ϕ_3 phase. The dual VCOs perform phase-domain integration of the residue from the SAR stage. The VCO differential output is differentiated digitally before being combined with SAR output to generate the final ADC output.

To reduce switching power in the capacitive DAC, the bidirectional single-sided switching scheme of [2] has been adopted. The switching scheme is illustrated for a 3-bit capacitive DAC in Fig. 4.3. The SAR DAC has no redundancy as the second-stage VCO can absorb decision errors in the SAR stage as long as the error is not so large as to cause phase overflow in the VCO stage. The VCO stage relaxes the

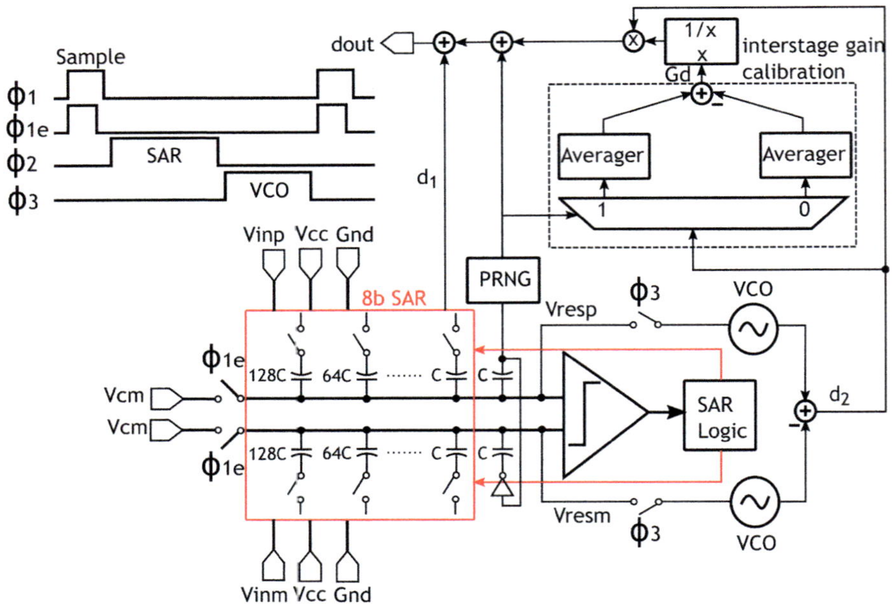

Fig. 4.2 Circuit diagram of the proposed SAR + VCO ADC

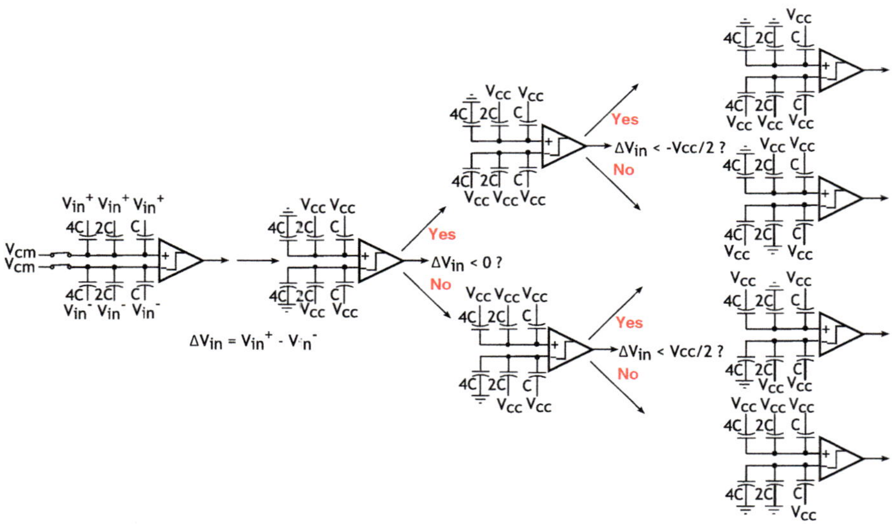

Fig. 4.3 SAR switching technique

precision requirement of the SAR stage and thus a low power comparator can be used in the SAR stage. The SAR in turn reduces the VCO swing and obviates the need for any VCO nonlinearity calibration.

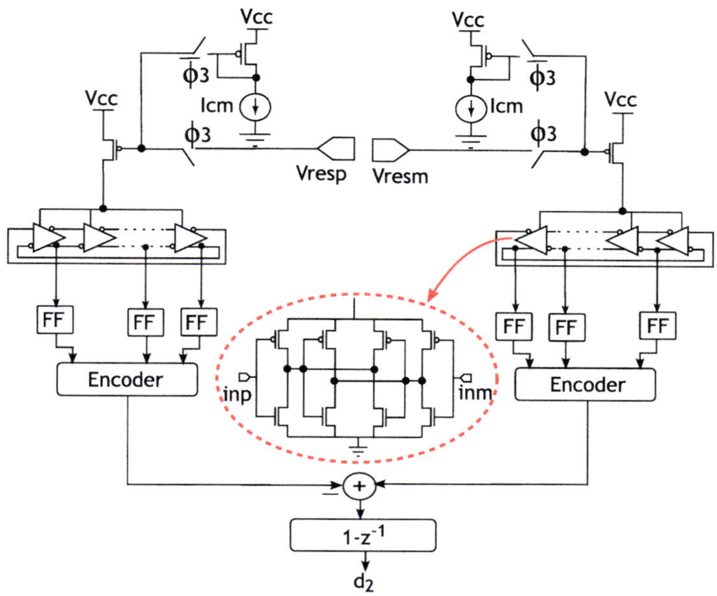

Fig. 4.4 Circuit diagram showing the second-stage VCO

The circuit diagram of the second stage is shown in Fig. 4.4. Each VCO consists of a seven-stage pseudo-differential ring inverter chain. The VCO performs phase-domain integration during the clock phase ϕ_3, but the VCO is not stopped during the clock phases ϕ_1 and ϕ_2. This is to prevent charge leakage which can corrupt the phase information held by the VCOs. Instead, the VCOs are biased with I_{cm} and run at a fixed frequency during $\overline{\phi_3}$. I_{cm} is kept low to save power and reduce phase noise during the VCO idle phase ($\overline{\phi_3}$). The digital logic for the second-stage runs at the ADC sampling frequency. There is no phase-overflow counter running at high VCO frequency, which significantly lowers the power consumption of the second stage compared to [11].

4.2.2 *ADC Model*

The ADC model is shown in Fig. 4.5. G represents the SAR residue voltage attenuation due to parasitic capacitors, K_{vco} is the VCO tuning gain, and G_d is the digital gain used to scale the VCO output d_2 before combining with SAR output d_1. R_n is the pseudo-random sequence used to dither the VCO stage. The ADC output can be written as

$$d_{\text{out}} = V_{\text{in}} + (q_1 - R_n)\left(1 - \frac{GK_{\text{vco}}}{G_d}\right) + \frac{q_2(1 - z^{-1})}{G_d} \tag{4.1}$$

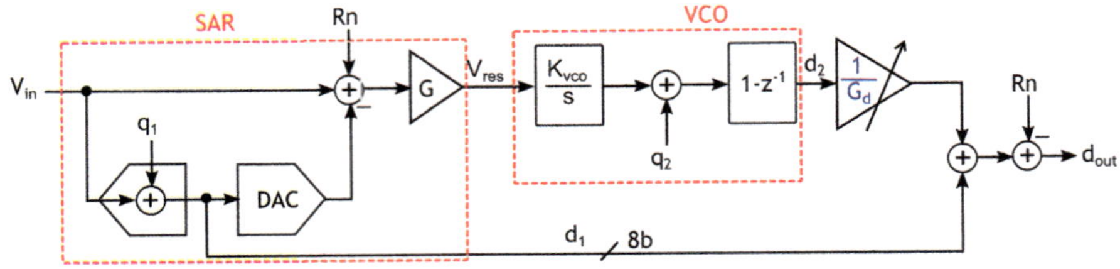

Fig. 4.5 Block diagram of the proposed ADC

If the digital gain G_d is set equal to the analog interstage gain $G_A \equiv GK_{\text{vco}}$, both R_n and SAR conversion error q_1 can be canceled at the output. The SAR comparator thermal noise is also removed at the output. Thus, the second stage allows the use of a low power comparator in the first stage. SAR quantization and comparator thermal noise as well as R_n can only be canceled at the output if the VCO stage has sufficient linearity. The final quantization noise at the ADC output comes solely from the VCO, q_2, and is first-order shaped. Thermal noise at the ADC output comes from VCO phase noise and kT/C noise which is much smaller compared to VCO phase noise. Thus, as long as the VCO is linear, the ADC resolution depends only on the VCO stage and is independent of the SAR resolution. An 8-bit SAR is chosen to ensure that the VCO is sufficiently linear.

Any mismatch between G_A and G_D will result in SAR quantization noise, comparator thermal noise, and R_n leaking to the output which will raise the in-band floor and increase distortion. To ensure $G_A = G_D$, we digitally adjust G_D such that

$$G_D \equiv G_A = \overline{d_2(R_n = 1)} - \overline{d_2(R_n = 0)} \tag{4.2}$$

where

$$d_2 = -q_1 G_A + q_2(1 - z^{-1}) + G_A R_n \tag{4.3}$$

The averaging in (4.2) is done by an infinite-impulse response (IIR) filter. q_1 does not depend on R_n and, hence, has very low correlation with R_n. Thus, q_1 is canceled in the subtraction between d_2 for $R_n = 1$ and $R_n = 0$. q_2 is the quantization noise of multi-bit VCO quantizer and has low power. Thus q_2 can be easily averaged out by the IIR filter. The interstage gain extraction circuit consists of two averagers and a 1-to-2 DEMUX. Depending on the value of R_n, the VCO output d_2 is sent to one of the two averagers. The averagers are built using low-pass, first-order IIR filters as shown in Fig. 4.6. The IIR filter output can be written as

$$y = \frac{\mu x}{1 - (1 - \mu)z^{-1}} \tag{4.4}$$

The speed of convergence of the calibration algorithm depends on the IIR filter bandwidth μ. It can be seen from (4.2) and (4.3) that the dominant noise source

Fig. 4.6 IIR filter used for background calibration

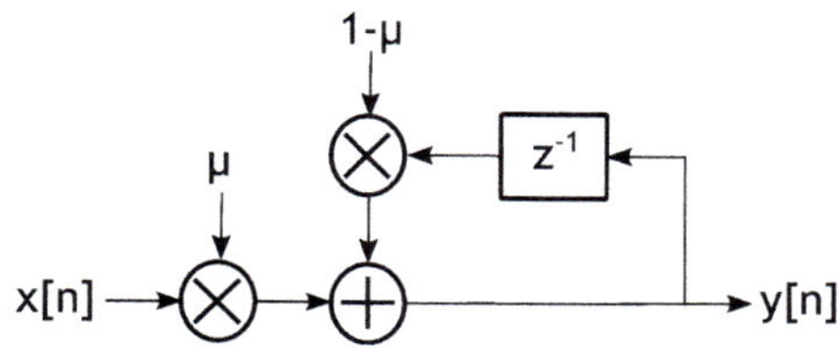

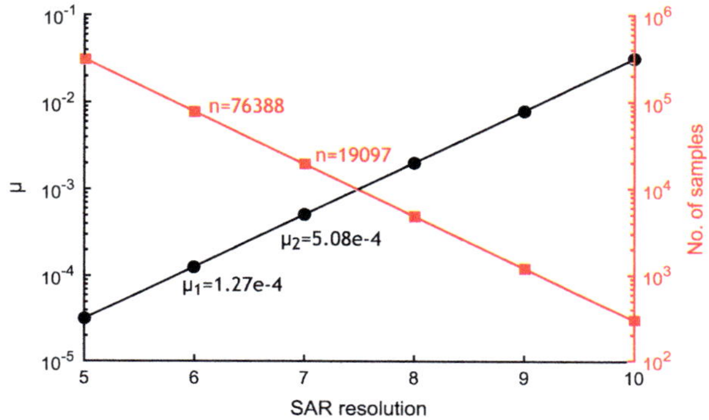

Fig. 4.7 IIR filter bandwidth versus SAR resolution

at the input to the IIR filter is the SAR quantization noise q_1. R_n is a deterministic signal and does not affect the IIR filter bandwidth. If the SAR resolution is increased by two times, its quantization noise power reduces by 4, and the IIR filter bandwidth can be increased by four times resulting in four times faster calibration convergence. This can also be seen from Fig. 4.7. As the SAR resolution is increased from 6 to 7 bits, the IIR filter bandwidth increases by 4, and this reduces the calibration convergence time by a factor of 4.

4.2.3 Measurement Results

The SAR + VCO ADC is implemented in 40 nm CMOS. The chip photo is shown in Fig. 4.8. The core circuit occupies an area of 0.03 mm^2. The prototype consumes 350 μW from 1.1 V supply while operating at a frequency of 36 MHz. Out of the 350 μW total power, 100 μW is consumed by the VCO, SAR comparator, and capacitive DAC switching, while the remaining 250 μW power is consumed by the digital circuitry including the pseudo-random number generator, SAR logic, clock generator, and VCO digital logic.

The measured spectrum with a 2.2 V differential input at 500 kHz frequency is shown in Fig. 4.9. The ADC quantization noise comes from the VCO stage and is first-order shaped. At an OSR of 9, the ADC SNDR without calibration is 64.5 dB, while calibration improves the SNDR to 74.3 dB. Background calibration also

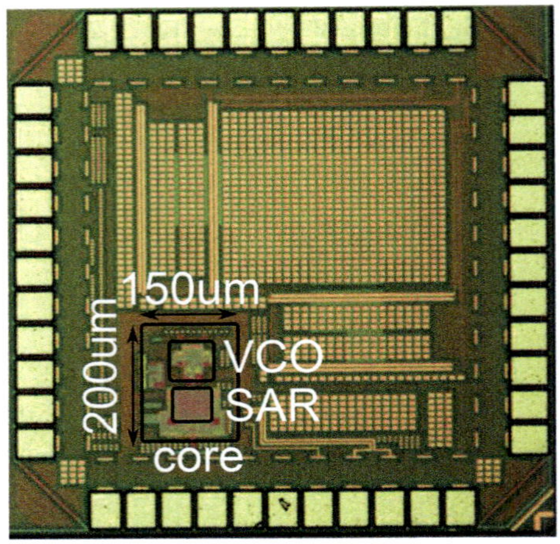

Fig. 4.8 Die photograph

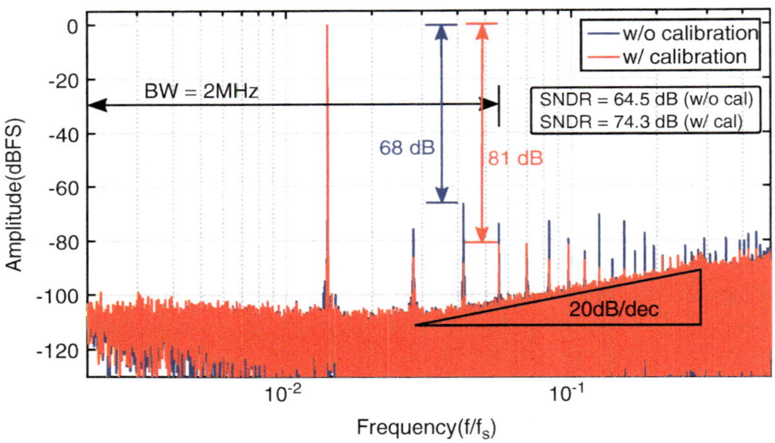

Fig. 4.9 Measured ADC spectrum with and without calibration

improves the SFDR from 68 to 81 dB. Even order distortion in the ADC spectrum after background calibration comes from mismatch in the capacitive DAC in the first-stage SAR.

Figure 4.10 shows the SNDR and SNR of the ADC versus input amplitude. The ADC has a dynamic range of 75.7 dB. The measured histogram of d_2 is shown in Fig. 4.11. The shift in d_2 distribution for $R_n = 1$ to $R_n = 0$ can be clearly seen from Fig. 4.11. The difference between $\overline{d_2(R_n = 1)}$ and $\overline{d_2(R_n = 0)}$ gives the interstage gain GK_{vco} as 1.3.

Figure 4.12 shows the calibration convergence speed of the proposed SAR + VCO ADC. The proposed background calibration has a very fast convergence and requires only 10^3 samples (or 25 μs) to converge. This is because the SAR quantization noise, which is a primary source of perturbation in the background calibration loop, is substantially attenuated by the first-stage 8-bit SAR [21].

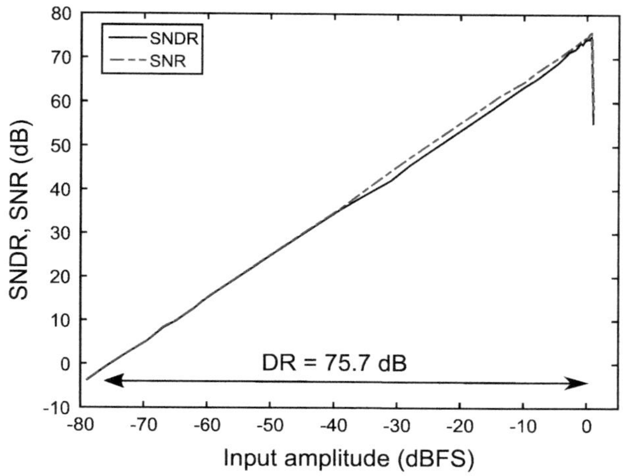

Fig. 4.10 Measured SNDR and SNR versus input amplitude

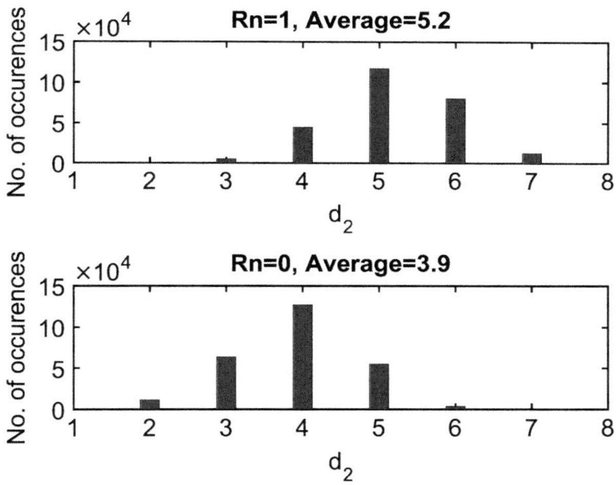

Fig. 4.11 Measured d_2 histogram for $R_n = 1$ and $R_n = 0$

The performance of the proposed ADC is compared with the state-of-the-art VCO-based ADCs in Table 4.1. It can be seen that the SAR + VCO architecture results in an FoM of 18.5 fJ/step which represents a significant improvement over the state-of-the-art.

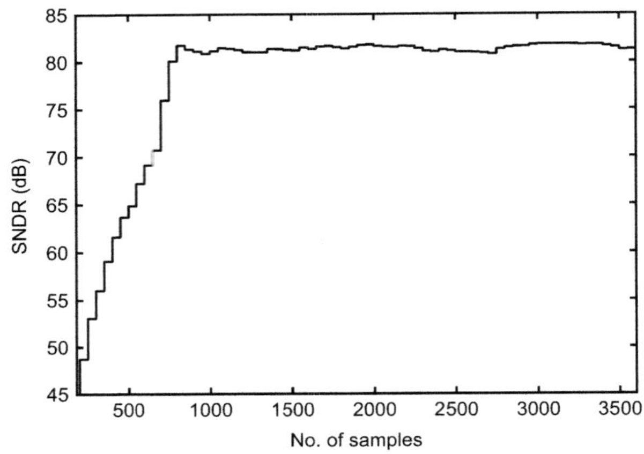

Fig. 4.12 Measured SNDR convergence curve

Table 4.1 Comparison with prior art

	[6]	[8]	[10]	**This work**	
Process (nm)	65	90	90	**40**	
Power (mW)	11.5	16	4.1	**0.35**	
Area (mm^2)	0.07	0.36	0.16	**0.03**	
Fs (MHz)	1300	600	640	**36**	
BW (MHz)	5.1	10	5	**3**	**2**
OSR	28	30	64	**6**	**9**
SNDR (dB)	75	78.3	74.7	**71.4**	**74.3**
FoM$_s$ (dB)	161	163	165	**171**	**172**
FoM$_w$ (fJ/step)	246	120	92	**18.5**	**21.3**

4.3 Hybrid ΔΣ SAR ADC

4.3.1 Circuit Architecture

Figure 4.13 shows the architecture of the proposed first-order NS SAR ADC. Compared to conventional SAR operations, two more clock cycles, ϕ_{ns_0} and ϕ_{ns_1} are added. Before ϕ_{ns_0} cycle, the SAR ADC does the normal conversion. Different from [19], there is no capacitor connected to V_{res} node during normal conversion, and, thus, the signal attenuation problem is avoided. To realize first-order noise shaping, the key is to integrate the residual voltage V_{res} and feed it back to the comparator input. During ϕ_{ns_0} cycle, a small capacitor, $C_2 = C/3$ is merged with the DAC capacitor, $C_1 = C$, to get the residue voltage, V_{res}. At the end of ϕ_{ns_0} cycle, C_2 will carry 0.75 V_{res}. In the following ϕ_{ns_1} cycle, C_2 dumps its charge onto another capacitor, $C_3 = C$, effectively realizing a passive integration. The voltage integrated on C_3 is labeled as V_{int}, which is fed back to the comparator input. Now

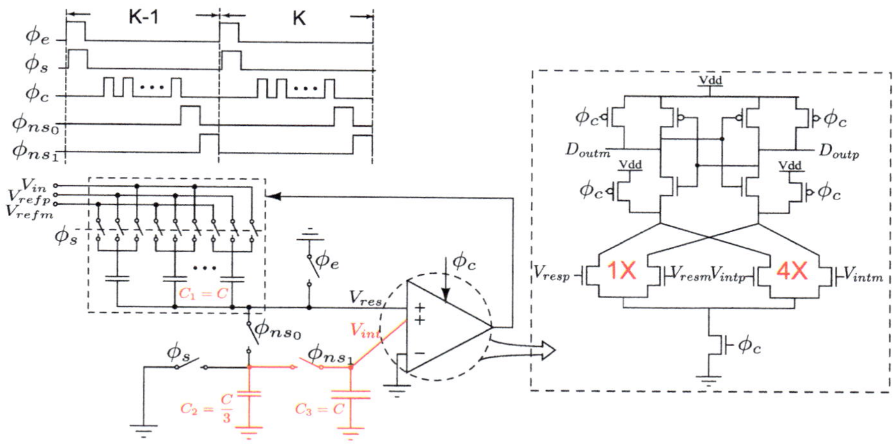

Fig. 4.13 Proposed NS SAR ADC architecture

the comparator has two path inputs, one of which is connected to V_{res}, while the other is connected to V_{int}. However, there is a limitation with passive integration that only a fraction of V_{res} is integrated, which degrades the noise-shaping performance. It seems that OTAs are still required to provide a gain to compensate the attenuation of V_{res}. Fortunately, as the comparators result is a 1-bit sign, what is required here is only a relative gain between V_{int} and V_{res}, which can be realized by simply sizing the comparator input transistors correspondingly. As shown in Fig. 4.13, to provide a gain of 4 on V_{int} path for a proper NTF, we size its corresponding input transistors 4 times larger than the V_{res} path. After ϕ_{ns_1} cycle, the charge on C_2 is cleared in next ϕ_s cycle to be ready for getting the new residual voltage. In real implementation, a *mode* signal is used to pull down V_{int} to ground so that the SAR ADC can be easily reconfigured to the conventional mode in case of Nyquist-rate applications. Additionally, foreground calibration on DAC mismatch can also be conducted in the Nyquist mode.

4.3.2 *Analysis*

To provide a better understanding of the proposed NS SAR architecture, Fig. 4.14 shows the general signal flow diagram assuming $C_1 = C_3 = C$, $C_2 = a/(1-a)C$, and the integration path gain of g. As can be seen from the derived NTF in D_{out} equation, there is a zero located at $(1-a)$ and a pole located at $(1-a)(1-ga)$. When $g = 1/a$, the pole is gone and the only left is the zero. In this design, we choose $a = 1/4$ and $g = 4$, realizing a NTF of $(1 - 0.75z^{-1})$. With $a = 1/4$, C_2 value is $C/3$, and consequently only 4/3 times more capacitors are required for the first-order NS. Note that the NTF is completely set by component ratios a and g,

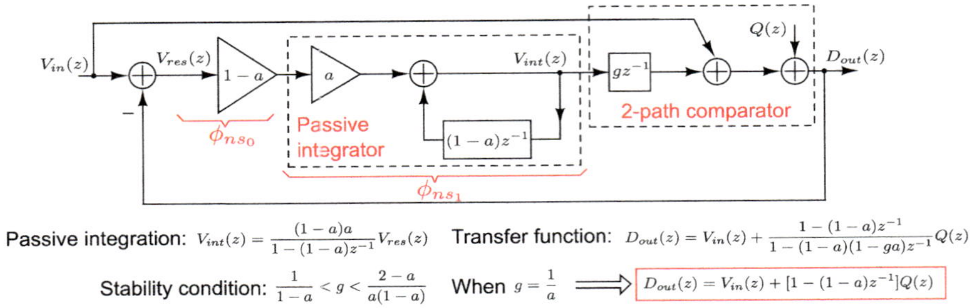

Fig. 4.14 General signal flow diagram of the proposed NS SAR ADC assuming $C_1 = C_3 = C$, $C_2 = a/(1-a)C$, and the integration path gain of g

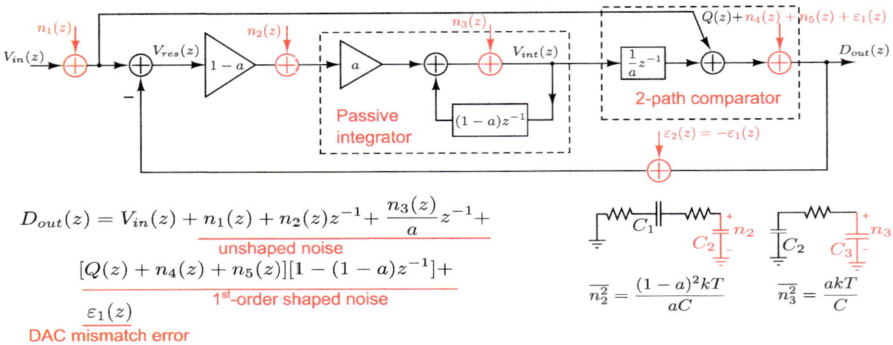

Fig. 4.15 Nonideal effects in the proposed NS SAR ADC with the integration path gain $g = 1/a$

and thus, is insensitive to PVT variations. To ensure stability, the pole needs to be within the unit circle. The stability condition is shown in Fig. 4.14. Given that the current stability condition is $4/3 < g < 28/3$, $g = 4$ determined by the comparator input transistor ratio is very far from the unstable boundary. Therefore, the proposed NS SAR architecture is highly robust.

With $g = 1/a$, Fig. 4.15 further investigates the nonideal effects including thermal noises and DAC mismatch errors in the flow. n_1 is the kT/C sampling noise which directly adds to the input signal. n_2 is the noise voltage on C_2 at the end of ϕ_{ns_0}, while n_3 is the noise voltage on C_3 at the end of ϕ_{ns_1}. Figure 4.15 also shows the noise power for n_2 and n_3. With $a = 1/4$, $\overline{n_2^2} = 9kT/4C$ and $\overline{n_3^2} = kT/4C$. As shown in the D_{out} equation, n_1, n_2, and n_3 directly pass through without being shaped, while the comparator noise n_4, the DAC noise n_5, and the quantization noise Q added at the same location are altogether shaped to the first order.

Another interesting merit of the proposed NS SAR ADC is its simplified digital DAC mismatch calibration. For conventional multi-bit $\Delta\Sigma$ ADCs, in order to completely remove the DAC mismatch error in the digital domain, we need to accurately extract not only the DAC mismatch percentage but also the DAC mismatch error transfer function (ETF), as the ETF may not be exactly 1 due to PVT variations. As a result, special techniques such as inserting a binary pseudo-random test signal [22] are required to measure the ETF. By contrast, the ETF in the NS SAR ADC is always 1 for any NTF under any PVT variation. The key reason is that the quantizer and the feedback DAC use the same capacitor array in a NS SAR ADC. It is different from conventional multi-bit ΔΣ ADCs whose DAC and quantizer are unrelated. As shown in Fig. 4.15, ε_1 represents the quantizer error due to capacitor mismatch, and ε_2 represents the feedback mismatch error. Since they are from the same origin in the NS SAR, it is easy to derive that $\varepsilon_2(z) = -\varepsilon_1(z)$. As a result, the ETF is 1 regardless of the values of a and g (see the equation in Fig. 4.15). Even though there exist capacitor mismatches, it is equivalent to a NS SAR that uses a nonbinary DAC array. As long as the capacitor mismatches are estimated, we can fully remove them in the digital domain. In this design, we reconfigure the NS ADC in the conventional Nyquist SAR mode at first and apply classic foreground calibration techniques [23] to estimate the DAC mismatch errors.

4.3.3 Chip Measurement Results

As a proof of concept, a prototype first-order NS SAR ADC is fabricated in a 0.13 μm CMOS process. Figure 4.16 shows its die photo and layout. The core area is 0.13 mm^2. The DAC array capacitor C_1 is 10-bit with a total capacitance of 2.1pF×2 and a unit MOM capacitor of 2fF. The passive integrator capacitors C_2 and C_3 are implemented using high-density MIM capacitors so that they consume much less area than C_1 (see Fig. 4.16). The sampling frequency is 2 MS/s. At 1.2 V supply, the chip consumes 60 μW power. 63% of the total power comes from the clock generation and the SAR logic, which are fully digital and can be greatly reduced with CMOS scaling. With a 95.37 KHz, −2 dBFS sinusoidal input, the measured output spectrum is shown in Fig. 4.17. At an OSR of 8, SNDR and SFDR are 74 and 95 dB, respectively, after calibration of DAC mismatch. Figure 4.18 shows the measured SNR/SNDR with different input amplitudes. Figure 4.19 shows the measured SNDR/Schreier FoM (FoM_S) trend with different OSRs. As shown in

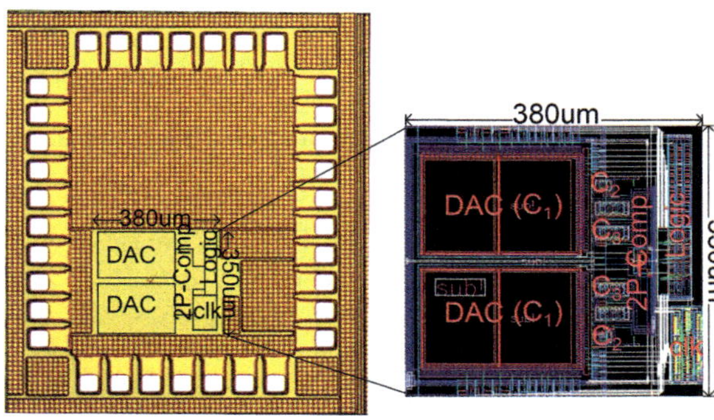

Fig. 4.16 Chip die photo and layout

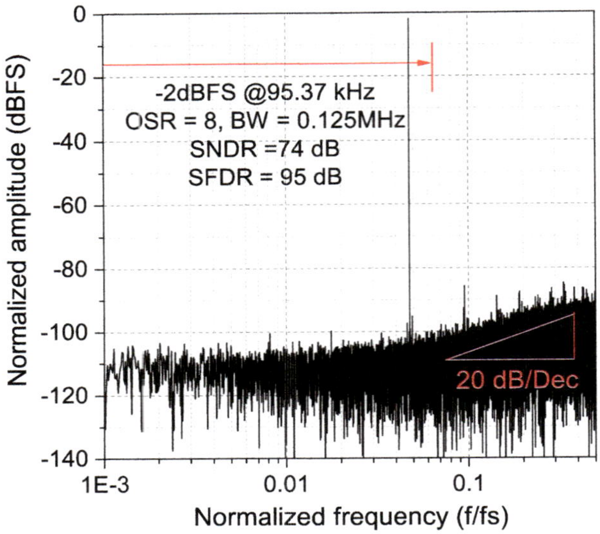

Fig. 4.17 Measured output spectrum with a 95.37 KHz, −2 dBFS sinusoidal input

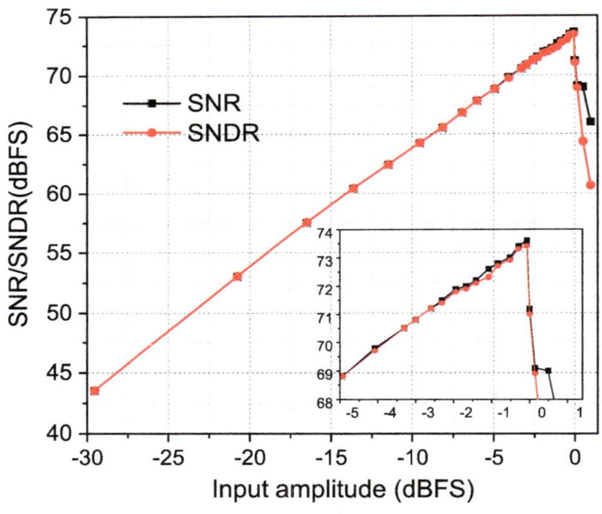

Fig. 4.18 Measured SNR/SNDR with different input amplitudes

Fig. 4.19a, with OSR doubled, SNDR increases by 6 dB which matches the NTF of $(1 - 0.75z^{-1})$. Therefore, according to $FoM_S = SNDR + 10\log_{10}(BW/Power)$, the FoM_S increases by 3 dB with OSR doubled. As shown in Fig. 4.19b, when OSR is 8, the chip achieves a FoM_S of 167 dB.

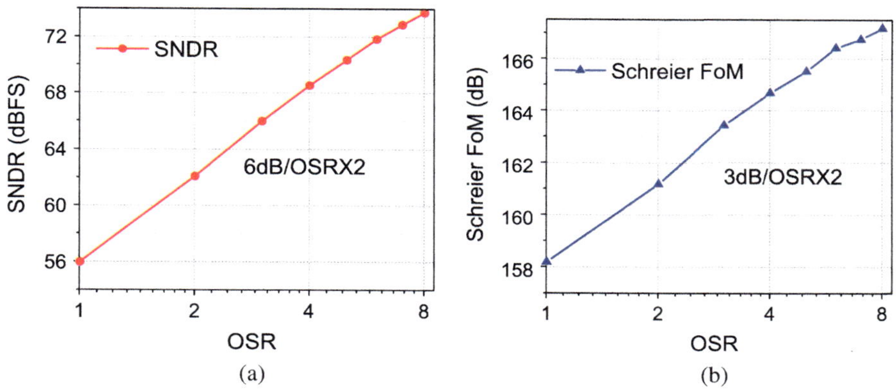

Fig. 4.19 With different OSRs: (**a**) measured SNDR and (**b**) Schreier FoM

Table 4.2 Comparison with previous NS SAR ADC works

Design	ISSCC12 [18]	VLSI15 [19]	**This work**
Architecture			
Need OTA	Yes	No	**No**
Need FIR filter	Yes	No	**No**
Signal attenuation	No	Yes	**No**
NTF zero location	0.64	0.5	**0.75**
Chip performance			
Technology (nm)	65	65	**130**
Supply (V)	1.2	0.8	**1.2**
Bandwidth (MHz)	11	6.25	**0.125**
Power (μW)	806	120.7	**61**
SNDR (dB)	62	58	**74**
ENOB (bit)	10	9.35	**12**
FoM_S (dB)	163	165	**167**

$FoM_S = SNDR + 10\log_{10}(BW/Power)$

Table 4.2 summarizes the architecture and chip performance comparisons between the proposed work and previous NS SAR ADC works. As can be seen, with a simple and power-efficient architecture, this work reaches the highest ENOB and the best Schreier FoM in an older process. Since the power is dominated by the digital portion, its power efficiency can be further improved with the CMOS scaling.

A simple and power-efficient NS SAR ADC architecture is proposed in this paper. Instead of using OTA-based active integrators, it uses one switch and two capacitors to realize passive integration. The capacitor ratio determines the NTF zero location, making the proposed architecture immune to PVT variations and highly robust. Compared to prior NS SAR ADC works, it gives the best noise-shaping performance with a zero closest to 1. A 0.13 μm CMOS prototype chip achieves the highest ENOB and the best Schreier FoM.

4.4 Conclusions

Two highly digital, hybrid $\Delta\Sigma$ ADCs are presented in this work. The SAR + VCO 0-1 MASH architecture performs time-domain quantization of the analog input signal and uses a simple, digital background calibration technique to address PVT sensitivity of VCO's tuning gain. The SAR + VCO achieves an energy efficiency of 18.5 fJ/conversion-step (Schreier FoM of 171 dB) which is the best among VCO-based ADCs. The hybrid, NS SAR ADC, achieves first-order noise shaping by using passive integrators. The NTF zero location is determined by the ratio of capacitors, thus, making the proposed architecture immune to PVT variations and highly robust. Compared to prior NS SAR ADC works, it gives the best noise-shaping performance with a zero closest to 1 and achieves a Schreier FoM of 167 dB. The SAR + VCO and NS SAR architectures achieve high resolution at low power by either canceling or first-order shaping the quantization and thermal noise of the SAR stage. The SAR + VCO architecture cancels noise from the SAR stage by using a fine-resolution VCO ADC but requires background calibration for accurate estimation of interstage gain. The NS ADC does not use a second quantizer but uses a passive integrator to first-order shape noise from the SAR stage. Thus, the NS SAR ADC has the advantage of much better PVT insensitivity over SAR + VCO architecture and does not require any background calibration. Compared to the SAR + VCO architecture, the NS SAR ADC introduces additional kT/C noise due to the capacitors C_2 and C_3 in the passive integrator. The order of noise shaping in the NS SAR ADC can be easily increased by increasing the order of the passive integrator and adding more paths to the comparator at the cost of increased kT/C noise and comparator thermal noise.

References

1. Sanyal, A., Sun, N.: An energy-efficient low frequency-dependence switching technique for SAR ADCs. IEEE Trans. Circuits Syst. Express Briefs **61**(5), 294–298 (2014)
2. Chen, L., Sanyal, A., Ma, J., Sun, N.: A 24-μW 11-bit 1-MS/s SAR ADC with a bidirectional single-side switching technique. In: IEEE European Solid-State Circuits Conference, Venice Lido, pp. 219–222 (2014)
3. Chen, L., Tang, X., Sanyal, A., Yoon, Y., Cong, J., Sun, N.: A 10.5-b ENOB 645 nW 100kS/s SAR ADC with statistical estimation based noise reduction. In: IEEE Custom Integrated Circuits Conference, San Jose, pp. 1–4 (2015)
4. Chen, L., Tang, X., Sanyal, A., Yoon, Y., Cong, J., Sun, N.: A 0.7V 0.6μW 100kS/s low-power SAR ADC with statistical estimation based noise reduction. IEEE J. Solid State Circuits **52**(5), 1388–1398 (2017)
5. Park, M., Perrott, M.: A 78 dB SNDR 87 mW 20 MHz bandwidth continuous-time $\Delta\Sigma$ ADC with VCO-based integrator and quantizer implemented in 0.13 μm CMOS. IEEE J. Solid State Circuits **44**(12), 3344–3358 (2009)
6. Taylor, G., Galton, I.: A mostly-digital variable-rate continuous-time delta-sigma modulator ADC. IEEE J. Solid State Circuits **45**(12), 2634–2646 (2010)

7. Straayer, M.Z., Perrott, M.H.: A 12-bit, 10-MHz bandwidth, continuous-time ADC with a 5-bit, 950-MS/s VCO-based quantizer. IEEE J Solid State Circuits **43**(4), 805–814 (2008)
8. Reddy, K., Rao, S., Inti, R., Young, B., Elshazly, A., Talegaonkar, M., Hanumolu, P.K.: A 16-mW 78-dB SNDR 10-MHz BW CT ADC using residue-cancelling VCO-based quantizer. IEEE J. Solid State Circuits **47**(12), 2916–2927 (2012)
9. Reddy, K., Dey, S., Rao, S., Young, B., Prabha, P., Hanumolu, P.K.: A 54mW 1.2 GS/s 71.5 dB SNDR 50MHz BW VCO-based CT ΔΣ ADC using dual phase/frequency feedback in 65 nm CMOS. In: IEEE Symposium on VLSI Circuits, Kyoto, pp. C256–C257 (2015)
10. Rao, S., Young, B., Elshazly, A., Yin, W., Sasidhar, N., Hanumolu, P.K.: A 71dB SFDR open loop VCO-based ADC using 2-level PWM modulation. In: IEEE Symposium on VLSI Circuits, Kyoto (2011)
11. Sanyal, A., Ragab, K., Chen, L., Viswanathan, T., Yan, S., Sun, N.: A hybrid SAR-VCO ΔΣ ADC with first-order noise shaping. In: IEEE Custom Integrated Circuits Conference, San Jose, pp. 1–4 (2014)
12. Ragab, K., Sun, N.: A 12b ENOB, 2.5 MHz-BW, 4.8 mW VCO-based 0-1 MASH ADC with direct digital background nonlinearity calibration. In: IEEE Custom Integrated Circuits Conference, San Jose, pp. 1–4 (2015)
13. Ragab, K., Sun, N.: A 12b ENOB, 2.5 MHz-BW, 4.8 mW VCO-based 0-1 MASH ADC with direct digital background nonlinearity calibration. IEEE J. Solid State Circuits **52**(2), 433–447 (2017)
14. Lee, K., Yoon, Y., Sun, N.: A scaling-friendly low-power small-area ΔΣ ADC with VCO-based integrator and intrinsic mismatch shaping capability. IEEE J. Emerg. Sel. Top. Circuits Syst. **5**(4), 561–573 (2015)
15. Yoon, Y., Lee, K., Hong, S., Tang, X., Chen, L., Sun, N.: A 0.04-mm 2 0.9-mW 71-dB SNDR distributed modula ΔΣ ADC with VCO-based integrator and digital DAC calibration. In: IEEE Custom Integrated Circuits Conference, San Jose, pp. 1–4 (2015)
16. Li, S., Sun, N.: A 174.3 dB FoM VCO-based CT ΔΣ modulator with a fully digital phase extended quantizer and tri-level resistor DAC in 130nm CMOS. In: IEEE European Solid-State Circuits Conference, Lausanne, pp. 241–244 (2016)
17. Sanyal, A., Sun, N.: A 18.5-fJ/step VCO-based 0-1 MASH ΔΣ ADC with digital background calibration. In: IEEE Symposium on VLSI Circuits, Honolulu, pp. 1–2 (2016)
18. Fredenburg, J., Flynn, M.: A 90-MS/s 11-MHz-bandwidth 62-dB SNDR noise-shaping SAR ADC. In: IEEE ISSCC Digest of Technical Papers, San Francisco, pp. 468–470, Feb 2012
19. Chen, Z., Miyahara, M., Matsuzawa, A.: A 9.35-ENOB, 14.8 fJ/conv.-step fully-passive noise-shaping SAR ADC. In: IEEE Symposium on VLSI Circuits Digest, Kyoto, pp. C64–C65, June 2015
20. Guo, W., Sun, N.: A 12b-ENOB 61μW noise-shaping SAR ADC with a passive integrator. In: IEEE European Solid-State Circuits Conference, Lausanne, pp. 405–408, 2016
21. Sun, N., Lee, H.-S., Ham, D.: A 2.9-mW 11-b 20-MS/s pipelined ADC with dual-mode-based digital background calibration. In: IEEE European Solid State Circuits Conference, Bordeaux, pp. 269–272, 2012
22. Kauffman, J., Witte, P., Lehmann, M., Becker, J., Manoli, Y., Ortmanns, M.: A 72 dB DR, CT ΔΣ modulator using digitally estimated, auxiliary DAC linearization achieving 88 fJ/conv-step in a 25 MHz BW. IEEE J. Solid State Circuits **49**(2), 392–404 (2014)
23. Lee, H., Hodges, D., Gray, P.: A self-calibrating 15 bit CMOS A/D converter. IEEE J. Solid State Circuits **19**(6), 813–819 (1984)

Chapter 5
A Hybrid Architecture for a Reconfigurable SAR ADC

Yun-Shiang Shu, Liang-Ting Kuo, and Tien-Yu Lo

5.1 Introduction

Reconfigurable analog-to-digital converters (ADCs) are commonly used in wireless communication systems, where a receiver chain usually needs to support multiple communication standards and digitize both narrow-band signals with high dynamic range (DR), such as for the GSM standard, and wide-band signals with moderate resolution, like in the LTE network. In sensor readout interfaces, the demands for configurable bandwidth and resolution are even more challenging when different physical signals in the real world are digitized by the same ADC working in a time-multiplexed manner. Usually, a successive-approximation-register (SAR) ADC is preferred because of its high power efficiency and flexible sampling rate, but for high-precision applications, a $\Delta\Sigma$ modulator-based ADC is still required. Therefore, in a multipurpose sensor platform, both SAR and $\Delta\Sigma$ ADCs are included to support various types of applications [1].

Figure 5.1 illustrates an example of a bio-signal acquisition system using electrodes to sense signals from a human body, where the performance of the ADC plays a key role in enabling different applications. When the electrodes are placed at proper locations on the human body, biopotentials, such as electroencephalography (EEG), electrocardiography (ECG), and electromyography (EMG), with different amplitude and frequency ranges can be detected. In Fig. 5.1, the amplitude of EEG signal may be only around several μV. In such cases, high dynamic range ADCs are required to distinguish a small signal from large interferers consisting of motion artifacts, DC offset, and 50/60 Hz coupling from the environment. In another bio-impedance case, the electrodes can be used to analyze body composition by

Y.-S. Shu (✉) • L.-T. Kuo • T.-Y. Lo
MediaTek Inc., Hsinchu, Taiwan
e-mail: yunshiang.shu@mediatek.com

P. Harpe et al. (eds.), *Hybrid ADCs, Smart Sensors for the IoT, and Sub-1V & Advanced Node Analog Circuit Design*, DOI 10.1007/978-3-319-61285-0_5

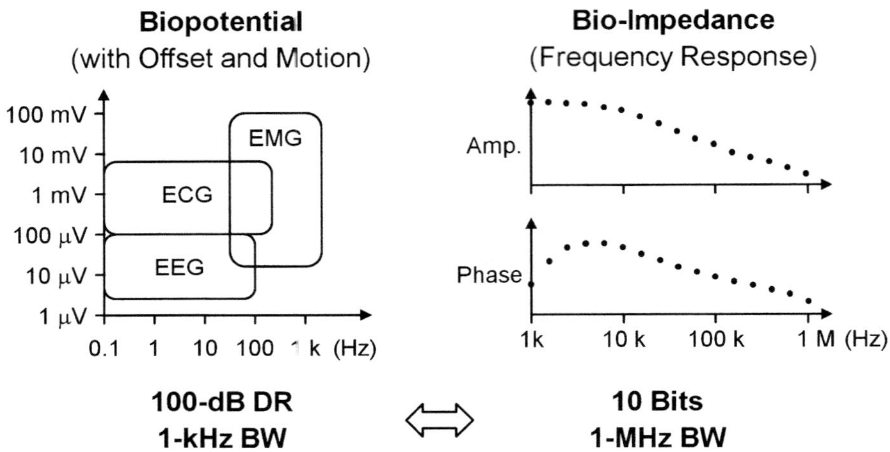

Fig. 5.1 Bio-signal acquisition with different ADC specifications

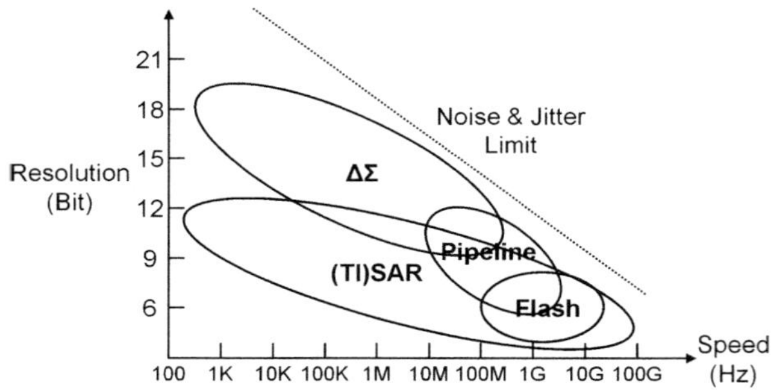

Fig. 5.2 Regimes of ADC topologies

measuring the amplitude and phase response of bio-impedance up to 1 MHz. To accommodate both cases with one ADC, the ADC needs to be configured between dynamic range performance of 100-dB within 1-kHz bandwidth and 10-bit resolution up to 1-MHz bandwidth.

Using one ADC to achieve both SAR- and $\Delta\Sigma$-type performance brings up a fundamental challenge in ADC design given that different ADC architectures are typically suitable for specific resolutions and conversion rates [2], as illustrated in Fig. 5.2. Flash ADCs can achieve high speed but at a low resolution, pipelined and SAR ADCs are suitable for medium-resolution applications, and $\Delta\Sigma$ modulators are capable of achieving high dynamic range but at a low conversion rate. Although it is possible for an ADC to achieve performance beyond its domain, it usually comes with a significant power and area penalty or extra cost for manufacturing. For example, SAR ADCs have been applied to high-resolution commercial products [3], but the requirements of trimming/calibration procedures and the use of high supply voltages to maximize signal-to-noise ratio (SNR) increase the production cost and power consumption.

Recently, the emerging hybrid architecture is blurring the boundaries between the different ADC architectures. This trend is being driven by both the low-power demands in mobile devices and also by the advance of CMOS technology. Designers have been trying to blend different architectures into one ADC design to achieve unprecedented power efficiency. Among the hybrid ADCs, SAR-assisted or SAR-based designs are the most popular due to the power-efficient nature of the SAR architecture and the fast switching behavior in advanced processes. Nowadays, SAR ADCs are capable of rivaling the performance of pipelined ADCs and, with time interleaving techniques, are even challenging the regime of high-speed flash ADCs [4]. Nevertheless, $\Delta\Sigma$-based designs are still dominant in the high-resolution space due to the strict noise and linearity requirements.

In the following sections, the fundamental ADC architectures are revisited to highlight ADC design challenges so as to further explain the principle, benefits, and limitations of hybrid architectures. Following this, a DAC mismatch error shaping technique is introduced and applied to a noise-shaping SAR ADC. This hybrid architecture allows a SAR ADC to be configured between SAR and $\Delta\Sigma$-type performance, as shown by the arrows in Fig. 5.2, and further blurs the boundaries between ADC topologies.

5.2 ADC Architectures and Fundamental Challenges

Hybrid ADCs are variations of the fundamental ADC architectures, including flash, subranging, pipelined, and $\Delta\Sigma$ modulator. The flash architecture is the most intuitive way to realize analog-to-digital conversion. This architecture is well known for its ability to achieve high speeds but has the drawback of exponentially increasing number of comparators as the resolution increases. However, another aspect which is seldom addressed is the requirements on comparator noise and offset. Theoretically, to gain every one extra bit, a 4× increase in device size is required to halve the mismatch error, and a 4× increase in power is needed in order to suppress noise by half. These imply that the most critical challenges in ADC design are device mismatch and circuit noise, which are related to the most important ADC characteristics of SFDR and SNR, respectively.

Different architectures have evolved to relax the burden on the comparators, as shown in Fig. 5.3. The subranging architecture divides the A/D conversion into coarse and fine steps in order to reduce the required number of high-performance comparators. It quantizes the input with a coarse ADC and uses an additional DAC to shift the signal into the input range of the fine ADC according to the coarse ADC output. Since $D_O = D_C + D_R$ and $V_I = V_C + V_R$, only an accurate DAC ($D_C = V_C$) and an accurate fine ADC ($V_R = D_R$) are required to guarantee the overall ADC performance. In such a case, the coarse ADC errors can be considered negligible as long as the residue, V_R, can be digitized by the fine ADC with sufficient input range. In other words, the mismatch and noise requirements of the coarse ADC are shifted to the additional DAC, which can achieve better SFDR and SNR with less power and area cost.

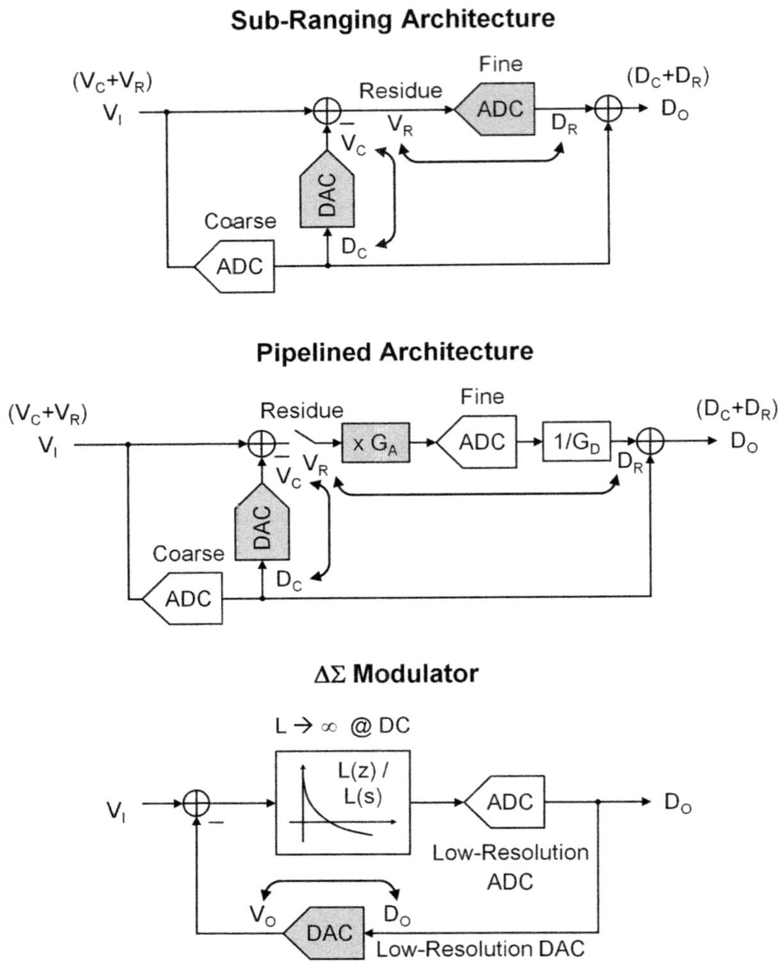

Fig. 5.3 Fundamental ADC architectures

A SAR ADC is an extreme case of subranging. It consists of multiple 1-bit stages, and the same comparator can be used in every stage to determine the polarities of the signal and residues. Consequently, there is only one comparator with a strict noise requirement, and the SFDR performance of the ADC is completely determined by the matching characteristic of the DAC cells.

Subranging and SAR ADCs impose a fundamental limitation on conversion rate due to the multistep operation and still need high-quality comparators in the fine ADCs. In contrast, a pipelined architecture, as shown in Fig. 5.3, solves these problems by sampling the residue of each stage and pipelining the coarse and fine operation. It also allows inserting a gain stage to amplify the residue before the fine quantization. With the interstage gain, G_A, the quantization step in the fine ADC can be enlarged to relax the comparator design. Therefore, only an accurate DAC ($D_C = V_C$) and an accurate gain stage ($V_R = D_R$) are required to keep $D_O = V_I$. This implies that the design requirement of the fine ADC is further shifted to the gain stage, where the analog gain, G_A, needs to match well with the ideal digital gain, G_D.

Once the circuit noise and device matching requirements are both shifted to the DAC arrays and gain stages, the ADC performance is now determined by the mismatch of the passive devices and the performance of the operational amplifiers in these blocks. If there is no strict power and speed constraint on the operational amplifiers, device matching eventually limits the ADC resolution to 10–12 bits, as exemplified by the regime of pipelined and SAR ADCs shown in Fig. 5.2.

A $\Delta\Sigma$ modulator with 1-bit quantizer is the only architecture which can achieve high resolution regardless of device mismatch. It is based on the concept of a negative feedback loop, where the feedback signal is equivalent to the input signal as long as the loop gain is high. If an ADC is put inside the loop, also shown in Fig. 5.3, its quantization and comparator errors will be suppressed by the high loop gain and can be ignored if the gain is high enough. This feature allows the use of a low-resolution ADC to digitize the signal, and therefore, either an inherently linear 1-bit DAC or a low-resolution DAC with dynamic element matching (DEM) can be adopted in the feedback path to realize an accurate DAC such that the digital output is equivalent to the feedback signal ($D_O = V_O$) and also the input signal, V_I. The high loop gain in $\Delta\Sigma$ modulators is usually realized by loop filters with poles at the band of interest. Since it is challenging to maintain a high loop gain over a wide bandwidth, the effective conversion rate of $\Delta\Sigma$ modulators is usually limited by the sampling rate achievable by the operational amplifiers.

Through this review of the fundamental ADC architectures, it can be observed that the requirements of device mismatch (SFDR) and circuit noise (SNR) on comparators are gradually shifted to DAC arrays and gain stages. Therefore, in modern ADC designs, comparator noise (in subranging ADCs), DAC linearity (in subranging and pipelined ADCs), and operational amplifier power consumption (in pipelined and $\Delta\Sigma$ ADCs) are the most critical challenges to be overcome by design innovations.

5.3 Hybrid Architecture with SAR ADCs

The principle of a hybrid architecture is to relax the aforementioned design challenges by choosing the best architecture for the sub-ADCs within the subranging, pipelined, and $\Delta\Sigma$ architectures in Fig. 5.3, depending on the particular design target and process. Although this is not a new concept, it has become compelling due to the recent trend of SAR-assisted designs. Traditionally, the sub-ADCs were realized with flash architectures for their high speeds, but with advances in

Sub-Ranging

	Coarse	Fine	
Chae, Y., ISSCC 2013 [6]	SAR	ΔΣ	Zoom ADC
Sanyal, A., CICC 2014 [7]	SAR	VCO	SAR-VCO ΔΣ
Liu, C.-C., ISSCC 2016 [8]	SAR	Digital Slope	SAR-Assisted Digital Slope
Dong, Y., ISSCC 2014 [9]	Flash	CT-ΔΣ	0-3 MASH

Pipelined

	Coarse	Fine	
Lee, C.C., VLSI 2010 [10]	SAR	SAR	SAR-Assisted Pipelined

ΔΣ Modulator

	Quantizer	
Tsai, H.-C., JSSC 2013 [11]	SAR	
Straayer, M.Z., VLSI 2007 [12]	VCO	

Fig. 5.4 Hybrid ADC examples

technology and design techniques, such as asynchronous timing control [5], SAR ADCs have been able to resolve more bits than flash ADCs with much lower power and area cost at an acceptable transport delay.

Figure 5.4 lists several examples to demonstrate the advantages of SAR-assisted hybrid architectures. For the subranging ADCs, resolving more bits in the coarse ADCs reduces the residue swing and enables more options to realize the fine ADCs. The zoom ADC [6] uses a 1-bit ΔΣ modulator as the fine ADC with relaxed requirement on operational amplifier bandwidth to achieve an ultralow power consumption. The SAR-VCO ΔΣ ADC [7] takes advantage of the small quantization steps of a VCO ADC without suffering from its limited linear input range. The SAR-assisted digital slope ADC [8] is able to achieve 100 MS/s since the time period needed to quantize the small residue in the linear slope stage is considerably reduced. In these subranging-based hybrid architectures, the demands for low-noise comparators and high-performance operational amplifiers are avoided.

For the pipelined architecture, additional bits in the coarse SAR ADC reduce the input swing of the gain stage. In the SAR-assisted pipelined ADC [10], the operational amplifier bandwidth and linearity requirements in the gain stage are greatly relaxed, and SAR ADCs can also be used as the fine ADC without strict comparator noise constraint. As to the ΔΣ architecture, a multi-bit SAR quantizer [11] can be used to reduce the quantization error and further lower the required loop gain and sampling rate, leading to relaxed operational amplifier design constraints. Figure 5.4 also includes two examples of hybrid ADCs [9, 12], which are not SAR-assisted, to demonstrate the idea of choosing different types of sub-ADCs depending on the particular design target.

A hybrid ADC can also be a SAR-based design. For example, the most significant bits (MSBs) of a SAR ADC can be resolved by a flash ADC [13] to enhance the conversion rate. In a noise-shaping SAR ADC [14], the residue after the SAR operation is sampled and used to shape both the comparator and quantization noise out of the band of interest, like $\Delta\Sigma$ modulators. This architecture relaxes the requirements on the comparator noise and reduces the number of required SAR cycles. It is especially interesting because the ADC can be switched between SAR and $\Delta\Sigma$ modes by disabling and enabling the noise-shaping function.

Although hybrid architectures can effectively overcome the challenges of comparator noise and operational amplifier design, the issue of DAC nonlinearity remains. Therefore, for high-resolution applications, $\Delta\Sigma$-based designs are still required to allow the use of low-resolution DACs. For example, the $\Delta\Sigma$ modulator with SAR quantizer [11] uses a digital modulator to reduce the number of bits for the feedback DAC. Also, although a noise-shaping SAR ADC can behave like a $\Delta\Sigma$ modulator, its achievable resolution is still limited by the DAC linearity.

5.4 DAC Linearization Techniques

To open up the application of hybrid ADCs for high-resolution A/D conversions, the issue of DAC nonlinearity must be taken care of properly.

Calibration is an effective method to remove device mismatch errors. It can be done in both analog and digital domains. Taking the subranging architecture in Fig. 5.3 as an example, since the target is to make $D_C = V_C$, analog calibration tunes the analog components so that V_C can approach the ideal D_C; similarly digital calibration adjusts the digital value of D_C to match the corresponding analog value, V_C, which is affected by device mismatch. However, calibration techniques raise several additional issues: how to measure the errors accurately, how to tune the analog values with sufficiently fine steps, how to track the long-term drift due to voltage and temperature variations, and the extra power and area cost for calibration circuitry and logic. Therefore, in most cases, it is still preferred either not to use calibration at all or to only calibrate the MSB cells for simplicity.

Dithering is one of the few options to improve DAC linearity without calibration. By injecting pseudorandom noise at the ADC input and removing it at the digital output, large periodic harmonic distortion can be spread over different frequencies. If a sinusoidal dither is applied, harmonics can also be modulated out of the band of interest [15]. However, the effectiveness of dithering depends on the amplitude of the injected dither, and the input signal range must be reduced to allow dither injection. Therefore, in practical cases, the improvement on linearity is limited because of the use of small dither to avoid excess dynamic range loss.

Dynamic element matching (DEM) is the most popular technique to linearize low-resolution DACs without calibration. This technique randomizes and shapes the mismatch errors out of the band of interest. However, due to the requirements of thermometer-coded DAC arrays and the exponentially increasing power and area

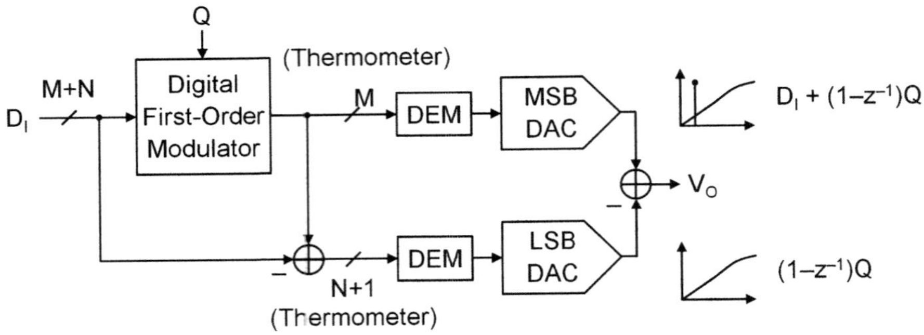

Fig. 5.5 Segmented noise-shaped scrambling scheme

cost with the number of bits, DEM is seldom used for DACs with >6-bit resolution. To extend the application of DEM, the segmented noise-shaped scrambling scheme [16], as shown in Fig. 5.5, is the only solution reported so far capable of linearizing high-resolution DACs without calibration. It uses digital $\Delta\Sigma$ modulation to separate the $(M + N)$-bit digital input, D_I, into the M-bit MSB of $D_\mathrm{I} + (1 - z^{-1})Q$ and the $(N + 1)$-bit least significant bit (LSB) of $(1 - z^{-1})Q$, where Q is the quantization error of the digital modulator. After conversion by two thermometer-coded DAC arrays with DEM, the first-order shaped quantization error is cancelled in the analog domain. By doing so, the mismatch between the two DAC arrays only results in a small residue of $(1 - z^{-1})Q$ within the band of interest.

The segmented noise-shaped scrambling scheme is very effective for high-resolution DACs. However, the digital $\Delta\Sigma$ modulator prohibits its application in SAR ADCs, since the modulator needs to receive all the digital bits before DAC switching, which is inconsistent with SAR operation. The transport delay of the digital modulator is also not suitable for the feedback loop in $\Delta\Sigma$ modulators. Therefore, this technique is seldom used in ADC designs, and how to linearize high-resolution DACs in ADCs (besides calibration) is still one of the fundamental challenges.

5.5 DAC Mismatch Error Shaping for SAR ADCs

In this section, a new technique, DAC mismatch error shaping (MES) [17], is introduced in detail to directly address the challenge of DAC linearity in SAR ADCs and to enable an over 100-dB SFDR without calibration. The idea of MES originates from the definition of DAC mismatch error. DAC mismatch error generally refers to the mismatch between the DAC cells, but more exactly, it is the mismatch between the DAC cells and their corresponding digital weights. Taking the subranging

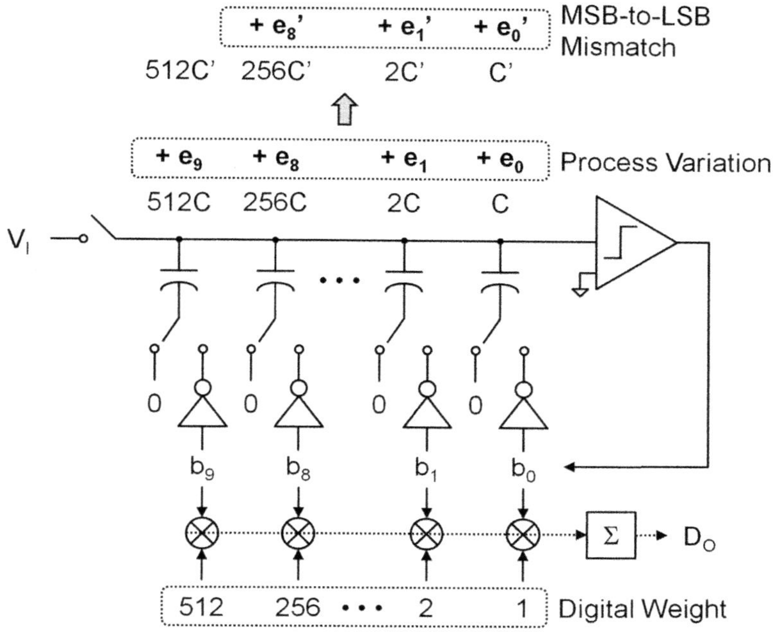

Fig. 5.6 Definition of DAC mismatch errors with MSB cell as reference

architecture in Fig. 5.3 as an example, DAC mismatch errors occur if the analog value, V_C, subtracted from the analog input, does not agree with the digital weights, D_C, added in the digital domain.

To precisely define DAC mismatch error, the relationship between the digital weights and the analog values should first be set by a reference. For example, in digital calibration, the digital weights of the MSB cells are measured and redefined by the following backend ADC [18]. That is, the LSB DAC is considered to have an ideal digital weight and is taken as a reference to define the digital weights of the MSBs. In DEM, the digital weight of a DAC cell corresponds to the average value of all the DAC cells, so that the summation of all the DAC mismatch errors is zero, leading to a notch at DC in the output spectrum.

In the MES technique, the MSB cell is taken as the reference to define the mismatch error, as illustrated in Fig. 5.6, where the 10-bit SAR ADC consists of a binary-weighted DAC array, 2^9C–2^0C, with corresponding digital weights, 2^9–2^0. In practice, each capacitor deviates from its ideal value due to process variation, as noted by the error terms of e_9–e_0. To further define mismatch error, the MSB cell is used as a reference to set $C' = (512C + e_9)/512$. Once C' is applied to all the DAC cells, the new error terms, e_8'–e_0', are the mismatches between the MSB cells and the LSB cells. That is, the MSB cell is considered perfectly matched with its digital weight, i.e., C'(analog value) $= 512$ (digital weight), and only the LSB cells contain mismatch errors relative to their digital weights.

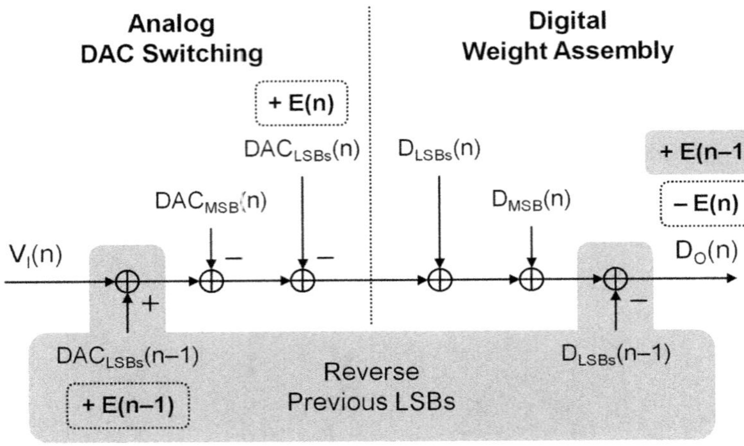

Fig. 5.7 Behavioral model of SAR ADC operation with DAC mismatch error shaping in gray area

Based on this definition, the SAR ADC operation can be modeled mathematically as depicted in Fig. 5.7, excluding the areas in gray for the moment. Ideally, the operations in the analog and digital domains should be perfectly matched such that the digital output equals the analog input exactly except for a small quantization error. In Fig. 5.7, the value of the MSB cell switching, $DAC_{MSB}(n)$, is first subtracted from the analog input. Then, the sum of the LSB cell switching, $DAC_{LSBs}(n)$, along with the LSB mismatch error, $E(n)$, is subtracted from the signal, where $E(n)$ represents a combination of $e_8'–e_0'$ depending on the LSB codes. After the corresponding digital weights of the LSBs, $D_{LSBs}(n)$ (=$DAC_{LSBs}(n)$), and the digital weight of the MSB, $D_{MSB}(n)$ (=$DAC_{MSB}(n)$), are added in the digital domain to reconstruct the signal, the LSB code-dependent error sequence, $E(n)$, appears at the digital output. This signal-dependent error results in harmonic distortions in the output spectrum.

To reduce the harmonic distortions, the principle of the MES technique is to duplicate the error incurred in the previous cycle, $E(n-1)$, and use it to cancel the error in the present cycle, $E(n)$. The cancellation is not exact but does generate a $(1 - z^{-1})$ high-pass filtering effect on the error sequence. In other words, if the error sequence, $E(n)$, is subtracted from itself with a one cycle delay, $E(n-1)$, the low-frequency components remain the same and are removed after subtraction. The gray area of Fig. 5.7 shows that an intuitive way to duplicate the previous error with opposite polarity is to reverse the operation of LSB in the previous cycle. By adding back the previous LSB value to the analog input and then removing it at the digital output, the error in the previous cycle, $E(n-1)$, is generated at the digital output to cancel $E(n)$.

The LSB reversion in the analog domain can be easily realized by a modified DAC switching scheme. Conventionally, all the DAC cells in a SAR ADC are reset in the sampling phase and switched to the references sequentially in the

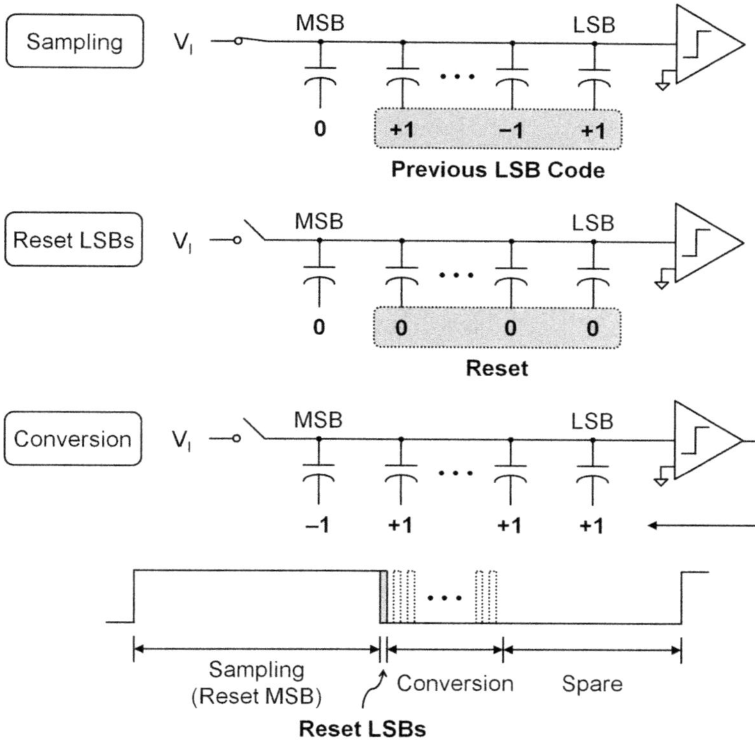

Fig. 5.8 SAR ADC operation with DAC mismatch error shaping

conversion phase. During this DAC switching process, the LSB mismatch error is injected. Figure 5.8 illustrates the operation of the modified switching scheme. In the sampling phase, the LSB cells are held on the references, which are set by the previous LSB code, instead of being reset to 0. After the sampling switch is disconnected, the LSB cells are switched from the references back to 0 to inject the previous LSB value, $DAC_{LSBs}(n-1)$, along with the LSB mismatch error, $E(n-1)$, into the sampled input. Once all the DAC cells are reset to 0, the normal conversion phase continues. The only difference in the timing control is an additional LSB reset phase. The time slot for this phase can be short since the incomplete settling of resetting the LSBs can be corrected for by redundancy in the LSBs [19].

Figure 5.9 shows the simulated output spectrums of a 12-bit SAR ADC before and after MES is applied to the 11-bit LSBs. It can be seen that low-frequency harmonics are greatly reduced and translated into high-frequency noise when MES is enabled. Once compared to an ideal first-order high-pass transfer function depicted in dashed line, it is interesting to observe that the harmonic suppression is actually better than the first-order shaping effect, as shown by the arrows. This phenomenon comes from the implicit dithering effect due to the LSB reversion because the LSB

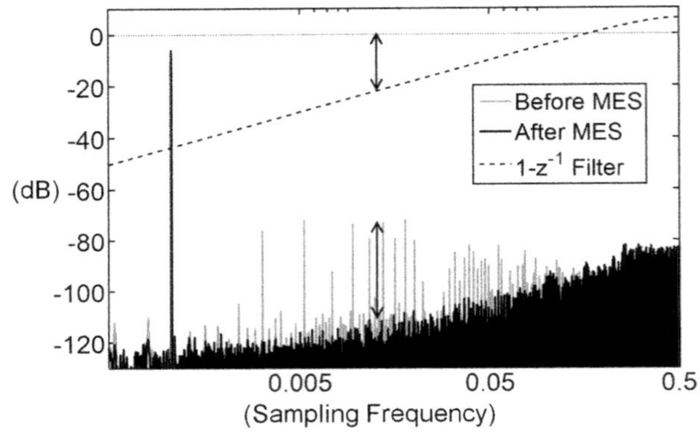

Fig. 5.9 Simulated output spectrum of 12-bit SAR ADC with DAC mismatch error shaping (MES)

values added to the analog input effectively represent the quantization noise after the MSB quantization. This dither-like signal is able to break large harmonics into different frequencies before shaping. Similar to most other dithering techniques, the injected dither occupies part of the signal range and sacrifices system dynamic range. In the example shown in Fig. 5.9, the signal must be reduced by 6 dB in order to allow the previous 11-bit LSBs to be injected at the ADC input.

An efficient way to mitigate the loss of dynamic range is to reduce the number of bits assigned to the LSB section. For example, if a 3-bit MSB is linearized using conventional data-weighted averaging (DWA) and MES is applied only to the LSB section to shape the MSB-to-LSB mismatch error, then the LSB section is reduced to one-eighth of full scale, resulting in less dynamic range loss due to MES. In this manner, the average value of the MSB cells becomes the reference to define the MSB-to-LSB mismatch error. This hybrid DAC linearization scheme comes with an additional benefit of mutual randomization effect because the MSB DWA sequence is randomized by the LSB dithering, and the MSB mismatch error shaped by DWA also randomizes the LSB codes. This effect results in less spurious tones before the first-order shaping takes effect.

A behavioral simulation with estimated circuit kT/C and comparator noise is applied to examine the effectiveness of the 3-bit DWA with MES and to compare it with the conventional DWA technique. Figure 5.10a is a histogram of SFDR from a 100-point Monte Carlo simulation with −2-dBFS input and random DAC mismatch at 64x oversampling ratio (OSR). It shows that DWA improves linearity as more MSB bits are included, as expected. Once MES is applied, SFDR with only 3-bit DWA becomes greater than 105 dB. The highest SFDR is limited by the noise floor due to the finite number of samples for FFT analysis. Figure 5.10b is another comparison at different input amplitudes. It shows that 8-bit DWA can achieve a high SFDR with a near full-scale input, but the linearity decreases and results in SNDR degradation around the 8-bit boundary because the MSB-to-LSB DAC mismatch error is not random enough in those conditions. On the other hand,

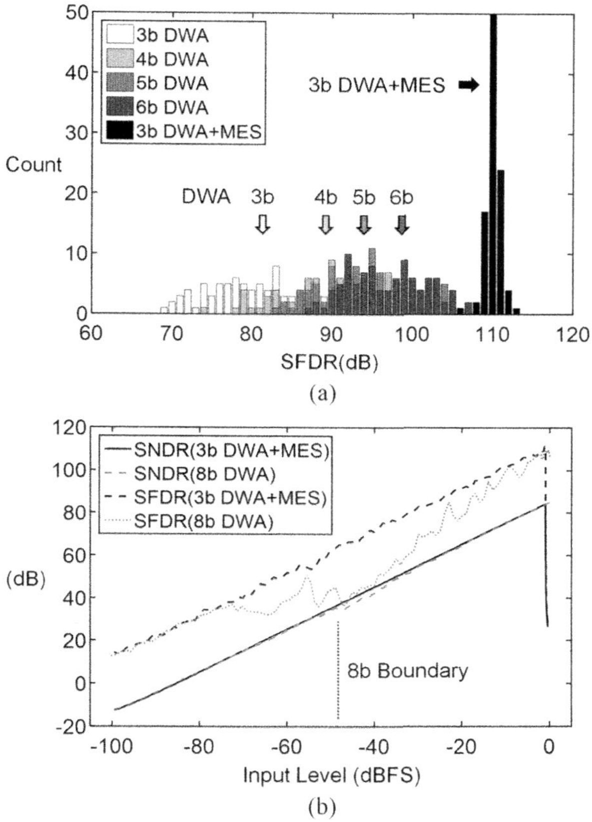

Fig. 5.10 Comparison of DWA and MES at 64x OSR: (**a**) histogram of simulated SFDR and (**b**) simulated SNDR/SFDR versus input level

the simulated results with the MES technique exhibit a constant SFDR/SNDR performance with typical circuit noise. The drop around 0 dB reflects the reduced dynamic range because of the LSB reversion.

5.6 General Form of DAC Mismatch Error Shaping Technique

The MES technique can be further extended to different types of error shaping other than first-order high-pass effect of $(1 - z^{-1})$. For example, if the previous LSB codes are inverted during the sampling phase, it results in a $(1 + z^{-1})$ filtering effect with a notch at $F_S/2$, where F_S is the sampling frequency. Similarly, if the inverted LSB codes from two cycles ago are applied, it generates a $(1 + z^{-2})$ transfer function with a pair of notches at $\pm F_S/4$ [17]. This feature is especially useful in band-pass ADCs, such as band-pass $\Delta\Sigma$ modulators.

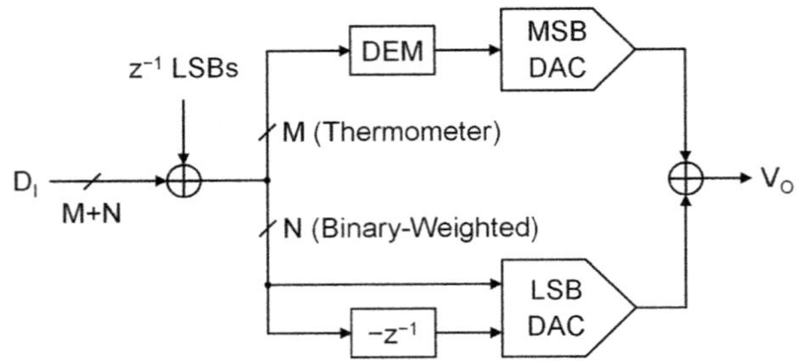

Fig. 5.11 Segmented DAC with mismatch error shaping

The hybrid DAC linearization scheme with DEM and MES can also be applied to typical high-resolution DACs, as shown in Fig. 5.11. Compared to the conventional segmented noise-shaped scrambling scheme in Fig. 5.5, this technique avoids the use of the digital modulator and the thermometer-coded LSB DAC. Similar to the DAC in the SAR ADC prototype, the $(M + N)$-bit digital input is separated into an M-bit thermometer-coded MSB DAC with DEM and an N-bit binary-weighted LSB DAC with MES. The LSB DAC needs to convert the present LSB codes and also performs the reverse operation of the previous LSB codes in the same cycle. This can be realized in switched-capacitor DACs, as in the SAR ADC example. For non-return-to-zero (NRZ) current DACs in continuous-time $\Delta\Sigma$ modulators, where each cell can only process one code in one cycle, two LSB DACs can be used to perform 1 and $-z^{-1}$ in alternate cycles to generate the $(1 - z^{-1})$ effect on the same DAC error. To compensate for the additional $- z^{-1}$LSBs in the analog domain, the z^{-1}LSBs are added to the digital input to keep the DAC output level, V_{O}, equivalent to the digital input value, D_{I}.

5.7 Reconfigurable Hybrid SAR ADC

The simulation results shown in Fig. 5.10 have demonstrated that a SAR ADC can achieve high linearity without calibration; together with noise shaping [14], both high SFDR and high SNR can be realized with a SAR ADC simultaneously.

Figure 5.12 shows the architecture of a 12-bit SAR ADC silicon prototype with MES and the first-order noise-shaping function. It is based on a coarse-fine architecture [20] with the concepts of hybrid ADC and hybrid DAC linearization. In this prototype, the three MSBs are resolved by a coarse flash ADC, whose thermometer-coded output passes through DWA logic and switches all the trilevel DAC cells simultaneously to avoid redundant charge transfers during the binary search process. The 11-bit LSB section, including 2-bit redundancy, is a nonbinary-weighted DAC array resolved by SAR operation with MES logic. The noise-shaping filter samples the residue after the SAR operation. The filtered residue is then used to shift the threshold voltage of the comparator to shape the comparator noise along with quantization noise.

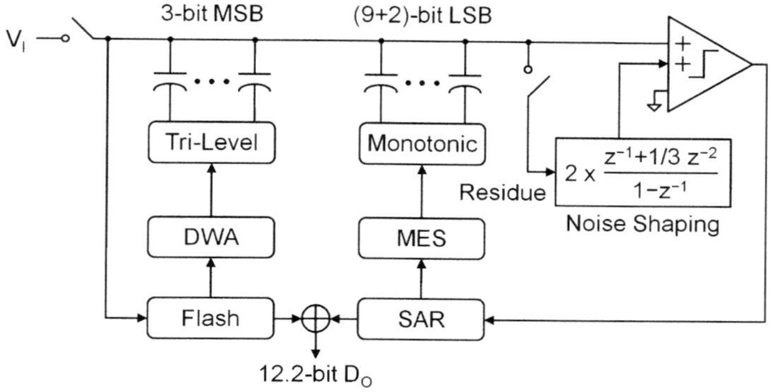

	ADC Architecture	DAC Linearization
MSB	Flash	Data Weighted Averaging
LSB	SAR	Mismatch Error Shaping
ADC	$\Delta\Sigma$	Dithering

Fig. 5.12 Architecture of 12-bit SAR prototype with hybrid ADC and DAC techniques

Figure 5.12 also indicates that this prototype includes the principles of three fundamental ADC architectures (flash, SAR, and $\Delta\Sigma$ modulation) as well as three DAC linearization techniques (DWA, MES, and dithering). Since the MES and noise-shaping functions can be enabled/disabled without interfering with the ADC's normal operation, this prototype can be switched from the conventional SAR mode to the oversampling $\Delta\Sigma$ modes by enabling the DWA + MES function and the noise-shaping filter. It is noted that this architecture can also be used as a simple flash ADC by disabling the DAC and the fine ADC functions.

The DAC resolution of the prototype is chosen to be 12 bits in order to minimize the residue at the noise-shaping filter input and to further save power. Since the comparator noise is around 10–11-bit level, the residue is mainly dominated by circuit noise instead of quantization error or large signal. Consequently, the switched-capacitor noise-shaping filter only needs to process small circuit noise with greatly relaxed gain, bandwidth, and linearity requirements on the operational amplifiers, resulting in an extremely power-efficient design.

5.8 Experimental Results

The 12-bit SAR prototype is fabricated in a 55-nm CMOS process. The oversampling $\Delta\Sigma$ mode operates at 1 MS/s and consumes 15.7 μW from a 1.2-V supply. When configured in the conventional SAR mode, the ADC consumes 22 μW at

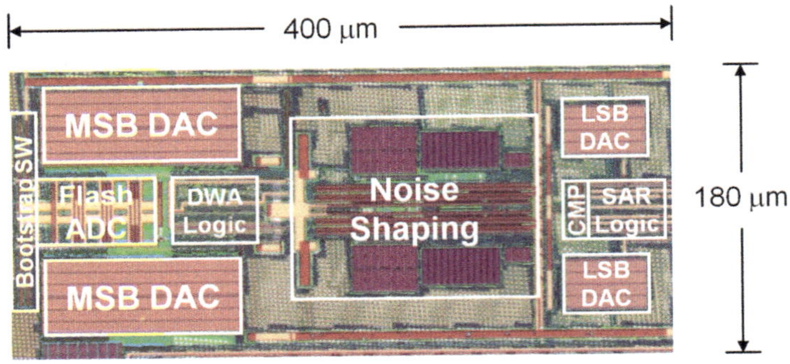

Fig. 5.13 Chip photograph

Fig. 5.14 Measured output spectrums in conventional SAR mode and oversampling $\Delta\Sigma$ mode at 1 MS/s

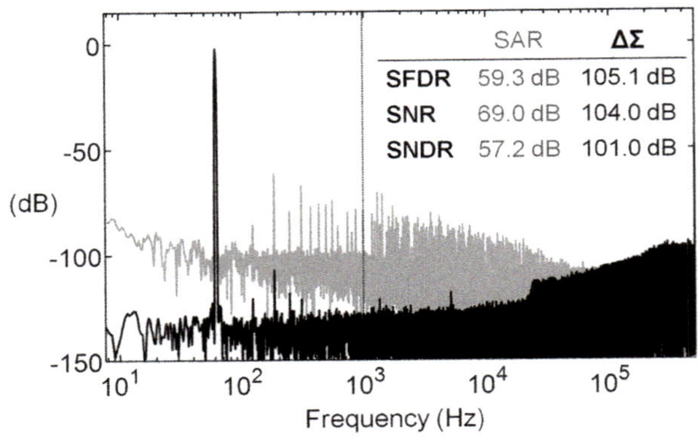

5 MS/s. A chip photo is shown in Fig. 5.13. The ADC occupies an active area of 0.072 mm^2 including decoupling capacitors. The noise-shaping filter is placed between the MSB and LSB DACs in order to exacerbate the MSB-to-LSB mismatch so as to verify the MES technique.

Figure 5.14 compares the measured output spectrums in the conventional SAR mode and the oversampling $\Delta\Sigma$ mode with a 60-Hz input. The spectrum of the conventional SAR mode shows a high noise floor with a clear trend of flicker noise and large harmonic distortions. The SNDR within the 1-kHz bandwidth is 57.3 dB limited by the harmonic distortions. Once the ADC is configured into the oversampling $\Delta\Sigma$ mode, all the in-band noise and the harmonics are significantly reduced with increased out-of-band noise, and the SNDR within 1-kHz bandwidth is improved to 101 dB. The 105-dB SFDR is limited by the third-order harmonic, which comes from the nonlinear parasitic capacitor on the top plates of the sampling DAC and is also observed from circuit simulation without DAC mismatch. The small bump around 20–30 kHz in the out-of-band noise is the remaining signal-dependent pattern of the DAC mismatch error, which exhibits much less tonal behavior due to the mutual randomization effects between DWA for MSBs and MES for LSBs.

Table 5.1 summarizes and compares the performance with state-of-the-art oversampling ADCs, including an oversampling SAR ADC [15], a continuous-time

Table 5.1 Summary of measured performance and comparison with state-of-the-art oversampling ADCs

	This ADC [17]		Harpe, P., ISSCC 2014 [15]		Sukumaran, A., JSSC 2014 [21]	Perez, A.P., ISSCC 2011 [22]
Architecture	Oversampling SAR		Oversampling SAR		CT-$\Delta\Sigma$	CT-$\Delta\Sigma$
Nyquist mode	9.5 bits up to 5 MS/s		11.3 bits up to 32 kS/s			
Process	55 nm		65 nm		0 18 μm	0.18 μm
Supply	1.2 V		0.8 V		1.8 V	1.5 V
Active area	0.072 mm^2		0.18 mm^2		0.24 mm^2	0.492 mm^2
Sampling rate	1 MHz		128 kHz		6.144 MHz	3.2 MHz
Order, bit	First-order, 12 bits		14 bits		Third-order, 1 bit	Third-order, 5 bits
Power	15.7 μW		1.37 μW		280 μW	140 μW
Bandwidth	1 kHz	4 kHz	4 kHz	16 kHz	24 kHz	50 kHz
SFDR	105.1 dB	105.1 dB	87.1 dB	86.9 dB	106 dB	96 dB
SNDR	101 dB	96.1 dB	79.1 dB	76.1 dB	98.2 dB	88.9 dB
FoM_s SNDR+10log(BW/P)	178.9 dB	180.0 dB	173.8 dB	176.8 dB	177.5 dB	174.4 dB

(CT) $\Delta\Sigma$ modulator [21], and a discrete-time (DT) $\Delta\Sigma$ modulator [22]. The ADC prototype achieves a 9.5-bit effective number of bits (ENOB) up to 5MS/s in the Nyquist mode. In the oversampling mode, the MES technique allows the 12-bit DAC to achieve 105-dB SFDR, which is comparable to the SFDR of the inherently linear 1-bit FIR DAC in the CT $\Delta\Sigma$ modulator [21]. This low distortion leads to over 100-dB SNDR at a 1.2-V supply, while higher voltages are used in the designs with high SNDR [21, 22]. The signal bandwidth of the prototype is the lowest because of the target applications. This ADC has the smallest area and is potentially more area efficient than CT $\Delta\Sigma$ modulators since large RC time constants and large devices for flicker noise are usually required in low-frequency designs. This architecture is also potentially more power efficient than DT $\Delta\Sigma$ modulators because of the relaxed operational amplifier design with small internal swings. Based on the measured SNDR over 4-kHz bandwidth, the prototype achieves the highest Schreier FoM of 180 dB compared with the state-of-the-art designs in [2] with >20-Hz signal bandwidth.

5.9 Conclusions

In this paper, the three fundamental challenges in modern ADC designs are revisited. While the emerging hybrid architectures can effectively relax the comparator noise and operational amplifier bandwidth requirements, a DAC mismatch error shaping (MES) technique is introduced to overcome the remaining challenge of DAC nonlinearity. A hybrid DAC linearization technique consisting of DEM, MES, and dithering is demonstrated in a SAR ADC prototype and achieves an over 100-dB SFDR without calibration. Once combined with noise shaping, the ADC can be configured between SAR and $\Delta\Sigma$ modes while maintaining the advantages of high power efficiency and flexible sampling rate in SAR ADCs. This hybrid architecture allows an ADC to be configured between flash, SAR, and $\Delta\Sigma$-type performance and further blurs the boundaries between different ADC architectures.

References

1. Van Helleputte, N. et al.: A multi-parameter signal-acquisition SoC for connected personal health applications. IEEE ISSCC Digest of Technical Papers, pp. 314–315 (2014)
2. Murmann, B.: ADC Performance survey 1997–.2016. Stanford Univ. [Online]. Available: http://www.stanford.edu/_murmann/adcsurvey.html
3. Analog Devices.: AD7485, 1 MSPS, serial 14-Bit SAR ADC datasheet [Online]. Available: http://www.analog.com/media/en/technical-documentation/data-sheets/AD7485.pdf
4. Kull, L. et al.: A 90GS/s 8b 657mW 64× interleaved SAR ADC in 32 nm digital SOI CMOS. IEEE ISSCC Digest of Technical Papers, pp. 378–379 (2014)
5. Chen, S.W.M., Brodersen, R.W.: A 6-bit 600-MS/s 5.3-mW Asynchronous ADC in 0.13-μm CMOS. IEEE JSSC. **41**(12), 2669–2680 (2006)

6. Chae, Y., Souri, K., Makinwa, K.A.A.: A 6.3 μW 20 bit incremental zoom-ADC with 6 ppm INL and 1 μV offset. IEEE JSSC. **48**(12), 3019–3027 (2013)
7. Sanyal, A. et al.: A hybrid SAR-VCO ΔΣ ADC with first-order noise shaping. Proceedings of the custom integrated circuits conference, IEEE, pp. 1–4 (2014)
8. Liu, C.C., Huang, M.C., Tu, Y.H.: A 12 bit 100 MS/s SAR-assisted digital-slope ADC. IEEE JSSC. **51**(12), 2941–2950 (2016)
9. Dong, Y., et al.: A continuous-time 0–3 MASH ADC achieving 88 dB DR with 53 MHz BW in 28 nm CMOS. IEEE JSSC. **49**(12), 2868–2877 (2014)
10. Lee, C.C., Flynn, M.P.: A SAR-assisted two-stage pipeline ADC. IEEE JSSC. **46**(4), 859–869 (2011)
11. Tsai, H.-C., et al.: A 64-fJ/conv.-step continuous-time ΣΔ modulator in 40-nm CMOS using asynchronous SAR quantizer and digital ΔΣ truncator. IEEE JSSC. **48**(11), 2637–2648 (2013)
12. Straayer, M.Z., Perrott, M.H.: A 12-bit, 10-MHz bandwidth, continuous-time ΣΔ ADC with a 5-bit, 950-MS/s VCO-based quantizer. IEEE JSSC. **43**(4), 805–814 (2008)
13. Lin, Y.Z. et al.: A 9-bit 150-MS/s 1.53-mW subranged SAR ADC in 90-nm CMOS. IEEE Symp. VLSI Circuits Digest of Technical Papers, pp. 243–244 (2010)
14. Fredenburg, J.A., Flynn, M.P.: A 90-MS/s 11-MHz-bandwidth 62-dB SNDR noise-shaping SAR ADC. IEEE JSSC. **47**(12), 2898–2904 (2012)
15. Harpe, P., Cantatore, E., van Roermund, A.: An oversampled 12/14b SAR ADC with noise reduction and linearity enhancements achieving up to 79.1 dB SNDR. IEEE ISSCC Digest of Technical Papers, pp. 194–195 (2014)
16. Adams, R., Nguyen, K.Q.: A 113-dB SNR oversampling DAC with segmented noise-shaped scrambling. IEEE JSSC. **33**(12), 1871–1878 (1998)
17. Shu, Y.-S., Kuo, L.-T., Lo, T.-Y.: An oversampling SAR ADC with DAC mismatch error shaping achieving 105 dB SFDR and 101 dB SNDR over 1 kHz BW in 55 nm CMOS. IEEE JSSC. **51**(12), 2928–2940 (2016)
18. Karanicolas, A.N., Lee, H.-S., Barcrania, K.L.: A 15-b 1-Msample/s digitally self-calibrated pipeline ADC. IEEE JSSC. **28**(12), 1207–1215 (1993)
19. Kuttner, F.: A 1.2V 10b 20MSample/s non-binary successive approximation ADC in 0.13μm CMOS. In: IEEE ISSCC digest of technical papers, pp. 176–177 (2002)
20. Tai, H.-Y. et al.: A 0.85fJ/conversion-step 10b 200kS/s subranging SAR ADC in 40nm CMOS. in IEEE ISSCC Digest of Technical Papers, pp. 196–197 (2014)
21. Sukumaran, A., Pavan, S.: Low power design techniques for single-bit audio continuous-time delta sigma ADCs using FIR feedback. IEEE JSSC. **49**(11), 2515–2525 (2014)
22. Perez, A.P., Bonizzoni, E., Maloberti, F.: A 84dB SNDR 100kHz bandwidth low-power single op-amp third-order ΔΣ modulator consuming 140μW. IEEE ISSCC Digest of Technical Papers, pp. 478–479 (2011)

Chapter 6
A Hybrid ADC for High Resolution: The Zoom ADC

Burak Gönen, Fabio Sebastiano, Robert van Veldhoven, and Kofi A. A. Makinwa

6.1 Introduction

Digital audio systems require high-resolution and high-linearity ADCs to digitize analog signals with high dynamic range. In today's system on chips (SoCs), such ADCs are typically integrated with digital signal processing blocks, thus forcing both to be realized in nanometer CMOS technology. Due to the relatively high silicon cost of such technologies, ADC area then becomes an important component of product cost. Moreover, the trend toward more mobile and wearable applications poses stringent constraints on the available power/energy. Thus, ADCs for digital audio systems should be both area and energy efficient.

Sigma-delta modulators (SDMs) are often employed in audio applications because they can achieve excellent linearity, even without calibration. However, compared to Nyquist-rate ADCs, which are typically less linear or require extensive calibration, SDMs are somewhat less energy efficient [13, 14]. To improve energy efficiency, recent ADCs have combined elements of Nyquist-rate and SDM architectures [1–6]. Such hybrid ADCs try to address the inability of both Nyquist-rate and SDM architectures to simultaneously achieve wide dynamic range, high resolution, and high accuracy, in an efficient manner. To do this, they typically split their input range into coarse and fine segments, each of which can be converted by different sub-ADCs that are optimized to handle the dynamic range of the different segments. The challenge then lies in combining the results of these conversions efficiently and accurately.

B. Gönen (✉) • F. Sebastiano • K.A.A. Makinwa
Delft University of Technology, Delft, The Netherlands
e-mail: B.Gonen@tudelft.nl

R. van Veldhoven
NXP Semiconductors, Eindhoven, The Netherlands

P. Harpe et al. (eds.), *Hybrid ADCs, Smart Sensors for the IoT, and Sub-1V & Advanced Node Analog Circuit Design*, DOI 10.1007/978-3-319-61285-0_6

In this paper, we describe a dynamic zoom ADC [10, 11], i.e., a hybrid ADC that consists of a compact and efficient coarse SAR ADC and an accurate and high-resolution fine discrete-time SDM (DT-SDM). The hybrid ADC achieves 109-dB dynamic range (DR), 106-dB signal-to-noise ratio (SNR), and 103-dB signal-to-noise-and-distortion ratio (SNDR) in a 20-kHz bandwidth, while dissipating 1.12 mW and occupying only 0.16 mm^2 in a 0.16-μm CMOS process.

The paper is organized as follows: first, the energy and area efficiency of high-resolution high-linearity ADCs is discussed (Sects. 6.2 and 6.3). This is followed by an overview of hybrid ADC architectures (Sect. 6.4). The zoom ADC and its system-level design are then introduced (Sect. 6.5), followed by its circuit design (Sect. 6.6). Finally, experimental results are presented (Sect. 6.7), followed by conclusions.

6.2 Energy Efficiency of High-Resolution DT-SDMs

The energy efficiency of an ADC is measured in terms of its energy per conversion E_{conv}, i.e., the energy spent by the ADC to produce an output sample. Figure 6.1 shows E_{conv} for ADCs published in recent years [14]. For low-resolution ADCs with N output bits, the energy per conversion is often limited by the energy required to compute the N output bits, thus E_{conv} scales with the number of conversion steps, i.e., $E_{\mathrm{conv}} \propto 2^N$. For high-resolution ADCs, i.e., ADCs with >75-dB DR, the energy per conversion is limited by the need to achieve sufficiently low thermal noise, which requires a quadratic increase of energy for each additional quantization step ($E_{\mathrm{conv}} \propto 2^{2N}$) [13]. Consequently, ADC power consumption will scale with DR. This consideration leads to the definition of the Schreier figure of merit (FoM$_S$) [12]:

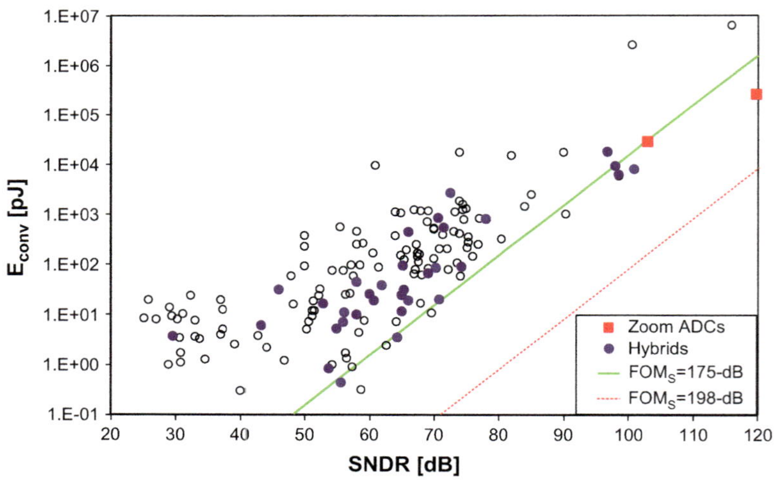

Fig. 6.1 SNDR vs energy per conversion of ADCs (2012–2017) [14]

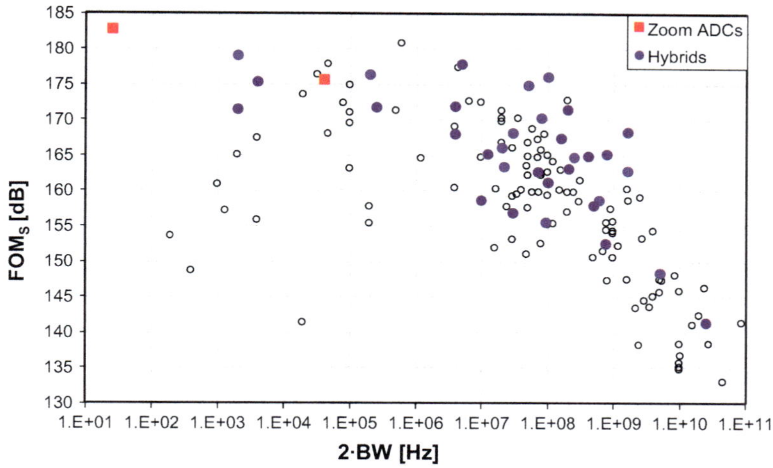

Fig. 6.2 Signal bandwidth vs $\mathrm{FoM_s}$ of ADCs (2012–2017) [14]

$$\mathrm{FoM_s} = DR + 10 \cdot \log_{10}\frac{f_{bw}}{P} \ [dB] \tag{6.1}$$

where DR is the dynamic range in dB, f_{bw} is the ADC bandwidth, and P is the ADC power consumption. Sometimes $\mathrm{SNDR_{max}}$ is used instead of DR, e.g., as in [14], because the former is usually worse than the latter. Energy efficiency is also difficult to combine with high speed, as shown in Fig. 6.2, which reports $\mathrm{FoM_S}$ vs input bandwidth for ADCs published in recent years [14]. $\mathrm{FoM_S}$ is higher for low and moderate bandwidths (0–10 MHz). It can be shown that a 198-dB $\mathrm{FoM_S}$ is theoretically achievable [15], which means that there is still an approximately 20-dB gap between this limit and the current state of the art (Fig. 6.1).

The reason for this gap lies in the implicit assumption behind the definition of $\mathrm{FoM_S}$, i.e., that most of an ADC's power consumption is used in its input stage to reduce thermal noise. In practice, however, this is not the case. First, other ADC subblocks consume a non-negligible amount of power. In a DT-SDM, for example, such subblocks will include the integrators that follow the input stage, the quantizer, the biasing circuits, and the digital back end. Second, even the power consumption of the input stage is often not limited by thermal noise but by other requirements such as linearity, slew rate, and settling time.

Bearing this in mind, it is then clear that to maximize $\mathrm{FOM_S}$, a number of different design strategies can be adopted. First, the power consumption of all ADC subblocks, especially that of the critical input stage, should be reduced. Several recent works have targeted improvements in the efficiency of the amplifiers used in the SDM loop filters. Various inverter-based amplifiers have been proposed [1, 10, 16–22] which double efficiency by summing the transconductances of NMOS and PMOS transistors biased by the same current. A further improvement is achieved in

[19] by stacking multiple inverters and adopting a high-supply-voltage technology. As a second design strategy, the chosen ADC architecture must maximally relax all requirements on the input stage apart from those related to thermal noise. For example, the use of multi-bit quantization reduces the signal swing processed by the loop filter and thus relaxes the input stage's slewing and settling time. However, to avoid excess loop delay, this often involves the use of quantizers based on power-hungry flash ADCs. Furthermore, fast and power-hungry digital logic is needed to implement the dynamic element matching techniques needed to guarantee linearity. Thus, efficient multi-bit quantization schemes are required.

6.3 Area Efficiency of High-Resolution ADCs

A similar approach can be used to analyze ADCs from an area efficiency perspective. Most of their silicon area should then be used to ensure low enough thermal noise. However, since the matching of integrated components scales with the square root of their area, i.e., two times more accuracy requires four times larger area, the accuracy requirements on active and passive components also impose a lower limit on the silicon area [20]. This implies that in a DT-SDM the total area should ideally be dominated by thermal-noise-critical and matching-critical components, such as sampling capacitors, the first integrator, and the DAC.[1] It should also be noted that over-sampling effectively reduces the in-band thermal noise in an ADC, hence relaxing the area requirement of the noise-critical capacitors for the same DR. Although good for area efficiency, over-sampling comes at the expense of increased power consumption in the quantizer and in the digital back end.

However, components not limiting noise or accuracy will also occupy a non-negligible chip area. For example, SDMs with multi-bit quantizers usually suffer an area penalty due to the quantizer's exponentially increasing area [20]. Furthermore, not all passives are sized for thermal noise or accuracy requirements. For example, the size of the integration capacitors in the switched-capacitor (SC) integrators of a DT-SDM is determined by the choice of loop-filter coefficients and the desired integrator output swing.

Figure 6.3 shows an area vs DR comparison of state-of-the-art audio ADCs [11]. It shows that higher DR indeed corresponds to higher chip area. It should be noted that both device matching and capacitor density (capacitance per unit area, in F/μm^2) and, hence, the resulting chip area are strongly technology dependent. Technology scaling also helps to reduce the power consumption of the digital logic and the quantizer, thus facilitating, for example, the use of multi-bit SDMs. System-level design should then include a careful choice of the technology in order to exploit the possible presence of high-density passives.

[1] While it is trivial that lower thermal noise requires larger capacitors in DT circuits, this is also true for continuous-time circuits: lower thermal noise implies lower resistances and, consequently, larger capacitors for the same total bandwidth.

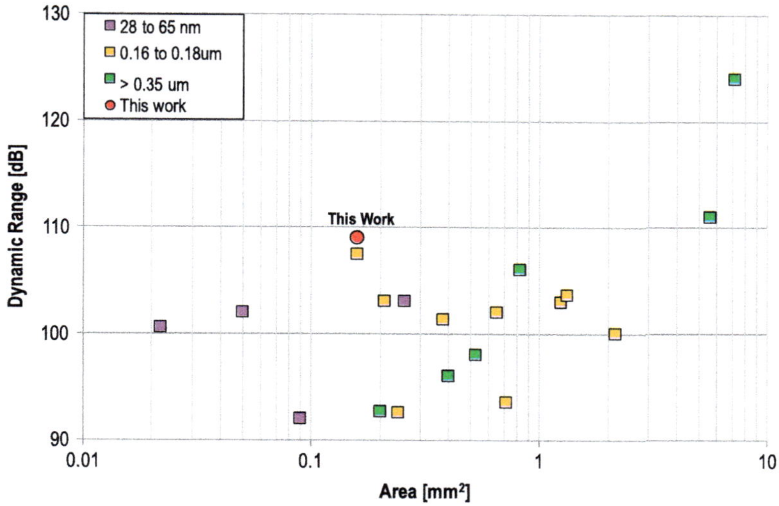

Fig. 6.3 Chip area vs DR for state-of-the-art audio ADCs [11]

To improve the area efficiency of DT-SDMs for a given technology, the following design flow should be followed. First, the over-sampling ratio (OSR) should be increased until the power consumed by the digital back end and the quantizer becomes significant. Second, the architecture should be chosen to reduce the area of system blocks that do not directly determine resolution and accuracy, such as the integration capacitors not in the first integrator, the quantizer, and the digital logic.

6.4 Hybrid ADCs

As mentioned before, most ADC architectures are not energy efficient when high resolution and high linearity are both required. Even conventional SAR ADCs, known for their excellent energy efficiency, suffer under high-resolution requirements due to the increased power consumption of the comparator. Furthermore, some ADC architectures, such as VCO-based converters, exhibit excessive nonlinearity for large input signals, thus limiting their DR [8]. Thus, it is often beneficial to divide the input dynamic range into "manageable" subranges, i.e., coarse and fine ranges, to decouple the problems associated with large signals, low noise, and high accuracy levels. In this way, the challenges of each design space, i.e., "subrange", can be addressed with an appropriately tailored ADC architecture. ADCs based on this approach are called *hybrid ADCs*. It is beneficial to have a close look into a subset of the hybrid ADCs, subranging ADCs, to understand their architectural motivation. Subranging ADCs were originally used to improve the efficiency of flash ADCs for high resolution by dividing the input range into multiple

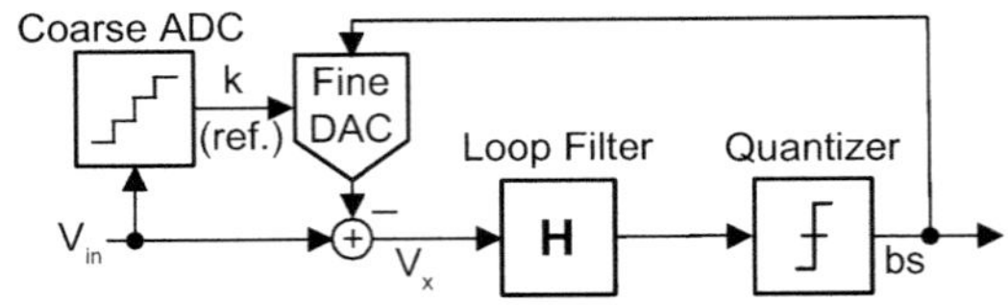

Fig. 6.4 Block diagram of a dynamic zoom ADC

subranges, most commonly into two coarse and fine ranges. However, they suffered from interstage matching, i.e., the coarse converter and the fine converter should have a perfectly matched range. This results in tough requirements on the thermal noise and accuracy of the coarse converter, which, in turn, lead to degraded energy efficiency. For this reason, back-end correction techniques such as redundancy (over-ranging) and/or calibration are often employed to relax the coarse converter's accuracy requirements [15], leading to very efficient designs.

By dividing the full input range into two or more subranges, hybrid two-step architectures in the form of SAR + SAR pipeline [9], SAR + single slope [7], SAR + SDM [1–4, 8], and flash + SDM [5, 6] achieve state-of-the-art energy efficiency, as shown in Figs. 6.1 and 6.2. In addition, their linearity is often improved thanks to the reduction of the signal swing at the input of the linearity-critical fine converter. It is observed that the architectures of the coarse and fine converters are tailored to the desired performance. For low-to-moderate input bandwidths, i.e., not close to the speed limits of the technology used, the coarse converter is often a SAR ADC [1–3, 7–9] due to their compactness and superior energy efficiency. When the speed of the coarse conversion is important, a flash ADC is preferred [5, 6]. In very efficient high-resolution (DR > 75 dB) hybrids, the fine converter is either a SDM [1, 2, 5], an over-sampling SAR [4], or a single-slope ADC [7], to achieve high resolution with maximum efficiency.

The zoom ADC architecture has been proposed for high-resolution and high-linearity applications, in which it simultaneously achieves excellent energy efficiency and small die area [1]. The system block diagram of a zoom ADC is shown in Fig. 6.4. It consists of a coarse ADC and a fine SDM. The coarse ADC's output (k) corresponds to an analog range $k \cdot V_{LSB,C} < V_{in} < (k+1) \cdot V_{LSB,C}$ where $V_{LSB,C}$ is its quantization step or least significant bit (LSB). The digital value k is then used to adjust, i.e., "zoom in," the references of the SDM's DAC such that $V_{REF-} = k \cdot V_{LSB,C}$ and $V_{REF+} = (k+1) \cdot V_{LSB,C}$. These reference voltages straddle the input signal V_{in}, thus ensuring that it lies in the input range of the fine $\Delta\Sigma$M. In contrast to other Nyquist-rate ADC + SDM hybrids, there is no computation of an analog residue signal resulting from the coarse conversion. Instead, only the digital result of the coarse conversion is used to "zoom in" on the signal level. By using a wider fine input range (i.e., over-ranging), the coarse converter's linearity and accuracy can be considerably relaxed. The overall linearity is determined by the fine SDM, in particular by its DAC, whose linearity is then improved by using dynamic element matching techniques.

Like a multi-bit SDM, zooming reduces the signal swing at the input of the SDM, thus relaxing the slewing requirements of the first SDM stage. Its performance will also be similar to that of a multi-bit SDM with the same OSR, despite the fact that there its multi-bit quantizer is *outside* the SDM loop. Consequently, higher quantizer delays can be tolerated, which means that the coarse converter can be implemented as a compact and efficient SAR ADC.

6.5 Incremental Zoom ADC

The first zoom ADCs were implemented as incremental converters, in which the coarse and fine conversions were performed sequentially [1, 2, 23]. The time-domain operation of an incremental zoom ADC is shown in Fig. 6.5. The conversion starts with a SAR phase to quickly determine the correct zoom range, followed by a fine 1-bit SDM phase that uses the nearest two reference levels to accurately determine the final digital value. This approach works well for quasi-static signals, such as those encountered in sensor readout [2, 23], or instrumentation applications [1], but it does not work for dynamic signals.

The time-domain operation of an incremental zoom ADC with a dynamic (time-varying) signal is shown in Fig. 6.6. After the SAR period, the ADC will assume that the chosen reference values are valid throughout the whole fine conversion. However, this is not true for dynamic signals, leading to modulator overload. Thus, the maximum input signal frequency will be limited to when assuming a 1-LSB$_C$ (coarse LSB) of fine input range, i.e., no over-ranging:

$$f_{\text{in,max}} < \frac{f_s}{2\pi \cdot \text{OSR} \cdot 2^N} \tag{6.2}$$

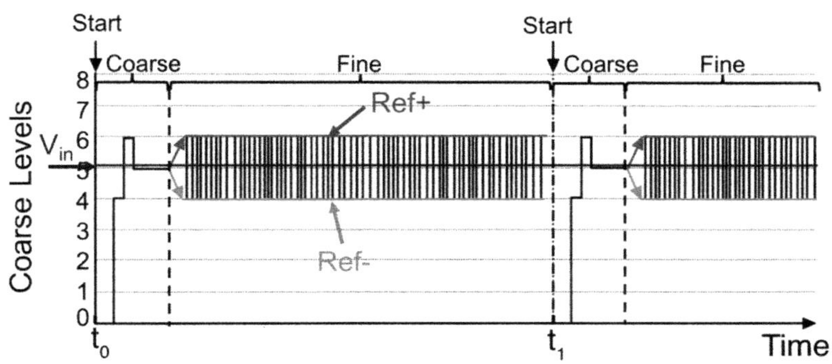

Fig. 6.5 Time-domain operation of an incremental zoom ADC with a static input. Showing the SAR ADC's comparator output during the coarse period, and the SDM's bitstream during the fine period

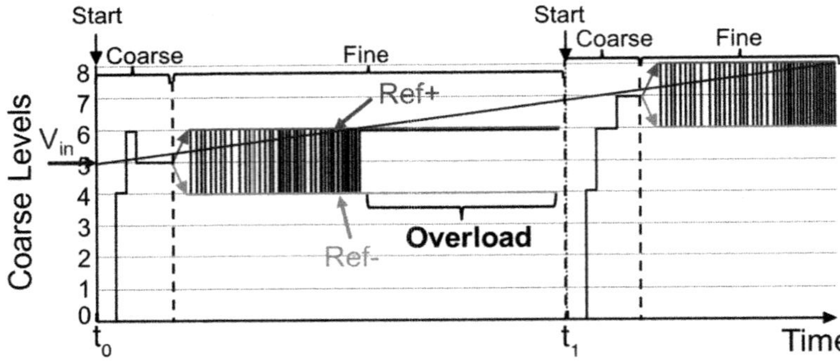

Fig. 6.6 Time-domain operation of an incremental zoom ADC with a dynamic input. Showing the SAR ADC's comparator output during the coarse period, and the SDM's bitstream during the fine period

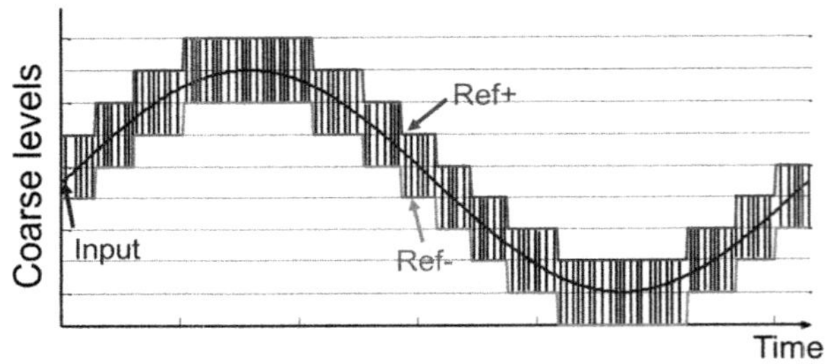

Fig. 6.7 Time-domain operation of a dynamic zoom ADC

where f_S is the sampling frequency, OSR is the over-sampling ratio of the zoom ADC, and N is the coarse ADC's number of bits. Thus, this architecture is only well suited to the conversion of quasi-static signals.

6.5.1 Dynamic Zoom ADC

In order to achieve greater input bandwidth, a *dynamic* zoom ADC is proposed in which the coarse and fine conversions are performed concurrently, i.e., in parallel [10, 11]. The time-domain operation of a dynamic zoom ADC with a sinusoidal input is shown in Fig. 6.7. The SDM references Ref + and Ref- then track the input signal fast enough to ensure that they always straddle it. The maximum input bandwidth for a dynamic zoom ADC is then

$$f_{\text{in,max}} < \frac{f_{\text{coarse}}}{2\pi \cdot 2^N} \tag{6.3}$$

where f_{coarse} is the coarse ADC sampling frequency, which is an integer fraction of f_S, i.e., f_S/N for an N-bit SAR ADC or f_S for a flash ADC. Compared to its incremental counterpart, the maximum input frequency of the dynamic zoom ADC is not a function of the SDM's OSR, thus allowing the use of a large OSR with the associated benefits in terms of resolution and area occupation.

6.5.2 A Dynamic Zoom ADC for Digital Audio

A prototype dynamic zoom ADC for digital audio has been designed as a proof of concept. The targeted specifications are 106-dB SNR and SNDR higher than 100 dB in the 20-kHz audio bandwidth with 1.25 V_{rms} input range. The chosen process technology is 0.16-μm CMOS. The system-level design starts with architectural choices. For a dynamic zoom ADC, these include the choice of the following parameters: F_s (OSR), coarse ADC resolution, coarse ADC redundancy (over-ranging), SDM loop-filter structure and order, and SDM quantizer resolution.

6.5.3 The SDM: OSR, Loop Filter, and the Quantizer

The efficiency of a SDM is, to first order, independent of its OSR. Increasing OSR is desirable to reduce the signal-to-quantization-noise ratio (SQNR) and the chip area and increase $f_{in,max}$ in (6.3). However, the power consumption of the digital sections (DEM, SAR controller, clocking) increases proportionally with the SDM sampling frequency f_s and, consequently, proportionally with OSR. For the chosen 0.16-μm CMOS technology, system simulations revealed that 11.2896 MHz (audio standard), corresponding to OSR = 282, is a good compromise between chip area, $f_{in,max}$, and digital power consumption.

For high energy efficiency, a thermal-noise limited SNR is desired, i.e., the quantization noise should be much less than the thermal noise. To achieve the targeted thermal-noise limited 110-dB SNR, SQNR = 130 dB is chosen. The zoom ADC's total SQNR is determined by the coarse resolution and the SDM's SQNR. The last depends on its loop-filter order, the quantizer resolution, and the OSR. Zooming relaxes the SQNR requirement of the SDM by reducing its input range. Thus, more than 1-bit quantization in the loop is not necessary.

To determine the loop-filter order of the SDM, Fig. 6.8 shows the $SQNR_{max}$ (for an ideal loop filter) as a function of the OSR for a zoom ADC with a 1-bit SDM quantizer and a coarse SAR ADC with 3–5 bits, for different loop-filter orders [12]. It is observed that for each increased bit in the coarse ADC, $SQNR_{max}$ increases by 6.02 dB similar to multi-bit SDMs. For the chosen OSR = 282, a second-order loop filter would be sufficient. However, for a robust design, a third-order SDM is chosen. Thanks to the noise scaling of the third stage, the power consumption of the third stage is expected to account for only 15% of the whole loop filter (simulated). For the implementation, a switched-capacitor (SC) loop filter is chosen for its robustness to clock jitter. The SC loop filter is chosen as a cascade of integrators with feed-forward (CIFF) for its superior linearity.

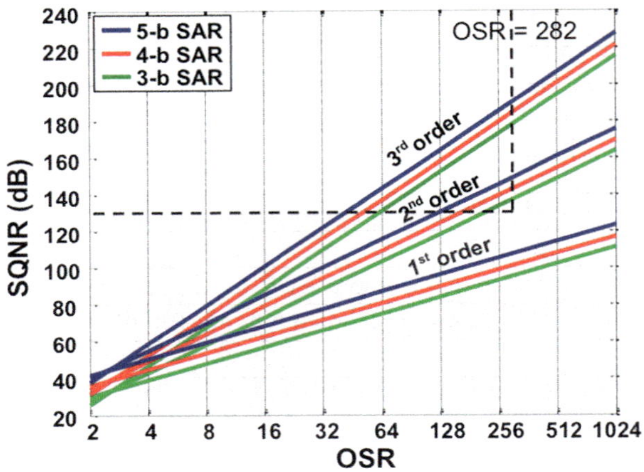

Fig. 6.8 OSR vs maximum SQNR of a zoom dynamic for different loop-filter orders and SAR resolutions

6.5.4 The SAR ADC and Over-ranging

As mentioned before, the fine DAC in Fig. 6.4 uses the digital result (k) of the SAR ADC to dynamically adjust its references. If according to the coarse converter the input signal satisfies $k \cdot V_{LSB,C} < V_{in} < (k + 1) \cdot V_{LSB,C}$ where $V_{LSB,C}$ is the coarse converter quantization step or least significant bit (LSB), the references of the fine DAC are set to $V_{REF-} = k \cdot V_{LSB,C}$ and $V_{REF+} = (k + 1) \cdot V_{LSB,C}$. However, using the zoomed-in references has several problems. Foremost, SDMs are not stable over the full range of their DACs. So, if V_{in} is close V_{REF+} or V_{REF-}, the SDM could be overloaded. Furthermore, any error in the coarse ADC due to mismatch or the coarse converter's thermal noise can lead to an error in k causing V_{in} to fall outside the SDM's input range. In that case, the SDM overloads and fine conversion becomes totally invalid, similar to the interstage mismatch problem of subranging converters [15]. To address this issue, the input range of the SDM can be widened by using *over-ranging*, so that the SDM DAC references are chosen as $V_{REF+} = (k + 1 + M/2) \cdot LSB_C$ and $V_{REF-} = (k - M/2) \cdot LSB_C$ where M is the over-ranging factor. Since the SDM DAC range is widened both at the low and at the high side, the DAC references symmetrically straddle the signal, i.e., V_{in} is approximately in the center of the SDM input range. Thus, even in the presence of a coarse conversion error smaller than $\pm M/2$, the SDM can operate without overloading. This allows for larger errors in the coarse ADC converter.

Figure 6.9 shows simulated maximum acceptable SAR ADC INL vs M for zoom ADCs with 4–6-bit SAR ADCs and third-order SDM with 1-bit quantization. For each data point, a 100-point Monte Carlo simulation has been run, and the maximum INL which causes less than 10-dB SQNR deviation is reported. The offset of the SAR ADC is not included for the sake of simplicity. The maximum tolerable INL ($\pm LSB_C$) is found independent of the coarse resolution; however, the relative matching of the unit elements increases quadratically for each coarse bit, i.e., from

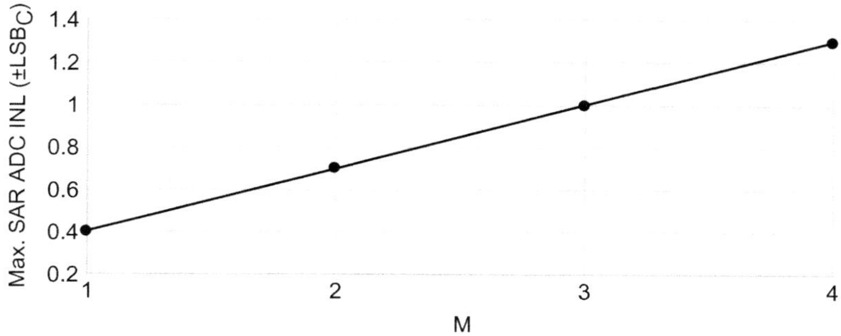

Fig. 6.9 Maximum simulated tolerable SAR ADC INL (LSB_C) vs M for zoom ADCs with a third-order SDM

5 bits to 6 bits, due to the smaller size of LSB_C. As it is seen from Fig. 6.9, the maximum acceptable INL error increases proportionally with M, thus dramatically relaxing the SAR ADC's accuracy requirements. Even missing codes are tolerated for $M \geq 3$.

Over-ranging comes at the cost of a lower SQNR, since doubling M results in a 1-bit less coarse resolution, i.e., 6-dB less SQNR, but it greatly simplifies the design of the SAR ADC and, consequently, its power consumption and area occupation. Thus, it makes an energy-efficient two-step conversion possible while keeping the SDM input range small ($= [M + 1]\cdot LSB_c$), avoiding strict matching requirements between two converters, and overloading in the SDM. Over-ranging also helps in increasing the maximum input signal bandwidth by modifying (6.3) into

$$f_{\text{in,max}} < \frac{(M + 1) \cdot f_{\text{coarse}}}{2\pi \cdot 2^N} \tag{6.4}$$

Figure 6.10 shows M vs $f_{\text{in,max}}$ for SAR ADCs with 4–6 bits and with $f_s = 11.29$ MHz. To allow for the 20-kHz signal bandwidth, viable options are both a 4-bit SAR ADC with $M = 2$ and a 5-bit SAR ADC with $M = 4$. However, the latter provides a better coarse resolution, i.e., a more precise reference range estimation, with negligible additional power and area. So, in this work a 5-bit SAR ADC with 4-LSB_c over-ranging is used.

6.5.5 *The Overall System*

A system-level block diagram of the proposed dynamic zoom ADC is shown in Fig. 6.11. The CIFF loop filter is optimized for area as explained in the following. In a CIFF SDM, the first integrator's gain coefficient a_1 (Fig. 6.11) is usually smaller than one to utilize a large stable input range [12]. However, in our case, a larger a_1 is possible since zooming causes a first-integrator input much smaller than the zoom

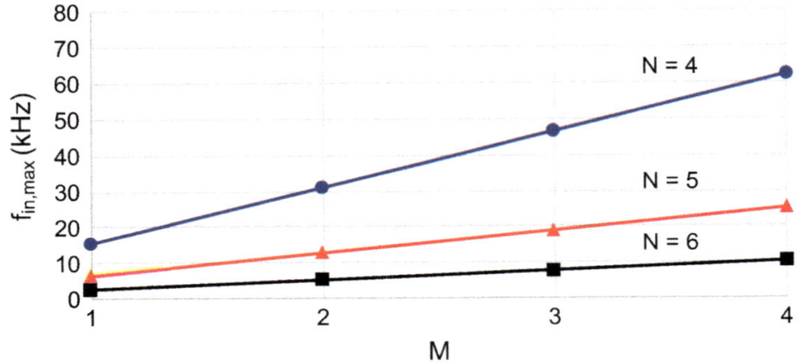

Fig. 6.10 $f_{in,max}$ vs M for 4–6-bit SAR ADCs clocked at 11.29 MHz f_S

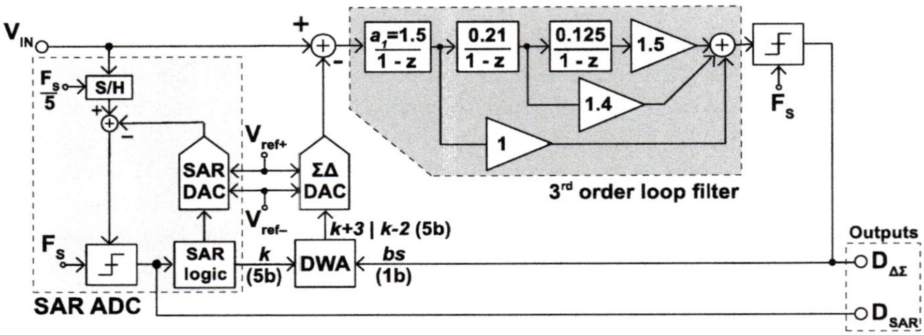

Fig. 6.11 System-level block diagram

ADC total full-scale input. A larger a_1 can be exploited to save a considerable silicon area, as explained in the following. Coefficient a_1 in the proposed implementation can be expressed as:

$$a_1 = \frac{C_S}{C_{\mathrm{int}\,1}} \tag{6.5}$$

where C_{int1} is the first integration capacitor and C_S is the sampling capacitor. The integrator is assumed to be non-inverting for the sake of simplicity. C_S is determined by the kT/C noise requirement, and it is fixed for a given OSR, thus resulting in C_{int1} being proportional to a_1. Since the area of a DT-SDM is dominated by the first-stage capacitors, $a_1 > 1$ allows for a large area saving. As a drawback, this increases the output swing of the first integrator, which however is quite small in a zoom ADC, and increasing it does not constitute an issue. Hence, $a_1 = 1.5$ is chosen (Fig. 6.11) corresponding to a first-integrator output swing of 27% of the full-scale output range. Such output swing is within the linear range of the inverter-based class-AB OTA used to implement the first integrator (see Sect. 6.5.5).

6.6 Circuit Design

A simplified circuit schematic of the proposed dynamic zoom ADC is depicted in Fig. 6.12. The 5-pF sampling capacitor is sized for kT/C noise and implemented by a parallel array of 31 capacitors that is also used as the fine DAC. The size of each integration capacitor $C_{\text{int},p-n}$ is 3.3 pF each. In the tracking phase Φ_1, all capacitors $C_{s[1..31]}$ are connected to the input. At the end of the tracking period, switch S_{13}, driven by an earlier phase clock Φ_{1e}, opens to sample the input on all capacitors while cancelling the input common-mode voltage. Input common-mode rejection is thus limited by the relative matching of the two sampling capacitor arrays, which was considered sufficient for the targeted application. In the integration phase Φ_2, m DAC elements ($m = k - 2$ or $m = k + 3$) are connected to $V_{\text{ref},p}$ in the positive DAC (to $V_{ref,n}$ in the negative DAC), while the others are connected to $V_{\text{ref},n}$ ($V_{\text{ref},p}$). Thus, a charge-domain zooming is effectively performed via charge redistribution. Accuracy of the zooming is improved by scrambling the units used in each period by using a data-weighted averaging (DWA) DEM algorithm. Switches $S_{S[1..31]}$ are bootstrapped to improve their linearity.

Simple energy-efficient CMOS inverters are used to implement the integrators. A dynamic biasing scheme similar to the one in [1] is employed to bias the inverter in a PVT robust manner. However, the OTA in [1] uses the large cascode transistors as switches. Since the gate capacitances of the cascodes are loading, the biasing circuit in each clock period and the biasing circuit power consumption would be too high for the high sampling frequency used in this dynamic zoom ADC. Thus, a new dynamic biasing scheme shown in Fig. 6.13a is proposed in the following: during the sampling phase Φ_1, the input transistors M_1 and M_2 are diode connected

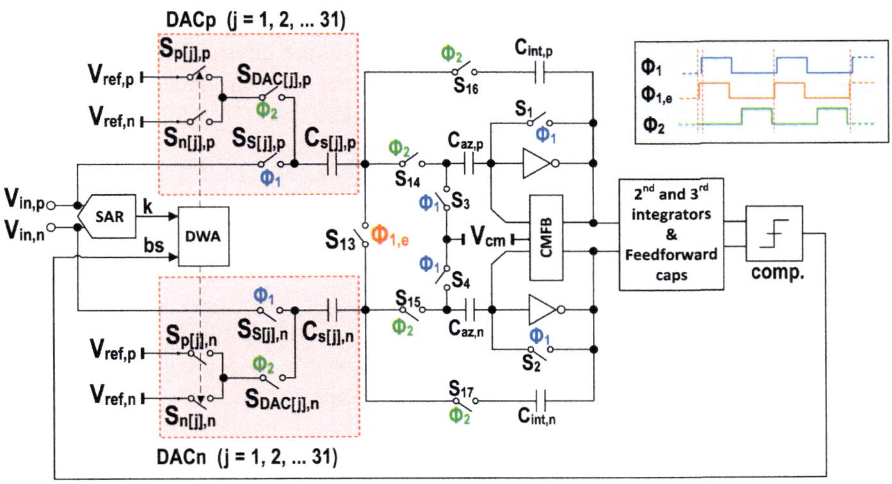

Fig. 6.12 A simplified circuit schematic of the proposed dynamic zoom ADC

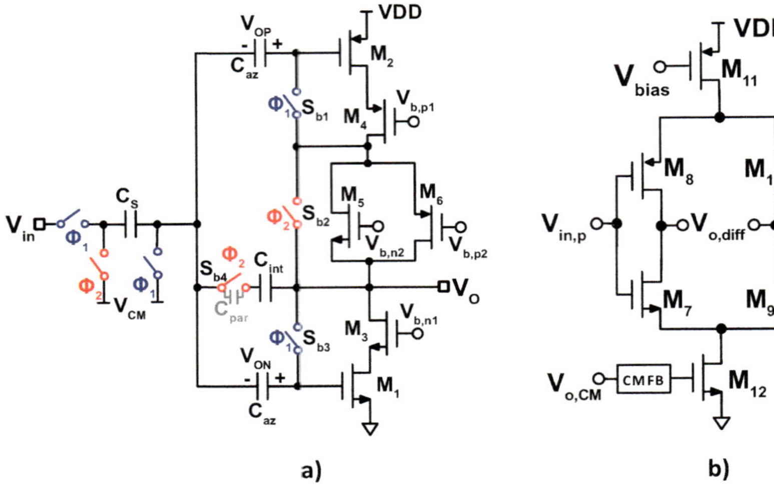

Fig. 6.13 (**a**) Proposed dynamic-biased inverter-based OTA. (**b**) Current-reuse OTA

by switches $S_{b1,3}$ and biased at 125 μA by a floating current source comprising M_5 and M_6. The bias voltages V_{OP} and V_{ON} are sampled together with of the offset and the 1/*f* noise of the OTA on the auto-zeroing capacitors C_{az} (2 pF each) effectively implementing auto-zeroing. In the integration phase Φ_2, the diode connections are broken, and the floating current source is simply bypassed by S_{b2}. A low-power biasing circuit can then be implemented, since it does not see any dynamic loading. Thanks to the cancellation of the input CM signal during sampling, the output CM drift of the OTA can be avoided by a simple SC common-mode feedback (CMFB) circuit [16]. Since the parasitic capacitance across S_{b4} (C_{par}) is discharged at every Φ_2, it might degrade the DC gain of the integrator. In the physical implementation, S_{b4}'s source and drain are shielded from each other by a grounded metal shield placed on top of the gate, so that C_{par} is reduced to less than 1fF, which is more than enough for a 65-dB DC OTA gain.

The second and third integrators are implemented by using fully differential current-reuse inverter-based OTAs biased at 50 μA each (Fig. 6.13b), and their capacitors were also scaled per their input-referred noise contribution scaling. The 1-bit quantizer is designed as a regenerative latch preceded by a static preamplifier and consumes 3.5 μA.

The implemented SAR ADC consists of a conventional synchronous logic, a charge redistribution capacitive DAC, and a comparator, as depicted in Fig. 6.14. The 11-fF unit capacitors are sized to ensure that coarse conversion errors are less than 1 LSB_C. The SAR ADC operates with the same sampling frequency of the SDM (f_s = 11.29 MHz), and it takes five cycles to make a conversion. The same comparator as in SDM is used.

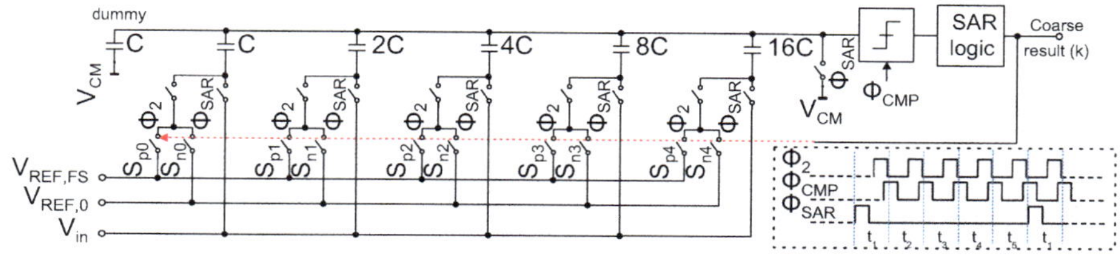

Fig. 6.14 The SAR ADC

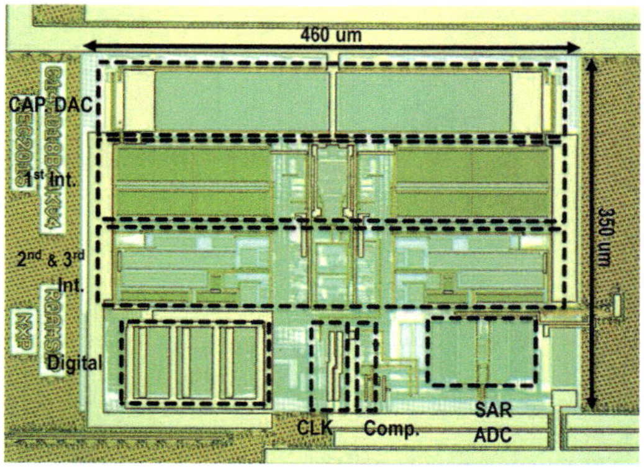

Fig. 6.15 Chip micrograph

6.7 Measurement Results

The prototype dynamic zoom ADC has been fabricated in a 0.16-μm CMOS technology [11]. It occupies an area of 0.16 mm^2 as shown in the chip micrograph (Fig. 6.15). Its total power consumption is 1.12 mW with the digital circuitry consuming 29% of the power (including DWA, SAR logic, and the nonoverlapping clock generator and excluding the digital decimator). The analog power consumption is dominated by the first integrator (56%, simulated). In contrast, the SAR ADC's analog section draws only 7 μW (measured).

The digital outputs of the ADC were the SAR ADC's comparator output, the SDM bit stream, and a clock synchronized to the data. Since the outputs were single-ended and full-CMOS level (0 V–1.8 V), their interference with the external voltage reference on the test PCB limited the measured SNDR to 98.3 dB in 20-kHz BW in the first experimental characterization [10]. After lowering the supply of the digital output drivers from 1.8 V to 0.9 V, the interference is reduced (Figs. 6.16 and 6.17) so that the maximum measured SNDR is 103 dB.

The ADC's peak SNR and DR were 106 dB and 109 dB, respectively, with DWA active (Fig. 6.17). Peak SNDR is limited to 72 dB with DWA off due to the fine DAC mismatch. Thanks to the input common-mode cancellation scheme, the CMRR is greater than 62 dB from DC up to 1 MHz for full-scale common-mode inputs. The ADC's 1/*f* corner measured to be below 20 Hz, proving the effectiveness of the auto-zeroing employed in the first OTA.

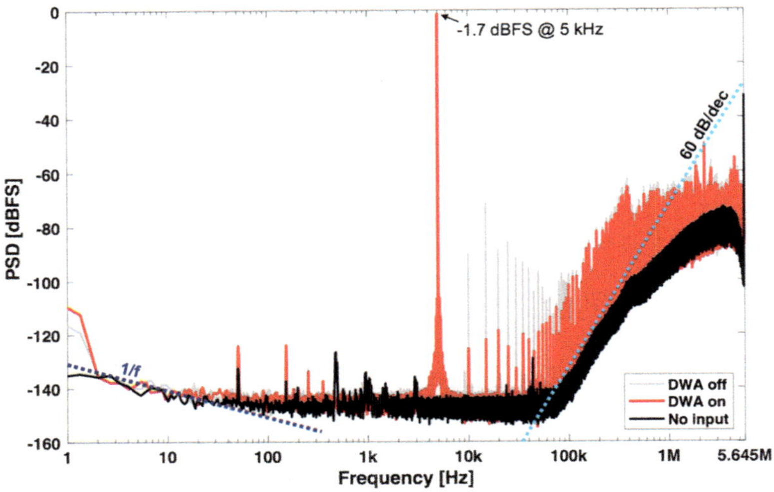

Fig. 6.16 Measured output spectra for DWA off, DWA on, and no input. Inputs are connected to V_{CM} for no input case, with DWA on

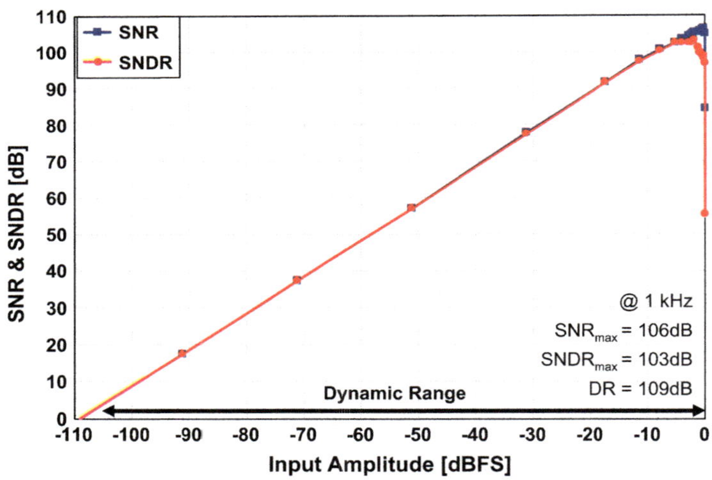

Fig. 6.17 Measured SNR/SNDR vs input amplitude (DWA on)

In order to test the overloading of the SDM with full-scale out-of-band signals, a full-scale sine wave is applied to the ADC's input, and its frequency is swept from 10 Hz to 100 kHz. In-band noise is measured for each point to predict the achievable DR as shown in Fig. 6.18. The degradation of the DR is observed with full-scale signals above 27 kHz, as predicted in system-level simulations. A first-order RC low-pass filter (LPF) with 30-kHz corner frequency is inserted before the

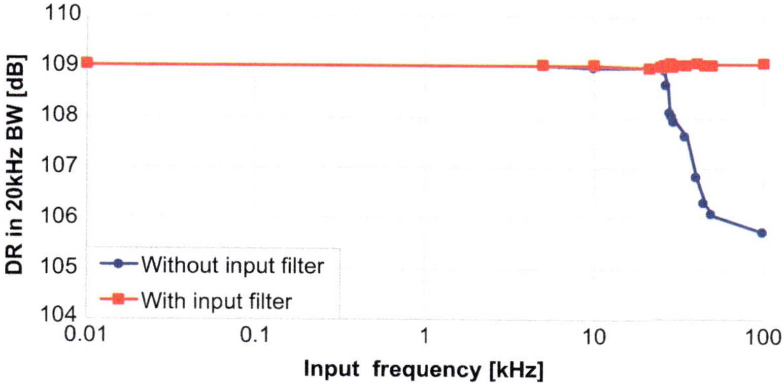

Fig. 6.18 DR in 20-kHz BW in the presence of in- and out-of-band full-scale inputs with and without an LPF at the input with 30-kHz corner frequency (DWA on)

Table 6.1 Performance summary and comparison with state-of-the-art audio ADCs

	Unit	This work	[24]	[25]	[26]	[27]	[18]
Year	–	2016	2016	2016	2016	2011	2016
Loop-filter type	–	DT	CT	CT	CT	CT+DT	DT
Technology	nm	160	160	130	65	40	130
Die area	mm^2	0.16	0.21	1.33	0.256	0.05	0.31
Power consumption	mW	1.12	0.39	0.28	0.8	0.5	0.3
Sampling frequency	MHz	11.29	3	6.144	6.4	6.5	6.1
Signal bandwidth	kHz	20	20	24	25	24	20
Peak SNR	dB	106	93.4	99.3	100.1	–	93.6
Peak SNDR	dB	103	91.3	98.5	95.2	90	97.7
DR	dB	109	103.1	103.6	103	102	100.5
FOMs[a]	dB	181.5	180.2	182.9	177.9	179	178.7

[a]FOMs = DR+10 log(signal bandwidth/Power)

ADC input, which ensures that the DR is constant up to at least 100 kHz (the max. measurement frequency is limited by the low-noise audio signal generator).

A performance comparison with the ADCs with similar resolution (>100-dB DR) and bandwidth is presented in Table 6.1. Although the proposed zoom ADC is a discrete-time design, it shows state-of-the-art 181.5-dB FoM_S. It is also considerably more area efficient than the previous designs implemented in similar technology nodes. As discussed before, the ADC's area is dominated by the capacitors defined by the kT/C noise required to obtain the 109-dB DR, so the area is used efficiently.

6.8 Conclusions

The dynamic zoom ADC is presented as a hybrid ADC suitable for high-resolution and high-linearity digital audio applications. The proposed zoom ADC employs a 5-bit SAR ADC working in parallel to assist a third-order SDM. This improved the overall energy efficiency by reducing the signal swing of the SDM and relaxed its nonthermal-noise-related power consumption. A 0.16-mm^2 prototype chip is implemented in 0.16-μm CMOS technology, achieving 109-dB DR, 106-dB peak SNR, and 103-dB peak SNDR while having an excellent FoM_S of 181.5 dB.

References

1. Chae, Y., Souri, K., Makinwa, K.A.A.: A 6.3 μW 20 bit incremental zoom ADC with 6 ppm INL and 1 μV offset. IEEE J. Solid State Circuits. **48**(12), 3019–3027 (2013)
2. Sechang, Oh., Jung, W., Yang, K., Blaauw, D., Sylvester, D.: 15.4b incremental sigma-delta capacitance-to-digital converter with zoom-in 9b asynchronous SAR. 2014 Symposium on VLSI Circuits Digest of Technical Papers, Honolulu, pp. 1–2 (2014)
3. Venca, A., Ghittori, N., Bosi, A., Nani, C.: A 0.076 mm^2 12 b 26.5 mW 600 MS/s 4-way interleaved subranging SAR- $\Delta\Sigma$ ADC with on-chip buffer in 28 nm CMOS. IEEE J. Solid State Circuits. **51**(12), 2951–2962 (2016)
4. Shu, Y.S., Kuo, L.T., Lo, T.Y.: An oversampling SAR ADC with DAC mismatch error shaping achieving 105-dB SFDR and 101-dB SNDR over 1 kHz BW in 55 nm CMOS. IEEE J. Solid State Circuits. **51**(12), 2928–2940 (2016)
5. Dong, Y., Yang, W., Schreier, R., Sheikholeslami, A., Korrapati, S.: A continuous-time 0–3 MASH ADC achieving 88-dB DR with 53 MHz BW in 28 nm CMOS. IEEE J. Solid State Circuits. **49**(12), 2868–2877 (2014)
6. Gharbiya, A., Johns, D.A.: A 12-bit 3.125 MHz bandwidth 0–3 MASH Delta-sigma modulator. IEEE J. Solid State Circuits. **44**(7), 2010–2018 (2009)
7. Liu, C.C., Huang, M.C., Tu, Y.H.: A 12 bit 100 MS/s SAR-assisted digital-slope ADC. IEEE J. Solid State Circuits. **51**(12), 2941–2950 (2016)
8. Sanyal A., Sun N.: A 18.5-fJ/step VCO-based 0–1 MASH $\Delta\Sigma$ ADC with digital background calibration. 2016 IEEE Symposium on VLSI Circuits (VLSI-Circuits), Honolulu, pp. 1–2 (2016)
9. van der Goes, F., Ward, C., Astgimath, S., Yan, H., Riley, J., Mulder, J., Wang, S., Bult, K.: 11.4 A 1.5-mW 68-dB SNDR 80-MS/s 2x interleaved SAR assisted pipelined ADC in 28nm CMOS. Proc. IEEE International Solid-State Circuits Conference, San Francisco, pp. 200–201 (2014)
10. Gönen, B., Sebastiano, F., van Veldhoven, R., Makinwa, K.A.A.: A 1.65mW 0.16mm^2 dynamic zoom ADC with 107.5dB DR in 20kHz BW. 2016 IEEE International Solid-State Circuits Conference (ISSCC), San Francisco, pp. 282–283 (2016)
11. Gönen, B., Sebastiano, F., Quan, R., van Veldhoven, R., Makinwa, K.A.A.: A dynamic zoom ADC with 109-dB DR for audio applications. IEEE J. Solid-State Circuits, (accepted for publication)
12. Schreier, R., Temes, G.C.: Understanding Delta-Sigma Data Converters. John Wiley and Sons, Hoboken (2005)
13. Murmann, B.: A/D converter trends: power dissipation, scaling and digitally assisted architectures. 2008 IEEE Custom Integrated Circuits Conference, San Jose, pp. 105–112 (2008)

14. Murmann, B.: ADC Performance survey 1997–2016. [Online]. Available: http://web.stanford.edu/~murmann/adcsurvey.html
15. Pelgrom, M.J.M.: Analog-to-Digital Conversion. Springer, Cham (2017)
16. Chae, Y., Han, G.: Low voltage, low power, inverter-based switched-capacitor delta-sigma modulator. IEEE J. Solid State Circuits. **44**(2), 458–472 (2009)
17. Christen, T.: A 15-bit 140-μW scalable-bandwidth inverter-based $\Delta\Sigma$ modulator for a MEMS microphone with digital output. IEEE J. of Solid-State Circuits. **48**(7), 1605–1614 (2013)
18. Lee, S., Jo, W., Song, S., Chae, Y.: A 300-μW audio $\Delta\Sigma$ modulator with 100.5-dB DR using dynamic bias inverter. IEEE Trans. Circuits Syst. I, Reg Papers. **63**(11), 1866–1875 (Nov. 2016)
19. Steiner, M., Greer, N.: 15.8 A 22.3b 1kHz 12.7mW switched-capacitor $\Delta\Sigma$ modulator with stacked split-steering amplifiers. 2016 IEEE International Solid-State Circuits Conference (ISSCC), San Francisco, pp. 284–286 (2016)
20. van Veldhoven, R.H.M., van Roermund, A.H.M.: Robust Sigma Delta Converters. Springer, Dordrecht (2017)
21. van Veldhoven, R.H.M., Rutten, R., Breems, L.J.: An inverter-based hybrid $\Delta\Sigma$ modulator. 2008 IEEE International Solid-State Circuits Conference – Digest of Technical Papers, San Francisco, pp. 492–630 (2008)
22. van Veldhoven, R.H.M, Nizza, N., Breems, L.J.: Technology portable, 0.04mm^2, Ghz-rate $\Delta\Sigma$ modulators in 65nm and 45nm CMOS. 2009 Symposium on VLSI Circuits, Kyoto, pp. 72–73 (2009)
23. Souri, K., Chae, Y., Makinwa, K.A.A.: A CMOS temperature sensor with a voltage-calibrated inaccuracy of $\pm$ 0.15°C (3σ) from -55°C to 125°C. IEEE J. Solid State Circuits. **48**(1), 292–301 (2013)
24. De Berti, C., Malcovati, P., Crespi, L., Baschirotto, A.: A 106-dB A-weighted DR low-power continuous-time $\Delta\Sigma$ modulator for MEMS microphones. IEEE J. Solid State Circuits. **51**(7), 1607–1618 (2016)
25. Billa, S., Sukumaran, A., Pavan, S.: A280-μW 24-kHz-BW 98.5dB-SNDR chopped single-bit CT $\Delta\Sigma$M achieving <10-Hz 1/f noise corner without chopping artifacts. 2016 IEEE International Solid-State Circuits Conference (ISSCC), San Francisco, pp. 276–277 (2016)
26. Leow, Y.H., Tang, H., Sun, Z.C., Siek, L.: A 1 V 103-dB 3rd-order audio continuous-time $\Delta\Sigma$ ADC with enhanced noise shaping in 65 nm CMOS. IEEE J. Solid State Circuits. **51**(11), 2625–2638 (2016)
27. Lo, T.Y.: A 102dB dynamic range audio sigma-delta modulator in 40nm CMOS. 2011 IEEE Asian Solid State Circuits Conference (A-SSCC), Jeju, pp. 257–260 (2011)

Part II
Smart Sensors for the IoT

The second part of this book is dedicated to recent advances in the field of sensors, interfaces, and references intended for use in wearable and IoT applications. Since such applications are usually battery powered, their key requirement is for high energy efficiency, which can then be combined with aggressive duty cycling to achieve extremely low (nW) levels of average power.

The first chapter, by Nick van Helleputte et al., discusses advancements in the design of analog circuits intended for use in wearable healthcare applications. A number of general trends, e.g., toward multimodal sensing, are discussed. Circuit topologies for the most relevant sensing modalities, e.g., ExG, bio-impedance, and photoplethysmogram (PPG), are presented, as well as some recent state-of-the-art implementations.

The second chapter, by Rajesh Pamula, Chris van Hoof, and Marian Verhelst, presents an ultra-low power PPG readout circuit that exploits various mixed-signal processing techniques. In particular, the use of compressive sampling (CS) allows the power consumption of its LED driver to be reduced by 30×. Heart rate information is then extracted in the compressed domain, thus avoiding the use of complex signal reconstruction techniques.

The third chapter, by David Ruffieux et al., describes an ultra-low power (240 nA) real-time clock module that achieves a typical accuracy of ±1 ppm at 1 Hz over the industrial temperature range (−40–85 °C). It combines a miniature 32 kHz quartz crystal and an ASIC in a miniature 8-pin ceramic package. An all-digital interpolation scheme allows its 1 Hz output to be trimmed with a resolution of 0.1 ppm, resulting in significant savings in both circuit area and power consumption.

The fourth chapter, by Sining Pan and Kofi Makinwa, presents two resistor-based CMOS temperature sensors. One is based on a Wien bridge RC filter, which has a temperature-dependent phase shift; while the other is based on a Wheatstone bridge, which outputs a temperature-dependent current. In both cases, the bridge outputs are digitized by continuous-time delta-sigma modulators. This results in sensors with state-of-the-art energy efficiency as well as low power dissipation (<200 μW).

In the fifth chapter, by Javier Perez Sanjurjo et al., an integrating dual-slope (DS) capacitance-to-digital converter (CDC) is presented. This is used to digitize the output of a pressure-sensing capacitive Wheatstone bridge. The proposed CDC generates a multi-bit output with the help of time domain rather than amplitude domain techniques. It also employs quantization noise shaping to reduce measurement time. The CDC achieves a capacitive sensing resolution of 5.4 aF (17-bits) while consuming only 146 μA from a 1.5 V power supply.

The sixth chapter, by Shikhar Tewari and Aatmesh Shrivastava, presents a 48 nW bandgap reference that can operate from supply voltages as low as 500 mV. This is achieved by using a switched capacitor charge pump, rather than resistors or current sources, to bias a pair of BJTs. The use of a charge pump also reduces the circuit's minimum supply voltage. At an output voltage of 500 mV, it achieves a temperature coefficient of 45 ppm/°C together with a PSRR that is well over 60 dB.

Chapter 7
Advances in Biomedical Sensor Systems for Wearable Health

Nick Van Helleputte, Jiawei Xu, Hyunsoo Ha, Roland Van Wegberg, Shuang Song, Stefano Stanzione, Samira Zaliasl, Richard van den Hoven, Wenting Qiu, Haoming Xin, Chris Van Hoof, and Mario Konijnenburg

7.1 Introduction

Global demographic trends, like increased access to healthcare and aging population, impose tremendous pressure on the traditional healthcare system. There is a clear ongoing paradigm shift toward preventive healthcare, which aims to avoid people getting sick in the first place or detect the onset of illness as soon as possible. The motivation is obvious, as such a preventive system will increase the quality of life, while reducing the cost of medical care. A critical requirement for such a preventive healthcare system is the ability to monitor the health status in an unobtrusive and permanent manner. Most medically relevant diagnostic equipment today still is expensive and bulky and requires trained professionals to operate them and interpret the data. Recent years have seen a significant rise in research toward bringing automated, reliable, high-quality diagnostic health assessment to a wearable platform.

Figure 7.1 shows a general block diagram for a wearable healthcare device. There is an analog front-end (AFE) which records (multiple) physiological signals and converts them to digital. There are numerous parameters that can be monitored that have clinical relevance for preventive healthcare systems. These include the

N. Van Helleputte (✉)
imec, Heverlee, Belgium
e-mail: Nick.VanHelleputte@imec.be

J. Xu • H. Ha • R. Van Wegberg • S. Song • S. Stanzione • S. Zaliasl • R. van den Hoven • W. Qiu • H. Xin • M. Konijnenburg
imec-Holst Centre, Eindhoven, The Netherlands

C. Van Hoof
imec, Heverlee, Belgium

imec-Holst Centre, Eindhoven, The Netherlands

P. Harpe et al. (eds.), *Hybrid ADCs, Smart Sensors for the IoT, and Sub-1V & Advanced Node Analog Circuit Design*, DOI 10.1007/978-3-319-61285-0_7

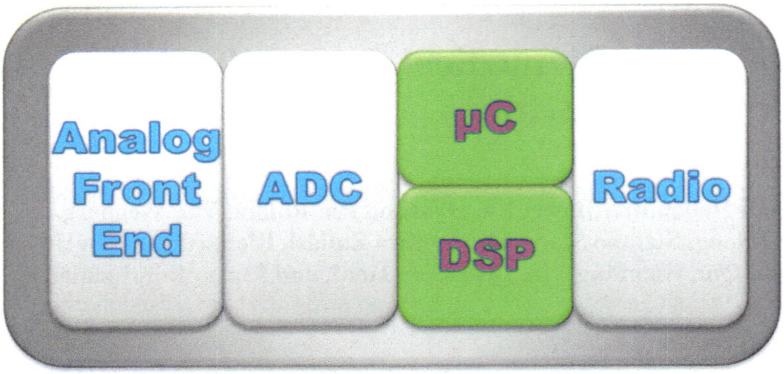

Fig. 7.1 General block diagram for a wearable healthcare device

electrocardiogram (ECG), electromyography (EMG), electroencephalogram (EEG), and bio-impedance (BIOZ) to monitor multiple physiological features such as heartbeat, respiration, hydration, body fat mass, and lung fluid accumulation. Also optical recordings like a photoplethysmogram (PPG) or functional near-infrared spectroscopy (fNIRS) are gaining interest. The on-chip microcontroller and/or dedicated DSP processes all these signals, which are then transmitted off-chip via RF for further processing and user notification in case abnormalities are detected.

This paper will discuss some recent state-of-the-art advancements in analog circuit design for wearable healthcare applications. In Sects. 7.2, 7.3, and 7.4, we will focus on building blocks for some of the most relevant signal modalities, ExG, bio-impedance, and PPG. Section 7.5 will discuss data convertor circuits.

7.2 ExG Readout Circuits

ExG (ECG, EMG, EEG) signals are measured on the surface of the skin by electrodes connecting to a high input impedance instrumentation amplifier (IA). This IA usually dominates the overall performance, such as noise, input impedance, CMRR, and power.

7.2.1 Ultralow-Power (ULP) IAs

Sub-μW ExG readout ICs with low supply voltages not only increase the continuous operation time of biomedical sensors but also reduce the size of the battery. This makes the ULP ExG readouts very popular for wireless biosensor nodes. However, minimizing power dissipation by scaling down the supply voltage is not

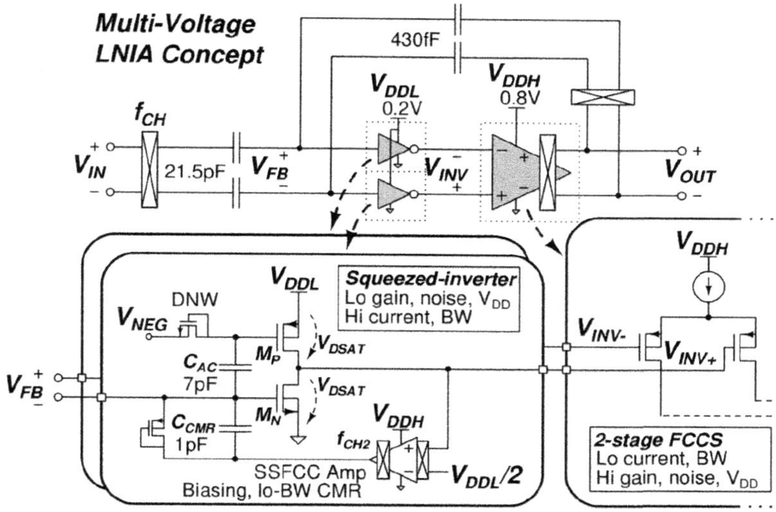

Fig. 7.2 Multi-voltage IA using local 0.2 V supply in its first stage [2]

straightforward, because aggressors (e.g., electrode offset, main interferences, and motion artifact) enforce a large dynamic range for the readout circuity. To solve this issue, [1–3] proposed the capacitively coupled IA architecture (Fig. 7.2), in which the input stage does not require a large voltage swing at the virtual ground. Therefore, the power of these IA architectures can be significantly reduced. In [2] the authors implement a 790 nW amplifier based on a 0.2 V inverter-based input stage. Nevertheless, a common drawback of the capacitively coupled IAs is that they suffer from low input impedance, especially when chopping [4] is performed prior to the input capacitors to mitigate IA's intrinsic $1/f$ noise. Another drawback is the implementation of a large resistor in the feedback loop. The conventional way of using a pseudo-resistor [5] has poor accuracy to PVT variation, while switched-capacitor resistor has limited resistance value. Switched-resistor implementation [6] solves these issues by switching on a medium-sized resistor only for a short period, so the effective resistance is boosted by the ratio of duty cycle of the pulse. Nevertheless, a high-frequency clock is required to generate the pulse.

7.2.2 High Input Impedance IAs

To improve the input impedance, [6, 7] proposed auxiliary buffers to drive the input coupling capacitance before the main input chopper is switching (see Fig. 7.3). Thus, the current to charge the input capacitance (C_{in}) is provided by the buffers instead of the input signal source. This technique ensures 300 MΩ input impedance over the ExG bandwidth [6] compared to the IAs utilizing a positive feedback

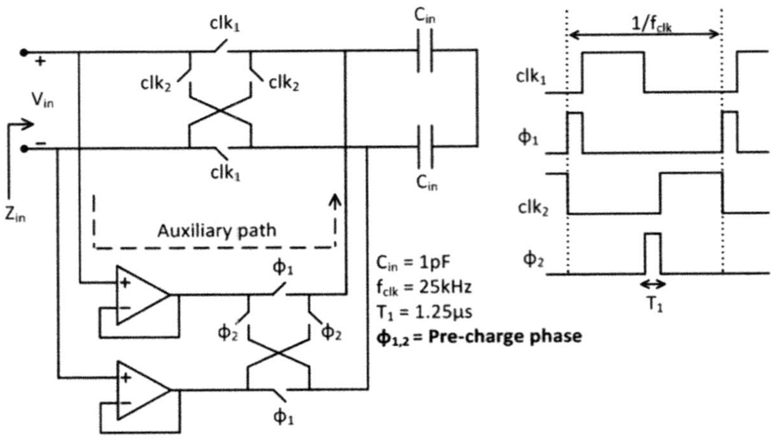

Fig. 7.3 Input impedance boosting by using two auxiliary buffers [6]

loop for impedance boosting (bootstrapping). The auxiliary buffers provide better stability and less dependence on IA's bandwidth. On the other hand, the analog buffers are power hungry, and their 1/*f* noise can degrade the IA's low-frequency noise performance, even if the buffers are duty cycled to be active only for a short period.

7.2.3 *Digitally Assisted IAs*

Apart from using the capacitively coupled IA, a low-supply voltage IA can be implemented in a time domain. A time-domain ECG amplifier (Fig. 7.4) is proposed in [8], in which the input biopotential signal is first converted into a 2b digital output through time-domain ADC and then decimated by a CIC digital filter. The large dynamic range requirement is moved to time domain and thus decoupled from VDD. In addition, the electrode offset is also compensated similar as in [9] via a digitally assisted servo loop. Nevertheless, time-domain IAs suffer from quantization noise of the comparator. Although the noise in ExG bandwidth can be reduced by using a higher-frequency clock, this also subjects to higher power consumption.

Table 7.1 shows a performance overview of some recently published instrumentation amplifiers. As is clear, the field spans various design nodes and architectures that focus on very high performance or ultralow power.

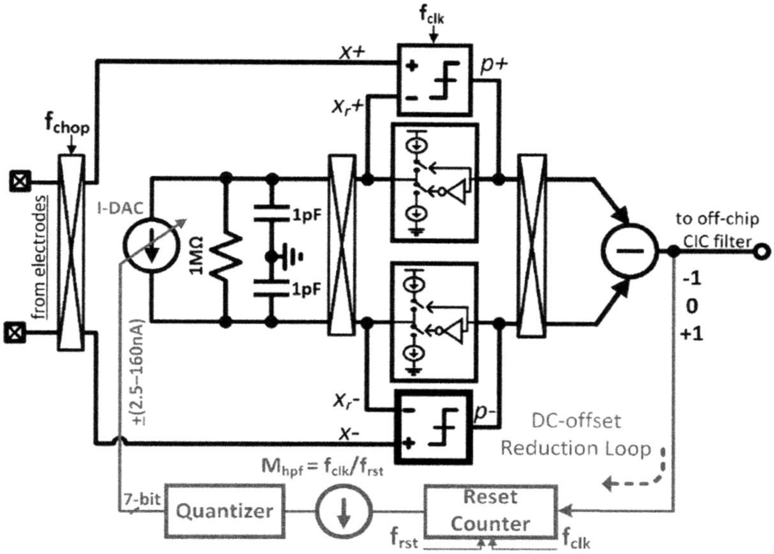

Fig. 7.4 Time-domain ECG amplifier

7.3 Bio-impedance Readout Circuits

7.3.1 Introduction

Typical BIOZ measurements require a fine resolution of mΩ range to achieve medical quality data while minimizing power consumption to ensure long battery life in wearable applications. The BIOZ channel consists of a current generator (CGEN) for applying differential current to the body and readout front-end to measure the resulting differential voltage. In two-point measurement setups (Fig. 7.5a), electrodes for current injection and voltage sensing can be shared. In this case, the measured signal and input-referred noise (expressed in Ω_{rms}) are given by the following two simplified equations:

$$V_{IN} = (Z_{BODY} + 2 \cdot Z_{ELEC}) \times I_{CG} \quad (7.1)$$

$$\overline{z_n} = \frac{\overline{v_{n,IN}}}{I_{CG}} = \frac{\sqrt{\overline{v_{n,CH}}^2 + \overline{i_{n,CG}}^2 \times (Z_{BODY} + 2 \times Z_{ELEC})^2}}{I_{CG}} \quad (7.2)$$

It is clear that the measured signal V_{IN} also contains the electrode impedance (Z_{ELEC}) which is usually much higher than the signal of interest (Z_{BODY}), and this limits its application to monitoring of only relative variation of BIOZ. Furthermore,

Table 7.1 Performance overview of recent IAs

	[10]	[2]	[3]	[8]	[9]	[6]	[11]
Technology	0.18 μm	0.18 μm	65 nm	40 nm	65 nm	40 nm	0.18 μm
Supply voltage	1.2 V	0.2/0.8 V	0.6 V	0.6 V	0.5 V	1.2 V	1.8 V
Input noise	0.61 μVrms (0.05–150 Hz)	0.94 μVrms	26 μVrms (1.5–370 Hz)	7.8 μVrms (1–150 Hz)	1.23 μVrms (1–150 Hz)	2 μVrms (1–150 Hz)	0.9 μVrms (0.5–100 Hz)
Electrode offset tolerance	400 mV	–	Rail to rail	150 mV	50 mV	<120 mV*	240 mV
Input impedance at 50 Hz	500 MΩ	–	1.6 GΩ*	50 MΩ	28 MΩ	300 MΩ	500 MΩ
CMRR at 50 Hz	110 dB	85 dB	60 dB	60 dB	88 dB	–	97 dB
Power	16 μW	0.79 μW	1 nW	3.3 μW	2.3 μW	2 μW	0.9 μW

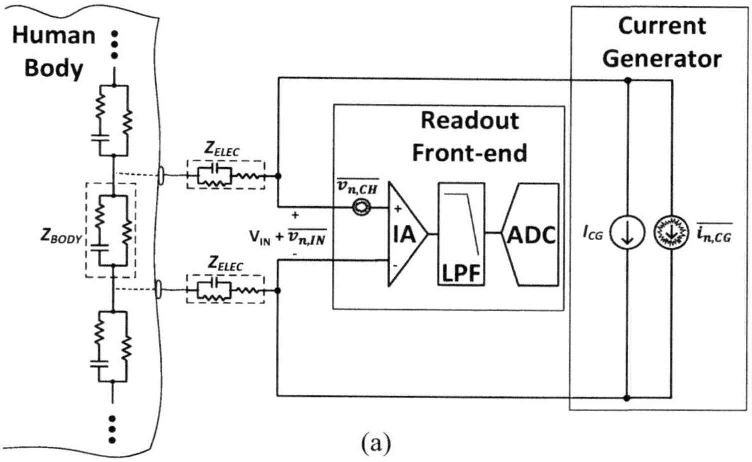

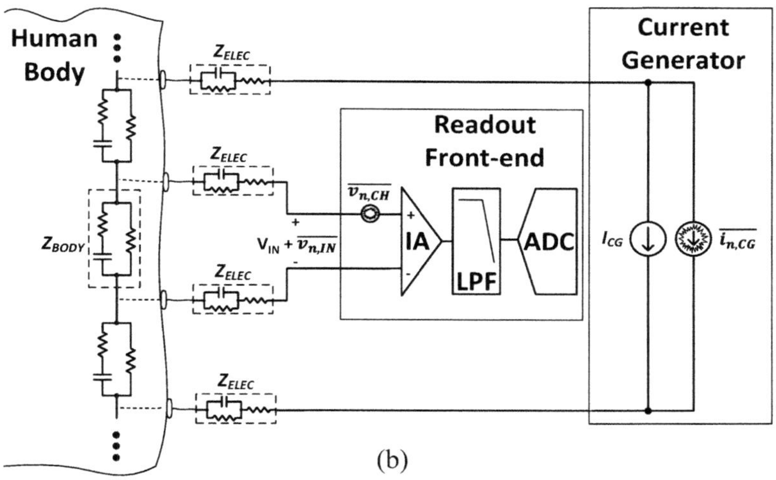

Fig. 7.5 Two (**a**) and four (**b**) electrode measurements of BIOZ

noise from the current generator $\left(\overline{i_{n,CG}^{2}}\right)$ becomes significant due to large Z_{ELEC} (few kΩ). For better accuracy, BIOZ can be measured with a tetrapolar method (Fig. 7.5b), where two electrodes are injecting current, while the remaining two are sensing the voltage. In this case, the equations become

$$V_{\text{IN}} = Z_{\text{BODY}} \times I_{CG} \tag{7.3}$$

$$\overline{z_n} = \frac{\overline{v_{n,\text{IN}}}}{I_{CG}} = \frac{\sqrt{\overline{v_{n,CH}^{2}} + \overline{i_{n,CG}^{2}} \times (Z_{\text{BODY}})^2}}{I_{CG}} \approx \frac{\sqrt{\overline{v_{n,CH}^{2}}}}{I_{CG}} \tag{7.4}$$

The input signal now only contains Z_{BODY} between the two sensing electrodes because there's no current flowing through the electrodes for voltage sensing. Of course this assumes an infinite input impedance, but in reality the input impedance tends to be high enough to make this approximation. However, it is important to understand how finite input impedance might result in a partial appearance of Z_{ELEC} into the formula also. The effect of the current noise in this case is negligible, thanks to small Z_{BODY} (<200 Ω) (Eq. 7.4). Therefore, four electrode measurements are preferred to achieve high accuracy as well as to obtain absolute value of Z_{BODY} [12–16].

The bio-impedance has a dependency on frequency because it has capacitive as well as resistive characteristic. Depending on the application (i.e., vital signs recording, respiration measurement, fluid retention analysis, etc.), one might be interested in either the resistive or capacitive parts, and one might be interested to characterize these at different frequencies. For continuous monitoring of respiration and heartbeat, BIOZ is often measured at a fixed single frequency (SF) of less than 100 kHz [12, 13, 17]. On the other hand, multifrequency (MF) measurement is preferred for in-depth analysis of body fluid such as electrical impedance tomography (EIT), lung fluid accumulation, and de-/overhydration detection, requiring a wide range of frequencies from 1 k to 1 MHz [14, 15]. In the following section, the state-of-the-art BIOZ readout circuits for SF and MF measurement will be shown.

7.3.2 Current Generator (CGEN)

Current generators usually come in the form of programmable differential current DACs. [13] proposed a pseudo-sine wave current generation for high-accuracy BIOZ recording (Fig. 7.6). However, it requires a large look-up table and complicates control and reliable demodulation. Hence, simple square-wave current injection is usually preferred [12, 14, 15]. However, even simple resistors tied to a fixed voltage source have been used also to implement rudimentary current generators, though for obvious reasons, these lack absolute accuracy. A good CGEN circuit needs to address several challenges. The IR drop over the electrodes impedance, especially if dry electrodes are used, can be significant requiring high compliance range designs. In [12], this was solved by running the CGEN on the high-voltage IO domain. Furthermore, applying a DC current to the human body is strictly prohibited for patient's safety, putting very strict requirements on the source and sink matching. An off-chip series-connected capacitor (C_{DC}) can easily block this; however, it imposes another challenge of DC biasing the CGEN output. The easiest way is using parallel resistors (R_{CG}) connected to V_{REF} (preferably VDD/2) [12], and the resistor will determine the output impedance of the CGEN. While a high output impedance is required for reliable BIOZ recording, the mismatch current between source and sink will create an IR drop over this bias resistor which should not push the CGEN outside the proper operating region. Hence, the CGEN designer must carefully balance the mismatch, compliance range, and output impedances.

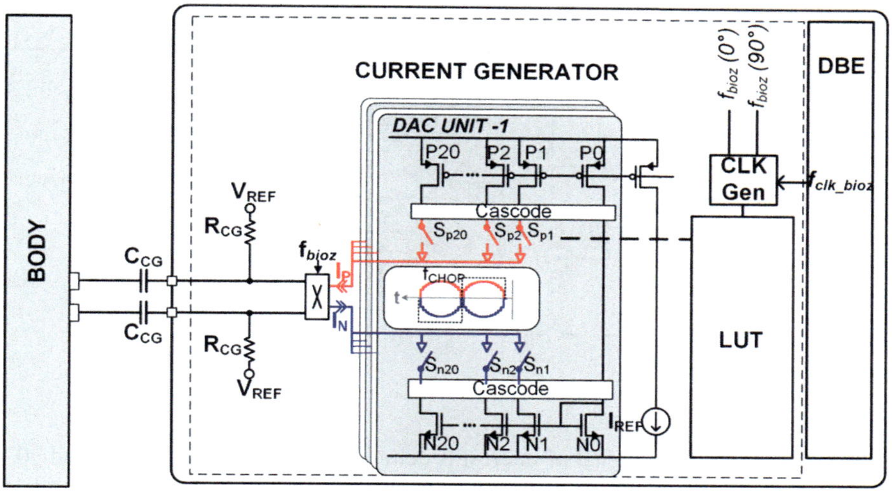

Fig. 7.6 Current generator schematic

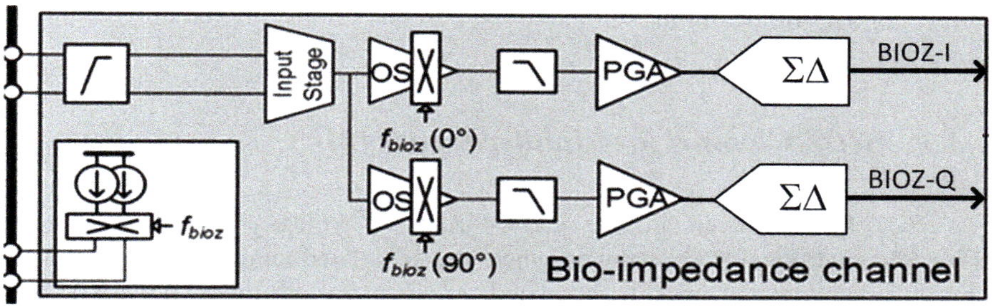

Fig. 7.7 BIOZ channel with current generator

7.3.3 BIOZ Channel for Single Frequency

High-performance and low-power BIOZ SF channels have been implemented with fully integrated CGEN [12]. Figure 7.7 shows a typical readout architecture, where the BIOZ readout adopts a chopper-demodulated IA, a programmable gain amplifier (PGA), and an ADC. While conceptually similar to an ExG readout, note that there's no modulator in front of the input stage. This is indeed not necessary because the signal of interest is already modulated to the BIOZ actuation frequency (f_{bioz}) by the CGEN. The work in [12] provides a means to extract real and imaginary components of the impedance with minimal power consumption overhead. The circuit diagram of the IA, consisting of an input stage and two output stages for the in (I) and quadrature (Q) phase, is shown in Fig. 7.8. The input stage converts the input voltage to current over a source degeneration resistor (R_{i}), and then the current is mirrored to output stages with mirroring ratio of N and flows through the output resistor (R_{o}).

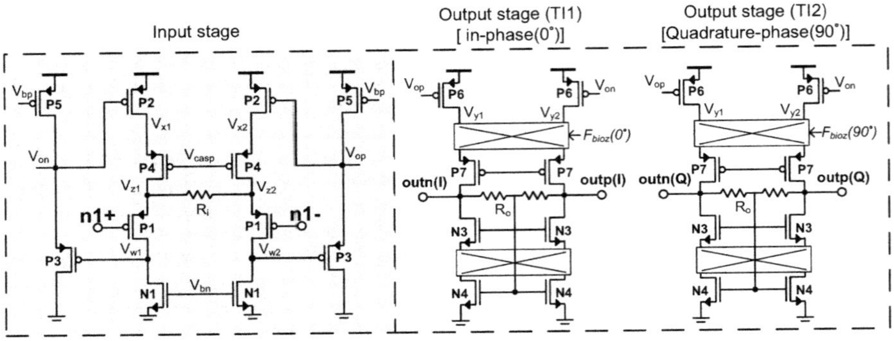

Fig. 7.8 Circuit diagram of IA

In the output stage, the signal of interest is demodulated to baseband by 0° and 90° phase clocks to extract the in-phase and quadrature outputs, respectively. Because the input stage determines the noise and input range, it consumes most of the power. The output stage consumes only 10 $\sim$ 20% of power. Hence, *I*/*Q* extraction comes at very little power overhead.

7.3.4 BIOZ Channel for Multifrequency (MF)

For BIOZ MF measurement, the readout front-end typically must handle larger BWs. The traditional BIOZ readout, where the modulated input signal is amplified first and demodulated back to baseband, requires power-hungry wide-BW IAs (BW > 1 MHz). A digital-intensive broadband approach using maximum length sequence (MLS) which has an equally distributed power spectrum was introduced, and it achieved a fast measurement (100 ms) with fine resolution (100 mΩ) (Fig. 7.9) [14]. However, this work also suffers from wide-BW requirements resulting in a limited frequency range up to 125 kHz.

A pre-modulation to intermediate frequency was introduced for low power consumption [15, 16] (Fig. 7.10). The high-frequency input signal is first demodulated to baseband and then modulated to an intermediate frequency ($\sim$kHz) which is low enough to be handled by low-power IAs (BW $\sim$ 30 kHz [15]) and also high enough to mitigate 1/*f* noise. This method also enables quadrature-phase measurement by using 90° shifted clock for demodulation. This work achieved wide frequency range up to 1.24 MHz while consuming only 52 μW.

To conclude our BIOZ recording overview chapter, Table 7.2 shows a state-of-the-art comparison.

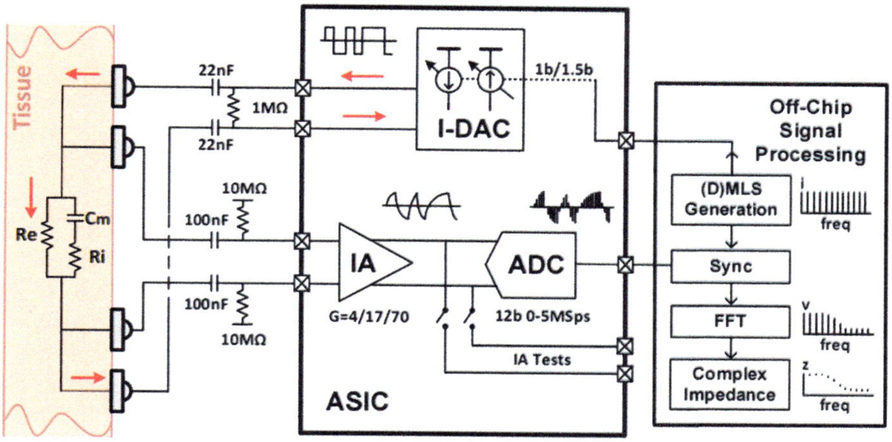

Fig. 7.9 Broadband BIOZ channel using MLS sequence

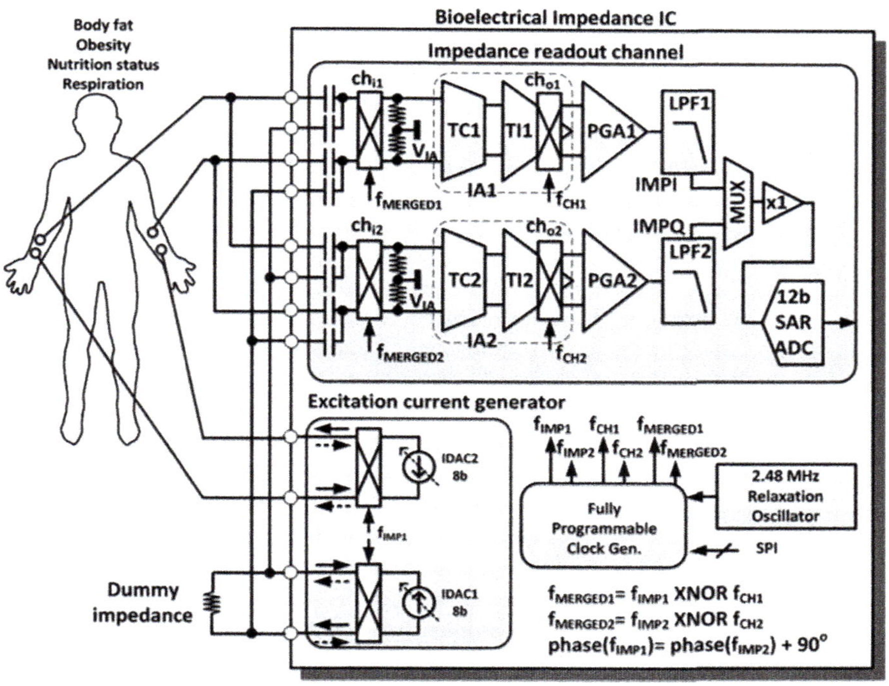

Fig. 7.10 BIOZ MF channel using pre-demodulation to intermediate frequency

Table 7.2 Performance overview of BIOZ recording ASICs

	[12]	[14]	[15]	[17]
Supply (V)	1.2: readout	1.8	1.2	2.0–3.6
	1.8: CG			
DC impedance range (Ω)	0–800	1–10 k	0–800	0–2.8 k
Current mag. (μApk)	25/50/75/100	0.5 50	0–338	1060
Frequency range (kHz)	20/40	1–125	0.25∼1240	1–150
Resolution (mΩrms)	6.6 at 25 μApk	100	8.68 at 166 μApk	100
Bandwidth	0.1–10 Hz	N/A	0.5–100 Hz	0.1–2 Hz
Multifrequency measurement	No	Yes	Yes	Yes
Power consumption (A)	47.1 μW, readout; 69.5 μW, CG at 25 μApk	155 μW	52 μW, readout; 298.8 μW, CG at 166 μApk	3.2 mW

7.4 Photoplethysmogram Readout Circuits

7.4.1 Introduction

A PPG signal is recorded by illuminating the skin and measuring the transmitted or reflected light that is modulated by the blood flow (heartbeat) [18]. While clinical PPG is a well-established field, reliable ambulatory PPG recording remains challenging because of motion artifacts, ambient light interference, and physiological differences among people. This necessitates the use of high dynamic range readouts with ambient light cancellation techniques. Besides, the power consumption of a PPG system (including the LED drivers and readout channels) is usually much higher than, for example, ECG, primarily because of the power required for the LED drivers. Low-power circuit solutions are thus paramount for wearable devices to reduce the battery size and improve user comfort.

A typical PPG recording circuit is shown in Fig. 7.11. It consists of an LED driver and LED and a photodetector together with appropriate readout channel. In the rest of this chapter, three design examples are discussed. The first example [12] supports various LEDs with different wavelengths together with multiple photo sensors for dynamically finding an optimized placement of LED/PD that functions well for all people. It also provides an ambient light cancellation technique which helps to partially remove the motion artifact and increase the dynamic range of the readout channel. The second example [20] provides one of the highest reported dynamic ranges, which improve the robustness of operation during ambulatory recording. The third example [19] focuses on the power minimization, in particular the power of the LED driver.

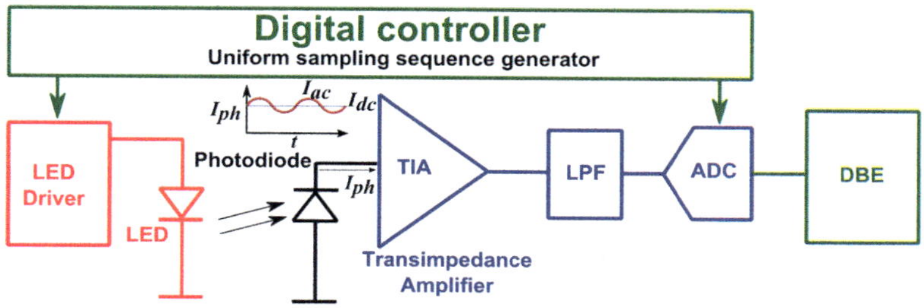

Fig. 7.11 A typical PPG readout system [19]

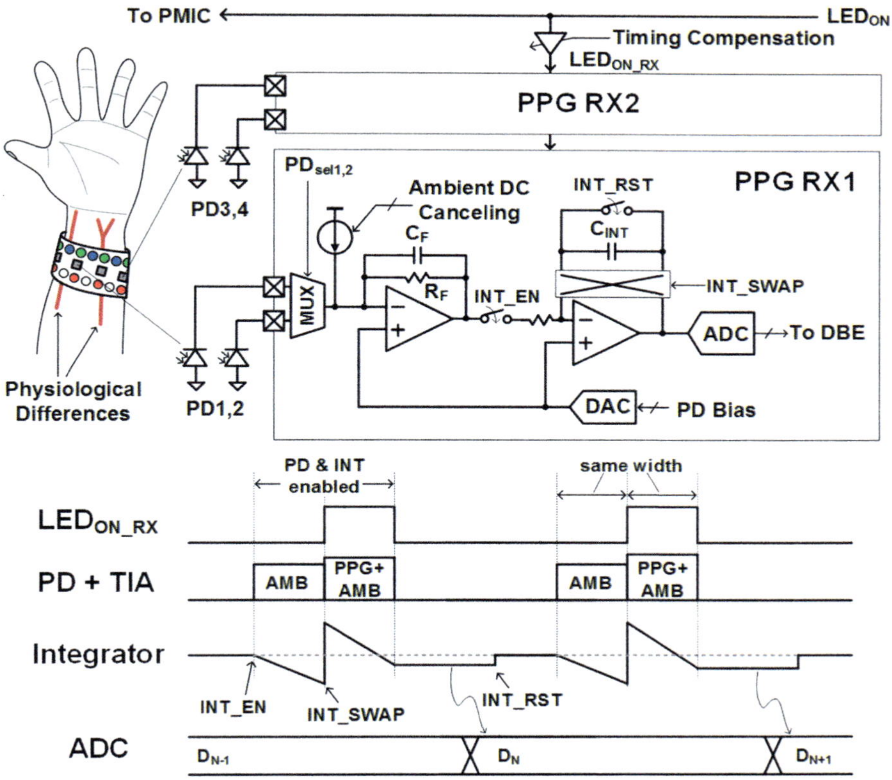

Fig. 7.12 System schematic of the PPG readout channel [12]

7.4.2 Ambient Light Cancellation

Figure 7.12 shows the low-power PPG readout channel with integrated ambient interferer removal from [12]. It consists of a TIA and an integrator, which amplifies the signal and removes the ambient component. Its operation is synchronized to the LED_{ON} (timing diagram). There are two phases of one integration cycle (PD&INT

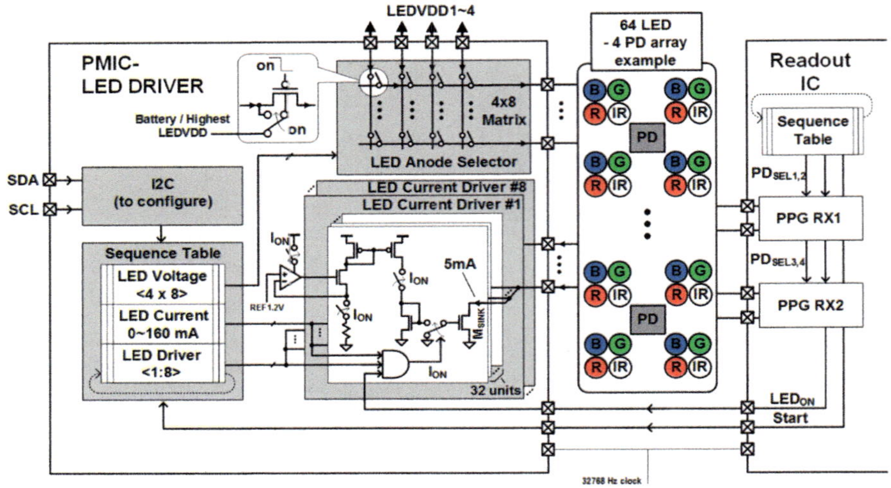

Fig. 7.13 System schematic of the LED drivers [12]

enabled period in timing diagram): in the first phase, the ambient light signal is integrated on C_{INT}; in the second phase, the C_{INT} is swapped, and then the ambient light together with the LED modulated PPG signal is integrated on the same capacitor. The pulse repetition frequency (PRF) is 4 kHz, and the pulse width is typically 10–20 μs. It is worth noting that the LED pulse width is narrow (10 μs); the ambient light can be regarded as constant during the whole integration period; thus, the signal from the ambient light is effectively cancelled. This can also be done by double sampling at the TIA output without the integrator and subtraction in digital domain [21]. Alternatively, course subtraction of ambient/DC current can be achieved via IDAC at the input of the TIA, to reduce the input DR range requirements of the TIA.

The LED drivers for pulse generation (Fig. 7.13) are organized in an 8×8 matrix with eight current drivers and eight driving voltage selectors, to control up to 64 LEDs in an orthogonal fashion. The LED drivers are controlled by a LED sequence table implemented in the digital controller, and the readout channel is kept synchronous. This feature provides the possibility for various settings, in terms of color, biasing current and voltage of the LEDs, to cope with physiological differences among people.

7.4.3 A High-DR PPG Readout

Figure 7.14 shows a high-DR PPG readout channel where there are two amplification stages (TIA + PGA) [20]. An ambient light measurement (obtained by reading

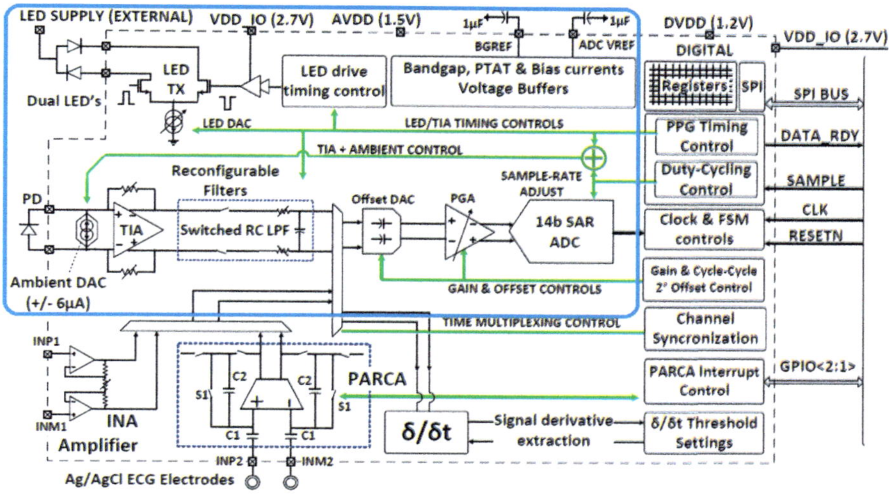

Fig. 7.14 System schematic of the PPG channel in [20]

out the PD without pulsing the LED) is inserted between LED pulse phases to enable system-level correlated double sampling. A fully differential topology is used for the TIA/PGA and the ADC, improving the DR over a single-ended solution by 6 dB. Moreover, an offset DAC is used to remove the DC component in the signal to allow further amplification. In a PPG signal, the AC component is usually less than 1% of the DC component; therefore, effective DC cancellation helps to increase dynamic range significantly. The obtained DR is 97 dB, which is among the highest reported.

7.4.4 Compressed Sampling

As previously mentioned, the dominant source of power consumption in a PPG recording is usually the LEDs. Hence, in order to save power, it is interesting to try to reduce the time the LEDs are active. [19] proposed an architecture for compressive sampling (CS)-based PPG acquisition to achieve exactly this. Nonuniform stimulation and sampling at sub-Nyquist rate can be used in CS-based PPG acquisition systems instead of conventional uniform stimulation and sampling at Nyquist rate, thereby reducing further the effective duty cycle of the LED. The example in [19] shows an LED power of 43 μwW, achieving a reduction of more than 30 times. However, it remains to be seen how well the accuracy can be maintained under true ambulatory scenarios.

7.4.5 Benchmarking and Conclusion

Table 7.3 shows an overview of PPG systems. The input-referred current noise is of significant importance, since it decides the sensitivity of the readout channel. A rms noise of <1 nA is typically achieved, while the noise can be reduced by increasing the gain of the channel at the cost of dynamic range. Providing a low-power readout channel with sub-100 pArms noise and high dynamic range at the same time is still challenging.

7.5 Biomedical Data Converter Circuits

7.5.1 Data Converters for Biomedical Applications

Biomedical signals are characterized by medium to high dynamic ranges but fairly low bandwidths. SAR ADCs (8b–12b) have been very popular in this field because of their minimal use of high-accuracy analog circuits [10, 24, 25]. However, their resolution is limited to about 12 bits due to matching requirements for the capacitors in the DAC. To accommodate higher dynamic ranges, $\Sigma\Delta$ ADCs are an attractive alternative making it possible to shift most of the processing to the digital domain and relax the requirements on the analog front-end. This also fits nicely with the trend toward more digital signal processing being integrated.

Konijnenburg et al. [12] propose a $\Sigma\Delta$ ADC for biomedical applications. To minimize power consumption, the $\Sigma\Delta$ ADC operates on a 32 kHz clock, avoiding power-hungry generation of high-speed accurate low jitter sampling clocks. The $\Sigma\Delta$ ADC is a single-loop second-order feed-forward SC $\Sigma\Delta$ ADC with 5-bit successive approximation (SA) in-loop quantization for the required ENOB of 15 bits. Figure 7.15 shows the measured output spectrum achieving a SNDR of 85.4 dB and a SNR of 87 dB with 3 μA current consumption.

It is worth noting that with a lot of $\Sigma\Delta$ ADCs, the modulator is not the power-limiting block, but the driver and reference generation are. In [12] the driver, which doubles as programmable gain amplifier (PGA) and anti-alias filter, consumes 10 μA. The input stage of the PGA consists of a differential difference amplifier (DDA) with two fully differential inputs [26, 27] and is implemented as a miller-compensated two-stage amplifier. The amplifier is chopper compensated to reduce flicker noise [4]. The first-stage amplifier is a symmetric amplifier with degenerated input differential pairs for handling the high signal swing in the input. The class A/AB output stage [28] helps to drive the $\Sigma\Delta$ ADC by enhancing the slew rate (SR).

While $\Sigma\Delta$ ADCs can achieve the required resolution, they don't specifically address multimodality, where multiple signals must be converted simultaneously. Indeed, [12] uses dedicated ADCs for different channels resulting in large silicon area. A $\Sigma\Delta$ ADC cannot easily be multiplexed due to the memory effect of the loop

Table 7.3 Benchmarking of PPG systems

	[12]	[19]	[20]	[21]	[22]	[23]
Technology	0.18 μm	0.18 μm	0.13 μm	0.18 μm	N. A.	N. A.
Supply voltage	1.2 V	1.2 V	1.5 V	1.8 V	2.0 V	1.8 V
Input noise	15.4 pVrms (1–64 Hz)	486 pArms (1–64 Hz)	4 pArms (0.1–20 Hz)	600 pArms (0.1–10 Hz)	36 pArms (0.1–20 Hz)	1–5 nArms (N. A.)
DC cancellation	10 μA	10 μA	12 μA and VDAC	100 μA	14 μA and VDAC	No.
Dynamic range	87 dB	N. A.	97 dB	91 dB	99 dB	72.3 dB
LED current	5–160 mA	5–160 mA	12 mA	0.1–25.6 mA	1–100 mA	8–250 mA
Channel power	130 μW	172 μW	69 μW	216 μW	400 μW	200 μW
System power	880 μW–24 mW	215–1372 μW	249 μW	336–1545 μW	N. A.	227–1280 μW

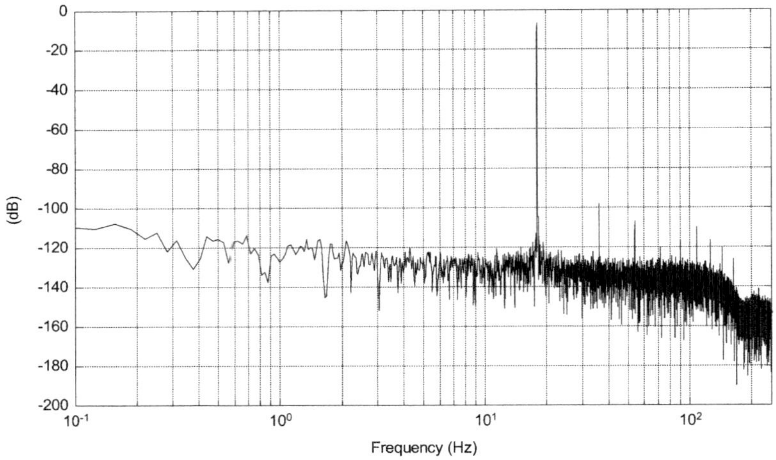

Fig. 7.15 Measured output spectrum of the ΣΔ ADC in [12]

filter integrators. In [29] a three-input multiplexed ΣΔ ADC is shown which allows sharing of active and passive components among the channels. For each input, a different integrator capacitor in the loop filter is switched to keep the charge and to solve the memory effect. The integrators of the loop filter of the ΣΔ ADC don't need to be reset, and a lower latency for the input channels is ensured. The ADC consumes 5.6 mW and achieves a SNDR of 74.7 dB in a 20 kHz BW. For higher-resolution time-multiplexing ADCs, SAR ADCs with in-band noise reduction and linearity enhancement techniques or incremental ΣΔ ADCs can be used.

7.5.2 High-Resolution SAR ADCs

Especially in advanced sensor platforms where higher-speed clocks and digital signal processing are readily available, oversampling and mismatch error shaping techniques as DWA can be introduced in a SAR ADCs [30, 31] to increase their effective resolution beyond 12b. In [30] the combination of dithering, chopping, differently encoded DAC array (4-bit thermometer plus 10-bit binary), and data-driven noise reduction (DDNR) techniques is used to achieve 79.1 dB SNR (87.1 dB SFDR) at 4 kHz BW consuming 1.37 uW.

In [31] a subranging design (Fig. 7.16) along with power-efficient switching schemes (trilevel/monotonic), DWA, segmented DAC array design, redundancy bits, mismatch error shaping (MES), and FIR-IIR filtering is presented. The design allows for various correction techniques because of the introduction of an extra sub-ADC. It achieves a good performance of 105.1 dB SFDR and 96.1 dB SNDR in a 4 kHz signal bandwidth. An ENOB of 15.65 bits and a power consumption of 15.7 μW is obtained for this design.

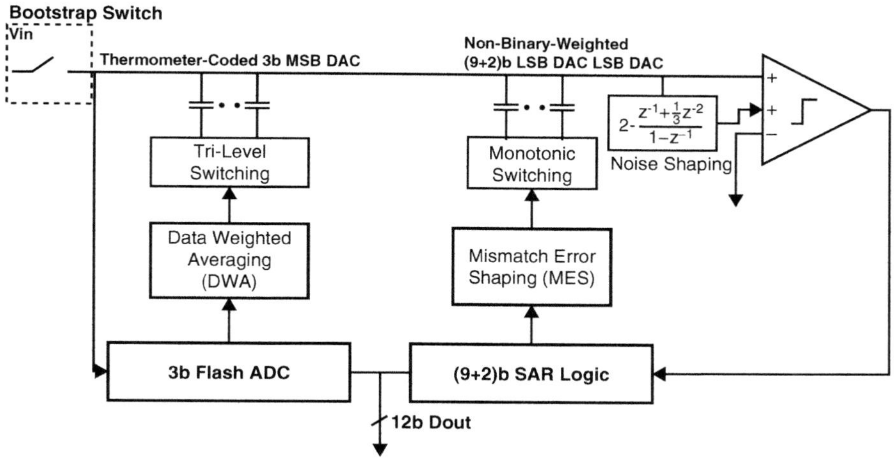

Fig. 7.16 A subranging ADC architecture [31]

7.5.3 Incremental SDMs

An incremental ΣΔ ADC [32–34] is a ΣΔ ADC used as a Nyquist ADC to convert several input channels. The incremental ΣΔ ADC requires a sample-and-hold driver circuit and decimation filter and is reset every time a new conversion starts. This new conversion runs for #N number of clock cycles to obtain a digital output code of the required resolution. For low power and a low conversion time per input channel, the number of clock cycles needs to be low. For first- or second-order ISDMs, the number of conversion cycles is quite large for achieving a high resolution [34], resulting in a very fast sampling clock. Higher-order or multi-bit architectures require less conversion cycles, but a more complex circuit is required, and moreover, the reported implementations still only achieve a moderate energy efficiency [33, 34]. There are in literature two efficient two-step ADC approaches to reduce the number of conversion cycles: adding an extended-range ADC (ER ADC) to the ISDM or a zoom ADC.

Figure 7.17 shows the concept of an ISDM with an ER ADC [35, 36]. The ER ADC is a Nyquist-rate ADC, and it is used to capture the "residue error" of the ISDM at the last cycle of each conversion. The input range of the ER ADC is assumed to be two times of the range of the "residue error" for introducing redundancy and at the same time relaxing the output swing of the integrator. By proper scaling of its output, the fine output Y_2 is obtained, which is then combined with the filtered output of the ISDM (Y_1) to produce Y_{out}. The decimation filter (DF) of the first-order ISDM is simply a counter, which converts the high-frequency data stream d_i to a single code. If the number of bits of the ER ADC is relatively high, e.g., 9 bits, the accuracy requirement for the ISDM is significantly relaxed, leading

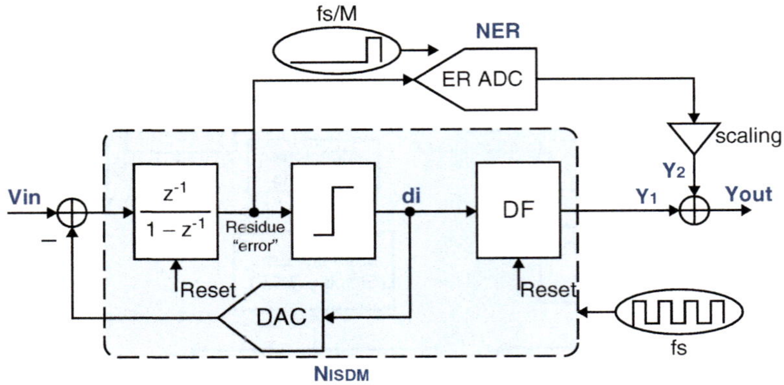

Fig. 7.17 Concept for an incremental SDM with an ER ADC [35, 36]

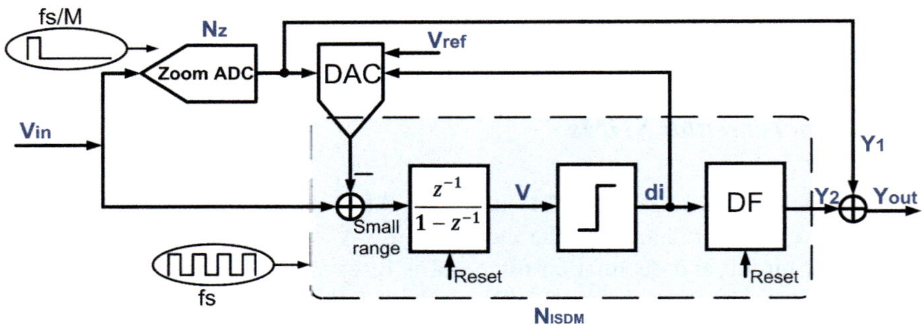

Fig. 7.18 Concept for an incremental SDM with a zoom ADC [37]

to a much shorter conversion time. A recent implementation achieves an 86.3 dB SNDR; this is an ENOB of 15 bits, with only 45 conversion cycles, and consumes 38.1 mW total analog and digital [36].

Figure 7.18 shows the architecture when a zoom ADC is added to the ISDM [38]. The zoom ADC is also a Nyquist-rate ADC, and it's operated at the beginning of each conversion to obtain a coarse result Y_1. Because the zoom ADC shares the same digital-to-analog converter with the ISDM, the reference voltage of the ISDM can be easily adjusted according to the coarse result. Therefore, its input is zoomed into a small range around the input signal. Since the input range of the ISDM is now quite small, it takes a much shorter time to obtain the fine output Y_2 for the system. In order to relax the requirement of the zoom ADC, 1b redundancy is implemented by increasing the input range of the ISDM to 2LSBs of the zoom ADC. Thus, the output of this architecture can be calculated with Y_1 as the scaled output of the zoom ADC and Y_2 as the output of the ISDM. The work in [37] achieves 119.8 dB SNR with 400 conversion cycles and a power consumption of 6.3 μW.

7.6 Conclusions

Wearable healthcare devices have entered our daily lives already in a very tangible way. They have the potential for even more disruptive fundamental societal impact by enabling true preventive healthcare. While research in these domains is advancing and great breakthrough results are being presented at a regular basis, power consumption remains prohibitively high for true, long-term monitoring. Hence, a continued quest for ever-lower power consumption remains a research challenge. An important general trend is multimodality readouts [12]. On the one hand, more signal modalities of course simply means that more relevant health parameters can be observed. But on the other hand, it also has the potential to increase the robustness of the recording or improve the power efficiency. For example, biopotential (ECG), bio-impedance, and optical (PPG) can all be used to measure heart rate. While all of these will suffer from motion artifacts, the underlying mechanisms are different. Hence, through sensor fusion algorithms, several recordings, each potentially of low quality and disrupted by heavy artifacts, can be combined to improve the overall quality. This paper focused on analog circuits for a number of common signal modalities for wearable healthcare and discussed a number of state-of-the-art implementations.

References

1. Song, S., et al.: A low-voltage chopper-stabilized amplifier for fetal ECG monitoring with a 1.41 power efficiency factor. IEEE Tran. Biomed. Circuits Syst. **9**(2), 237–247 (2015)
2. Yaul, F.M., Chandrakasan, A.P.: A sub-μW 36nV/$\sqrt{}$Hz chopper amplifier for sensors using a noise-efficient inverter-based 0.2V-supply input stage. IEEE ISSCC. 94–95 (2016)
3. Harpe, P., Gao, H., van Dommele, A.R., Cantatore, E., van Roermund, A.: 0.20 mm^2 3 nW signal acquisition IC for miniature sensor nodes in 65 nm CMOS. IEEE J. Solid State Circuits. **51**(1), 240–248 (2016)
4. Enz, C.C., Temes, G.C.: Circuit techniques for reducing the effects of op-amp imperfections: autozeroing, correlated double sampling, and chopper stabilization. Proc. IEEE. **84**(11), 1584–1614 (1996)
5. Harrison, R.R., Charles, C.: A low-power low-noise CMOS amplifier for neural recording applications. IEEE J. Solid State Circuits. **38**(6), 958–965 (2003)
6. Chandrakumar, H., Marković, D.: 5.5 a 2 μW 40 mVpp linear-input-range chopper- stabilized bio-signal amplifier with boosted input impedance of 300 MΩ and electrode-offset filtering. IEEE ISSCC. **59**, 96–97 (2016)
7. Birk, C., Mora-Puchalt, G.: A 60V capacitive gain 27nV/$\sqrt{}$Hz 137dB CMRR PGA with ±10V inputs. IEEE ISSCC. 376–377 (2012)
8. Mohan, R., Zaliasl, S., Gielen, G., Van Hoof, C., Van Helleputte, N., Yazicioglu, R.F.: A 0.6V 0.015mm2 time-based biomedical readout for ambulatory applications in 40nm CMOS. IEEE ISSCC. 482–483 (2016)
9. Muller, R., Gambini, S., Rabaey, J.M.: A 0.013mm2 5uW, DC-coupled neural signal acquisition IC with 0.5 V supply. IEEE J. Solid State Circuits. **47**(1), 232–243 (2012)
10. Van Helleputte, N., et al.: A 345 μW multi-sensor biomedical SoC with bio-impedance, 3-channel ECG, motion artifact reduction, and integrated DSP. IEEE J. Solid State Circuits. **50**(1), 230–244 (2015)

11. Bin Altaf, M.A., Zhang, C., Yoo, J.: A 16-channel patient-specific seizure onset and termination detection SoC with impedance-adaptive transcranial electrical stimulator. IEEE J. Solid State Circuits. **50**(11), 2728–2740 (2015)
12. Konijnenburg, M., et al.: A multi(bio)sensor acquisition system with integrated processor, power management, 8×8 LED drivers, and simultaneously synchronized ECG, BIO-Z, GSR, and two PPG readouts. IEEE J. Solid-State Circuits. **51**(11), 2584–2595 (2016)
13. Yan, L., et al.: A 13μA analog signal processing IC for accurate recognition of multiple intra-cardiac signals. IEEE Trans. Biomed Circuits Syst. **7**(6), 785–795 (2013)
14. Xu, J., et al.: A low power configurable Bio-Impedance Spectroscopy (BIS) ASIC with simultaneous ECG and respiration recording functionality. Proc. IEEE ESSCIRC, pp. 396–399 (2015)
15. Ko, H., et al.: Ultra low power Bioimpedance IC with intermediate frequency shifting chopper. IEEE Trans. Circuits Syst. II. **63**(3), 259–263 (2016)
16. Kassanos, P., et al.: An integrated analog readout for multi-frequency Bioimpedance measurements. IEEE Sensors J. **14**(8), 2792–2800 (2014)
17. TI AFE4300 Datasheet.
18. Reddy, K.A., George, B., Mohan, N.M., Kumar, V.J.: A novel calibration-free method of measurement of oxygen saturation in arterial blood. IEEE Trans. Instrum. Meas. **58**(5), 1699–1705 (2009)
19. Rajesh, P.V., et al.: 22.4 A 172uW compressive sampling photoplethysmographic readout with embedded direct heart-rate and variability extraction from compressively sampled data. 2016 IEEE International Solid-State Circuits Conference (ISSCC), San Francisco, pp. 386–387 (2016)
20. Sharma, A. et al.: Multi-modal smart bio-sensing SoC platform with >80dB SNR 35μA PPG RX chain. 2016 IEEE Symposium on VLSI Circuits (VLSI-Circuits), Honolulu, pp. 1–2 (2016)
21. Winokur, E.S., O'Dwyer, T., Sodini, C.G.: A low-power, dual-wavelength Photoplethysmogram (PPG) SoC with static and time-varying interferer removal. IEEE Trans. Biomed. Circuits Syst. **9**(4), 581–589 (2015)
22. TI AFE4404 datasheet
23. ADI ADPD103 datasheet
24. Yazicioglu, R.F., et al.: 200 μW eight-channel EEG acquisition ASIC for ambulatory EEG systems. IEEE ISSCC Digest of Technical Papers, pp. 164–165 (2008)
25. Xu, J., Yazicioglu, R.F., Harpe, P., Makinwa, K.A.A., Van Hoof, C.: A 160 μW 8-channel active electrode system for EEG monitoring. IEEE ISSCC Digest of Technical Papers, pp. 300–301 (2011)
26. Sackinger, D., Guggenbuhl, W.: A versatile building block: the CMOS differential difference amplifier. IEEE J. Solid State Circuits. **22**(2), 287–294 (1987)
27. Alzaher, H., Ismail, M.: A CMOS fully balanced differential difference amplifier and its applications. IEEE Trans. Circuits Syst. II, Analog Digi. Signal Process. **48**(6), 614–619 (2001)
28. Rabii, S., Wooley, B.A.: A 1.8-V digital-audio sigma-delta modulator in 0.8-μm CMOS. IEEE J. Solid State Circuits. **32**(6), 783–796 (1997)
29. Sebastiano, F., van Veldhoven, R.H.M.: A 0.1-mm2 3-channel area-optimized ADC in 0.16-um CMOS with 20-kHz BW and 86-dB DR. 2013 Proceedings of the ESSCIRC (ESSCIRC)
30. Harpe, P., Cantatore, E., van Roermund, A.: An oversampled 12/14b SAR ADC with noise reduction and linearity enhancements achieving up to 79.1dB SNDR. ISSCC Dig. Tech. Papers. 194–195 (2014)
31. Shu, Y., Kuo, L., Lo, T.: 27.2 an oversampling SAR ADC with DAC mismatch error shaping achieving 105dB SFDR and 101dB SNDR over 1kHz BW in 55nm CMOS. pp. 458–459
32. Markus, J., Silva, J., Temes, G.C.: Theory and applications of incremental delta sigma converters. IEEE Trans. Circuits Syst. I. **51**(4), 678–690 (2004)
33. Quiquempoix, V., Deval, P., Barreto, A., Bellini, G., Markus, J., Silva, J., Temes, G.C.: A low-power 22-bit incremental ADC. IEEE J. Solid State Circuits. **41**(7), 1562–1571 (2006)

34. Agnes, A., Bonizzoni, E., Maloberti, F.: High-resolution multi-bit second-order incremental converter with 1.5- V residual offset and 94-dB SFDR. Analog Integr. Circ. Sig. Process. **72**(3), 531–539 (2011)
35. Rombouts, P., De Wilde, W., Weyten, L.: A 13.5-b 1.2-V micropower extended counting A/D converter. IEEE J. Solid State Circuits. **36**(2), 176–183 (2001)
36. Agah, A., Vleugels, K., Griffin, P.B., Ronaghi, M., Plummer, J.D., Wooley, B.A.: A high-resolution low-power incremental ADC with extended range for biosensor arrays. IEEE J. Solid State Circuits. **45**(6), 1099–1110 (2010)
37. Chae, Y., Souri, K., Makinwa, K.: A 6.3 μW 20b incremental zoom-ADC with 6ppm INL and 1 μV offset. IEEE ISSCC Dig. Tech. Papers, pp. 276–277 (2013)
38. Kim, H., et al.: A configurable and low-power mixed signal SoC for portable ECG monitoring applications. IEEE Trans. Biomed. Circuits Syst. **8**(2), 257–267 (2014)

Chapter 8
An Ultra-low Power, Robust Photoplethysmographic Readout Exploiting Compressive Sampling, Artifact Reduction, and Sensor Fusion

Venkata Rajesh Pamula, Chris Van Hoof, and Marian Verhelst

8.1 Introduction

There is an ever-increasing interest in wearable medical devices as a reaction to the population's increased vulnerability to cardiovascular diseases (CVD) and mental disorders. Continuous monitoring of heart rate (HR) and heart rate variability (HRV) provide critical information about an individual's cardiovascular and mental health state [1]. Electrocardiography (ECG) is one of the standard signal acquisition modalities to continuously monitor HR and HRV. Despite being a popular technique, continuous monitoring of HR through ECG reduces patient comfort, particularly for long-term monitoring, due to the usage of electrodes and its requirements regarding skin preparation.

Photoplethysmography (PPG)-based continuous HR and HRV monitoring is emerging as an attractive alternative to ECG-based methods. Unlike ECG, PPG is a non-contact, single-point biosignal measurement technique, resulting in an increased patient comfort. PPG acquisition involves shining light (usually at 660 nm and/or 900–940 nm wavelengths, corresponding to red (R) and infrared (IR), respectively), through tissue and detecting the transmitted/reflected component of the light [2]. Light-emitting diodes (LEDs) are commonly used as light source, while the transmitted/reflected component is measured using photodiodes (PDs). Although PPG-based HR measurement is a well-understood concept, its usage in continuous monitoring is hindered due to its very large power consumption compared to ECG-

V.R. Pamula (✉) • C. Van Hoof
ESAT, KU Leuven and IMEC, Leuven, Belgium
e-mail: pamula@imec.be

M. Verhelst
ESAT, KU Leuven, Leuven, Belgium

P. Harpe et al. (eds.), *Hybrid ADCs, Smart Sensors for the IoT, and Sub-1V & Advanced Node Analog Circuit Design*, DOI 10.1007/978-3-319-61285-0_8

based methods. The power consumption of a typical PPG acquisition system ranges from few mWs to tens of mWs, dominated by the power consumption of the LED driver. Moreover, PPG acquisition is highly susceptible to motion artifacts, degrading its robustness and reliability.

In this chapter, a compressive sampling (CS)-based PPG readout is presented, which enables reduction of relative LED driver power consumption by up to a factor of 30x. The ASIC also integrates a digital back-end, which performs direct feature extraction from the CS signal to estimate average HR, without requiring complex reconstruction techniques. The possibility of artifact reduction, leveraging on sensor fusion and a spectral subtraction technique, is also presented.

8.2 Overview of Photoplethysmogram (PPG) Acquisition

As indicated in Sect. 8.1, PPG acquisition is an optical-sensing modality, which can be performed both in transmission and reflection modes. Figure 8.1 shows the principle of PPG acquisition, using transmission mode. The photocurrent (I_{ph}) acquired by the PD comprises of a small AC component, which arises due to increased flow of blood during systole. The AC component rides on top of a relatively large DC component, arising due to bone and the static blood in the tissue. Since the AC component is related to heart pulsation, its frequency is highly correlated to the HR. The peak-to-peak value of the AC component of the photocurrent depends on several factors including the intensity of the LED light, the quantum efficiency of the PD, the skin tone of the subject mode of PPG acquisition, and the location at which the acquisition is being performed [3]. In terms of relative values, the AC component is typically 1–4% of the DC component. In a voltage mode signal processing system, this photocurrent is converted into a voltage signal through a transimpedance amplifier (TIA), which is then processed further (Fig. 8.2).

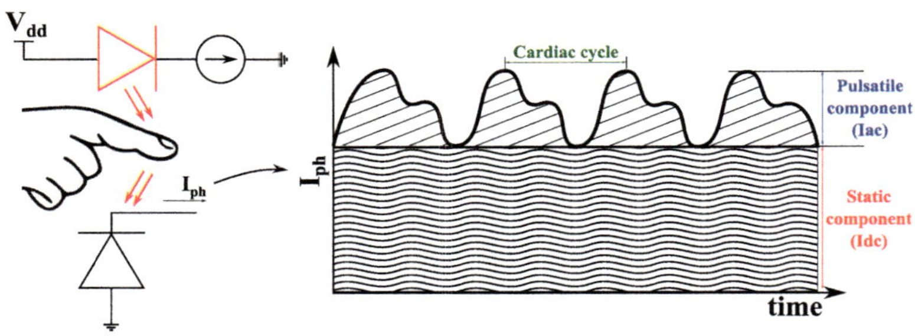

Fig. 8.1 Principle of transmission mode PPG acquisition

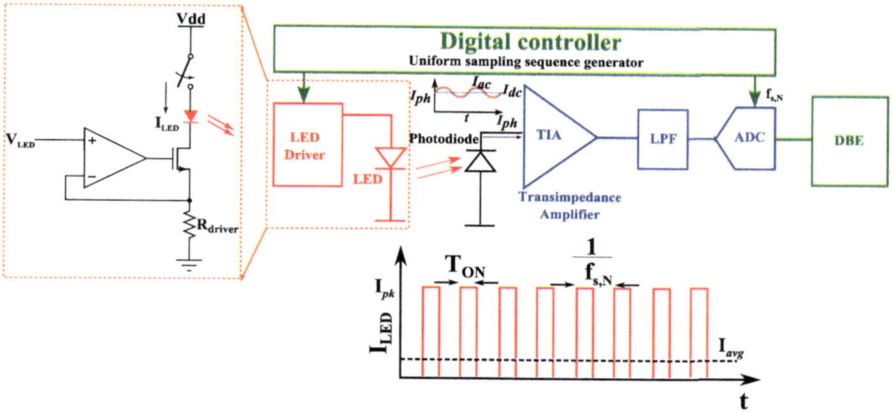

Fig. 8.2 Conventional PPG acquisition system employing uniform LED stimulation and sampling

The relatively small AC component in comparison to the DC component necessitates the need for a low-noise and large dynamic-range (DR) readout. Further, to conserve power in the LED driver, the LED stimulation is typically pulsed, with duty cycles in the range 0.25–2%, implying the need for fast settling of the TIA. The ON duration of the LED and hence the duty cycle (D) of the LED determine the average current (I_{avg}) flowing through the LED (I_{LED}) and hence its power consumption (Fig. 8.2). While the LED driver power can be reduced by further reducing D, it poses stringent requirements on the bandwidth of the TIA, thereby increasing its power consumption. In [4], the authors combined the LED driver power and readout power to formulate an optimization problem to arrive at optimum value for D which minimizes the entire system power. The implemented system is reported to consume 1.5 mW, of which 340 μWs is dissipated in the LED driver. However, [4] does not formulate the power consumption of TIA in terms of noise and assumes it is bandwidth limited. This assumption is not necessarily valid when high SNRs are required at low LED stimulation currents as the noise of the TIA dictates the readout power consumption in such cases rather than its bandwidth.

A highly sensitive analog front-end (AFE) for PPG acquisition using ambient light is proposed in [5]. The proposed AFE employs a wide range logarithmic digital-to-resistance converter (DRC) and a nonuniform quantizer based on laddered inverter quantizer (LIQAF). The AFE, implemented in a 0.18 μm CMOS process, consumes <4 μWs of power from a 0.5 V supply. While the possibility of measuring HR using ambient has been demonstrated with the proposed AFE, its efficacy is highly dependent on the ambient light levels. Moreover, the SNR achieved under nominal lighting conditions is not reported. This, therefore, limits the applicability of the proposed AFE in [5], particularly under low ambient light and low perfusion conditions, which necessitate the use of LED as light source. Therefore, it is desirable to explore techniques that can potentially reduce the relative LED driver power consumption. Enabling this objective, by employing the principles of CS, to reduce the relative LED driver power consumption, is the goal of the work presented in this paper.

8.3 Compressive Sampling (CS) for Photoplethysmography (PPG) Acquisition and Feature Extraction from Compressively Sampled Signal

8.3.1 Compressive Sampling (CS) for Photoplethysmography (PPG) Acquisition

Compressive sampling (CS) is an alternate signal acquisition paradigm which asserts that certain class of signals can be faithfully recovered from far fewer samples or measurements of the signal compared to traditional Nyquist-based sampling [6]. This acquisition protocol relies on the *sparsity* of the signal on a given basis and its *incoherence* to the sampling scheme. Several physiological signals such as ECG, EEG, etc. have been shown to be sparse on standard bases such as the wavelet basis and Gabor basis. PPG signals, in particular, have been shown to be sparse on frequency basis [7]. Since the frequency basis is maximally incoherent to canonical basis, CS for PPG signals is equivalent to randomly subsampling the signal in time domain. In mathematical terms, the signal acquisition protocol in CS framework can be expressed as

$$Y = \Phi X \tag{8.1}$$

where Φ is referred to as the measurement matrix and Y and X are the M-dimensional measurement vector and the N-dimensional signal vector, respectively ($M \ll N$).

As indicated in (8.1), CS-based acquisition of a signal acquires a *linearly transformed, lower-dimensional* representation of the signal, often referred to as *measurements* rather than the actual signal. This approach, therefore, results in *compression* (and hence the name compressive sampling), which is usually quantified through the metric, compression ratio (CR) defined as

$$CR = \frac{N}{M} \tag{8.2}$$

The equivalent (partial) measurement matrix for random subsampling is shown in Fig. 8.3, which is $M \times N$ reduced order identity matrix, formed by choosing M rows from the $N \times N$ identity matrix at random. The M rows chosen at random correspond to the M sampling instants in time domain (with the row index corresponding to the sample index). In practice, pseudorandom subsampling schemes are used, showing

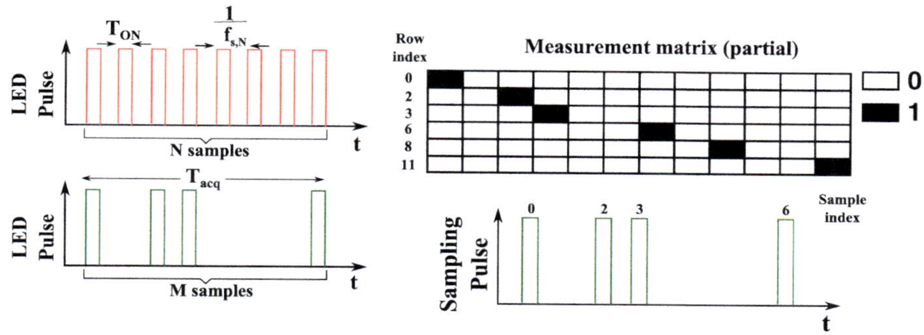

Fig. 8.3 Sampling sequence for CS PPG acquisition system and its equivalent measurement matrix structure

on par performance with fully random samples. The same pseudorandom sequence can be reused for every discrete window of length T_{acq} s (Fig. 8.3). Compared to the conventional PPG acquisition, based on uniform sampling, CS-based PPG acquisition acquires signal at an average sampling rate of $f_{s,CS}$ given by

$$f_{s,CS} = \frac{f_{s,N}}{CR}. \tag{8.3}$$

where $f_{s,N}$ is the uniform sampling rate.

Therefore, CS-based PPG acquisition systems have a LED driver duty cycle of $T_{ON} \times f_{s,CS}$ compared to $T_{ON} \times f_{s,N}$, as is the case for conventional uniform sampling PPG acquisition systems which hence enables reduction of LED driver power consumption by a factor of CR.

8.3.2 Feature Extraction from Compressively Sampled (CS) Photoplethysmographic (PPG) Signal

While the acquisition of a signal in the CS framework is relatively simple, recovering the signal back from the measurements, often referred to as the reconstruction process, is a very computationally intensive task. While several algorithms exist for signal recovery, with varying degrees of computational complexity [8], only a handful hardware implementations for the same exist [9–11]. Many CS implementations reported in the literature assume the presence of a powerful base station, to which the measurement data is off-loaded over a wireless link. The reconstruction is then performed at the base station, where the power constraints are relaxed (Fig. 8.4a). The advocates of this model [8, 12] argue that wireless transmission tends to dominate the sensor node power consumption in wireless sensor network (WSN) and body area network (BAN) applications, and hence transmitting the

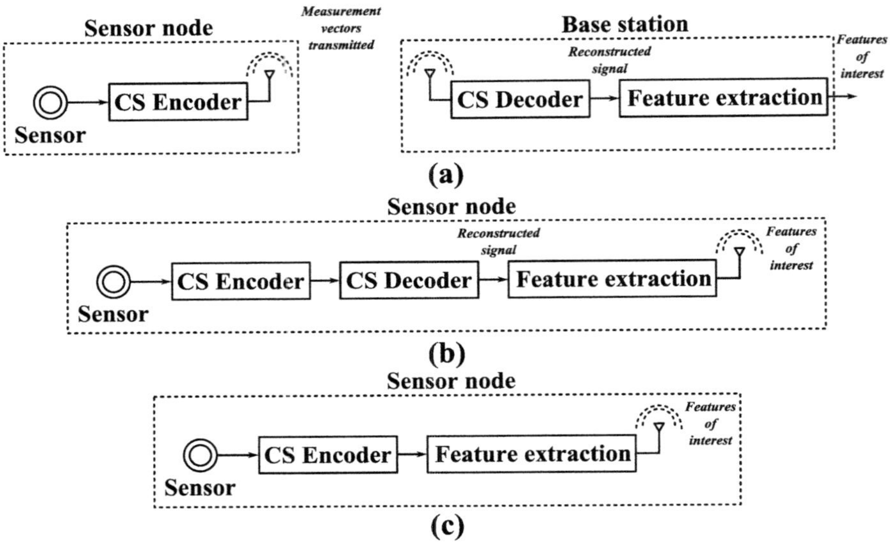

Fig. 8.4 Various possible CS-based acquisition systems in context of BAN. (**a**) Signal acquisition is performed at the sensor node, while reconstruction and feature extraction are performed at the base station. (**b**) Both CS encoding and decoding are performed on the signal node followed by feature extraction. (**c**) Feature extraction is performed on the sensor node directly from the CS data

compressed data rather than performing the reconstruction on the sensor node and then transmitting the data is efficient in terms of power consumption. This approach, therefore, shifts the problem of signal analysis and extracting the parameters of clinical interest to the base station. However, the need for analyzing the signal and extracting the relevant *features* locally on the sensor node is becoming important, particularly for privacy-and latency-sensitive biomedical applications [13]. Yet, locally reconstructing the signal on the sensor node (Fig. 8.4b) would consume power in the range of mWs [9], rendering feature extraction from the reconstructed signal on the sensor node infeasible for low-power sensing applications.[1]

Alternatively, the requirement of reconstructing the signal can be circumvented if the features of interest can be extracted directly from the CS data (Fig. 8.4c). This approach enables rapid signal analysis on energy-scarce BAN/WSN platforms directly from the CS data, without requiring complex reconstruction process. In this work, the use of least-squares spectral fitting techniques is explored for power

[1]The benefits of CS encoding and decoding, followed by feature extraction, all on the sensor node, over the conventional approach of performing feature extraction on the Nyquist rate sampled signal might not be obvious. CS-based approach can be useful in cases where high-power stimulation is involved, as in the case with PPG acquisition as well as in the cases where the maximum achievable sampling frequency of the ADC is limited [15].

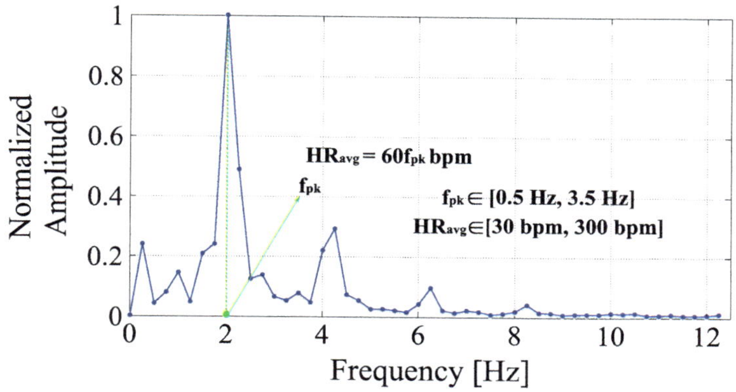

Fig. 8.5 Estimation of average HR from the frequency spectrum of PPG signal

spectral density (PSD) estimation directly from CS PPG signals. In particular, Lomb-Scargle periodogram (LSP) is used as the PSD estimator for randomly subsampled PPG signal [14]. Once the PSD is estimated, the average HR over time interval of T_{acq} can be estimated in the frequency domain by extracting the frequency corresponding to the peak in the PSD, f_{pk} as

$$HR_{\text{avg}} = 60 \cdot f_{pk}. \tag{8.4}$$

where HR_{avg} is the average HR in beats per minute (bpm) (Fig. 8.5). Once the average HR is estimated, HRV can be readily inferred from the variation of average HR across successive time intervals over which the spectrum is estimated.

The implementation details of a single-channel PPG readout ASIC [16], leveraging on the concepts of CS and feature extraction from CS domain discussed in this section, are presented in Sect. 8.4.

8.4 Compressive Sampling (CS) Photoplethysmographic (PPG) Readout Implementation

8.4.1 ASIC Implementation

The top-level architecture of the single-channel CS PPG acquisition ASIC is shown in Fig. 8.6. The ASIC embeds an AFE which performs a pseudorandom subsampled acquisition of the PPG signal and a digital back-end (DBE), which performs the HR estimation directly from the CS PPG signal. The AFE integrates a programmable gain TIA, the output of which is interfaced to a switched integrator (SI), which improves the SNR. The output of the SI is buffered and digitized through a 12-bit SAR ADC. A sub-1V bandgap reference is integrated on-chip to provide stable

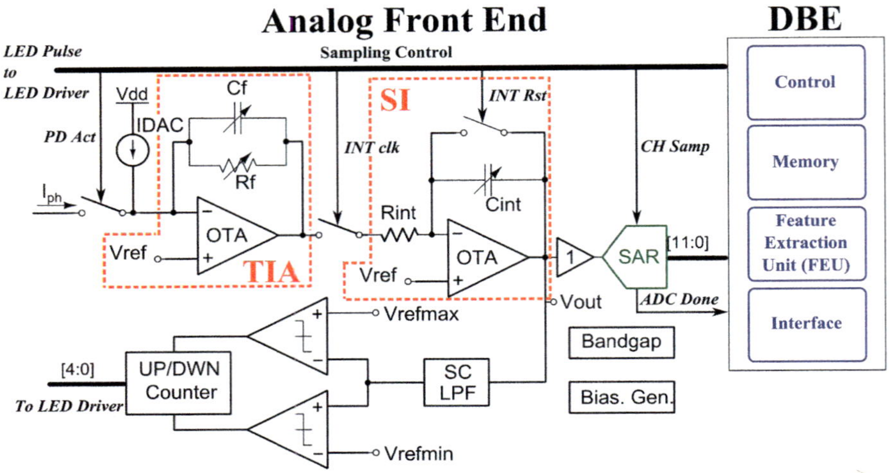

Fig. 8.6 The architecture of a single-channel CS PPG acquisition ASIC which embeds a DBE for feature extraction

on-chip bias and reference signals. The DBE comprises of a control unit (CU) that generates the necessary control signals required for the LED driver, AFE, and the ADC and also the required internal timing and synchronizing signals. Direct memory access (DMA) is integrated into the DBE which transfers the incoming data from the ADC into one of the data memory (DMEM) banks. The feature extraction unit (FEU), also part of the DBE, accelerates the process of LSP to enable extraction of HR directly from the CS PPG signal. The DBE is clocked through an external clock at 32 kHz. The ASIC also provides wide-scale programmability both for the gain and bandwidth settings of the AFE and CR, thereby extending its utility across a wide range of photocurrent amplitudes.

The first stage of the readout channel is a TIA that is interfaced to an off-chip photodiode (PD). The TIA converts the PPG signal that is acquired as a current signal at the output of the PD into a voltage signal, which is further processed by the signal processing chain in voltage domain. The TIA is realized by employing resistive feedback (R_f) around a two-stage Miller-compensated OTA. The large reverse bias junction capacitance of the PD, which manifests itself as a parasitic capacitance (C_p) at the inverting node, poses issues to the stability of the TIA. Hence, a compensation capacitor (C_f) is added in parallel to R_f to improve the stability margin of the TIA. As mentioned in Sect. 8.1, the relative large DC component of photocurrent necessitates the need for large DR for the readout. The channel DR requirements can however be relaxed if the DC component of the current is rejected early in the signal processing chain. This is achieved by interfacing a 5-bit current DAC (IDAC), capable of sourcing up to 10 μA of current at the input of the TIA.

The output of the TIA is fed into a switched integrator (SI), which is realized by incorporating a switched-capacitor (SC) in feedback around the OTA. The output of the TIA is converted into a current signal through R_{int}, which is then integrated onto C_{int} for a duration of T_{int}, thereby providing additional voltage amplification. The SI stage, apart from providing additional gain, also acts as a noise-limiting filter [17]. This is particularly important in pulsed PPG acquisition systems, where the thermal noise originating from the OTA of the TIA exhibits noise peaking at high frequencies. A mixed-signal feedback loop, comprised of a SC low-pass filter (SC LPF), comparators, and an up-down counter, tracks the output DC level of the SI. A 5-bit control code, to control the LED drive/IDAC current, is generated by the feedback loop such that the DC output of the SI stays within the threshold values (V_{refmin} and V_{refmax}), to ensure the proper utilization of the available channel DR.

The output of the SI is then digitized using a 12-bit SAR ADC, which comprises of a split capacitor DAC to reduce the area requirements, with a unit capacitance (C_u) of 800 fF. The pseudorandom subsampling instants of the ADC are controlled by the CU that forms part of the DBE. The digitized data, at the output of the ADC, is fed into the DBE for further processing to extract the HR. Interested readers are referred to [18], where the detailed description of the DBE architecture and implementation is presented.

8.4.2 Measurement Results

The ASIC is fabricated in a 0.18 μm process and occupies an area of $10\,\text{mm}^2$. To validate the functionality of the ASIC, an off-chip LED is modulated by a sinusoidal current with a frequency of 1.2 Hz (corresponding to 72 bpm HR), and the resulting PD current is read out for CRs of 8x and 30x as a voltage signal at the output of SI. As can been seen in Fig. 8.7, the acquired photocurrent and hence the voltage at the output of SI are (pseudo)random in nature and demonstrate the functionality of the AFE in CS acquisition mode. Figure 8.7 also shows an in vivo PPG acquisition, performed at 10x compression.

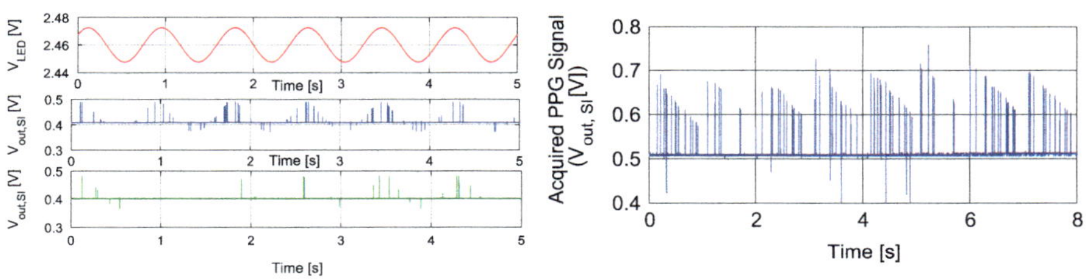

Fig. 8.7 (*Left*) Signal acquisition with CRs 8x and 30x when LED is stimulated with a sinusoidal current at 1.2 Hz. (*Right*) In vivo acquired PPG signal through the ASIC with a CR of 10x

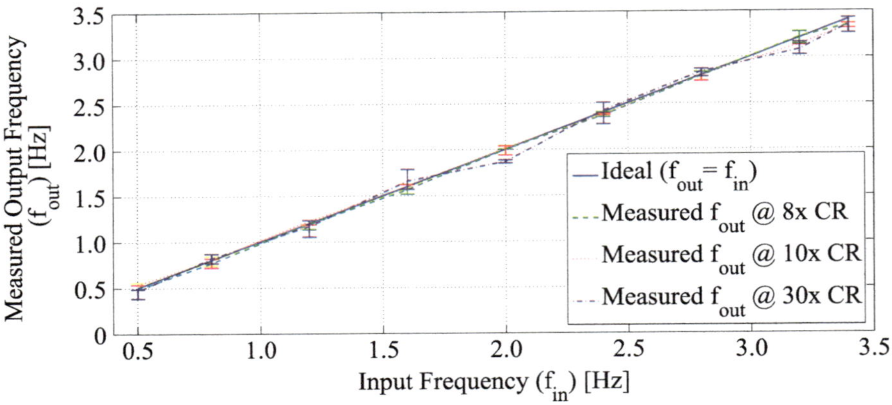

Fig. 8.8 Measured frequency corresponding to the peak in the PSD (f_{pk}) from the ASIC with LED modulated with a sinusoidal current whose frequency is swept from 0.5 to 3.4 Hz

The HR estimation performance is characterized by modulating the LED with a sinusoidal current, the frequency of which is swept from 0.5 to 3.4 Hz to cover the HR range of 30–204 bpm.[2] The LED modulation is carefully chosen so that the AC component of the photocurrent is approximately 20 nA_{pp}, which is comparable to the range of photocurrent measured while performing in vivo acquisition. The output of the readout is then compressively sampled with CRs 8x, 10x, and 30x, and feature extraction is performed on the acquired data. Since the feature extraction process estimates the frequency corresponding to the peak in the PSD, under ideal conditions, the estimated peak frequency (f_{pk}) should be identical to the input frequency. Figure 8.8 shows the extracted peak frequency for different CRs. The peak frequency serves as a proxy to estimate the HR using (8.4). The measured HR exhibits a worst-case error of 10 bpm at 30x compression for a nominal HR of 96 bpm. This error is still within the conformance specification provided by the ANSI-AAMI standards for heart rate meters [20].

Figure 8.9 shows the chip micrograph and the power consumption at different CRs. The ASIC consumes a total power of 172 μW from a supply of 1.2 V for the entire system of which the AFE consumes 158.8 μW, while the ADC and the DBE consume 6 μW and 7.2 μW, respectively. On the other hand, the LED driver power consumption scales from 1200 μW to 43 μW,[3] when scaling between uniform sampling mode (1x CR) and 30x CR, respectively, thanks to the compressive sampled acquisition paradigm. At lower CRs, LED driver continues to dominate the power consumption of the system, while at higher CRs, the AFE limits the power consumption due to fundamental noise limitations.

[2]Standard database [19] PPG signals lack annotations and hence sinusoidal modulation is chosen.

[3]The LED driver power consumption is measured while acquiring the PPG signal of a healthy individual. At the reported power levels, the resulting photocurrent is measured to have an AC component of 45 nA_{pp}, while the DC component is measured to be 1.6 μA.

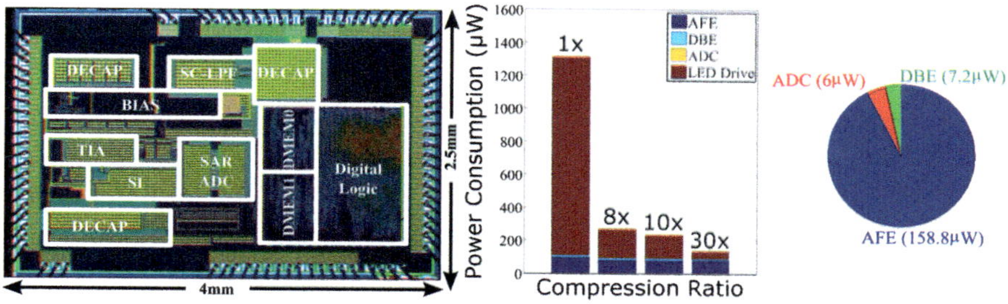

Fig. 8.9 The ASIC chip micrograph and measured power consumption breakdown of the ASIC and the off-chip LED driver for different CRs

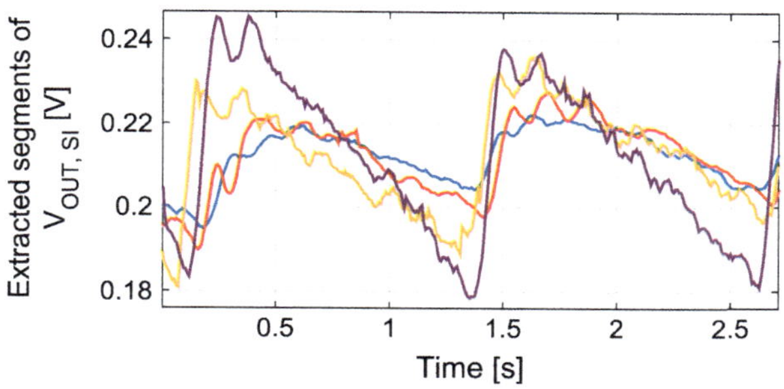

Fig. 8.10 In vivo acquired PPG signals under different SNR conditions. The corresponding values of acquired photocurrent and LED driver current are indicated in Table 8.1

Table 8.1 ASIC performance with different SNRs

AC component of PPG signal (Vpp)	**15 mV**	**25 mV**	**40 mV**	**60 mV**
AC component of photocurrent (I_{ac}(pp))	**3 nA**	**5 nA**	**8 nA**	**12 nA**
LED peak current	18 mA	56 mA	160 mA	314 mA
IDAC	312.5 nA	937.5 nA	2.1875 μA	2.5 μA
HR est.@CR = 10x	49	46	47	48
HR est.@CR = 1x(uni.)	48	48	48	48

The robustness of the ASIC under varying SNR conditions is demonstrated by performing in vivo acquisition of PPG under four different conditions, changing the LED driver current (drawn from a 5 V supply) while adjusting the IDAC setting to cancel most of the DC component out. The excerpts of the recorded PPG signals after being filtered are shown in Fig. 8.10. Table 8.1 shows the information of the different setups, the resulting AC component of the acquired signals, and the heart rate value calculated by the ASIC at a CR of 10x as well as with uniform sampling. The AC component of the photocurrent varies from 3 nA_{pp} for a LED driver peak current of 18 mA to 12 nA_{pp} when the LED driver peak current is increased to

314 mA. The HR, estimated from the uniformly sampled PPG signal using FFT, serves as the reference. The PPG signal is then compressively acquired at a CR of 10x, and the average HR estimated by the ASIC is compared against the reference. As can be seen in Table 8.1, the error in the average HR is estimated at 10x CR within 2 bpm under varying SNR conditions. The LED driver power consumption, on the other hand, scales proportional to the CR, from 6.1 mW to 615 μW for an acquired AC component of photocurrent of 12 nA_{pp}.

Finally, the performance of the CS PPG ASIC is summarized and compared against the state-of-the-art PPG acquisition systems in Table 8.2. Compared to the state-of-the-art, CS-based PPG, acquisition enables up to 30x reduction in the power consumption of the LED driver, thanks to the DBE, which accelerates LSP to enable feature extraction directly from CS data to accurately estimate HR with minimum power penalty.

8.5 Artifact Reduction and Electrocardiogram (ECG)-Assisted Photoplethysmogram (PPG) Acquisition

8.5.1 Artifact Reduction Through Spectral Subtraction

A wearable PPG acquisition system is usually subject to large amounts of motion. Depending on the location of the sensor and the nature of the motion, the artifacts in signals can manifest themselves in different ways, making it hard to derive HR and HRV from them. Unlike ECG, motion can potentially alter the flow of blood in the tissue, affecting the physiological signal in a rather fundamental way. An immediate consequence of large motion artifacts is the need for large DR for the PPG readout. Yet, even for PPG readouts with large DR, processing the signal infested with artifacts to extract relevant features such as HR is extremely challenging.

An attempt to mitigate motion artifacts in PPG acquisition systems was made by [4]. The authors in [4] rely on mechanically stabilizing the LED and PD pair housed in a double-ringed aluminum unit. This mechanical approach, however, is limited to PPG sensors that have a specific form factor – the ones that can be worn as a ring.

Figure 8.11 shows an alternate approach to mitigate motion artifacts in the digital domain. This process involves spectral estimation of PPG signals using the LSP approach. Simultaneously, a representative motion signal, such as the output of an accelerometer, is pseudorandomly subsampled, with the same sampling sequence as used for the PPG aignal. Subsequently, spectral estimation is performed using the LSP. The output spectra are then normalized and subtracted from each other to remove the motion component. Since this approach involves *denoising* through subtraction in the frequency domain, it is referred to as *spectral subtraction* technique and is popularly used to eliminate the background noise in speech signal processing [24].

Table 8.2 ASIC performance summary and comparison with the state-of-the-art

	This work	TBCAS'10 [21]	ISSCC'13 [5]	TBCAS'08 [22]	TBCAS'15 [23]
Tech. & supply	**0.18 μm CMOS 1.2 V**	1.5 μm BiCMOS 5 V	0.18 μm CMOS 0.5 V	0.35 μm CMOS 2.5 V	0.18 μm CMOS 1.8 V
Sampling frequency	**128, 16, 13 and 4 Hz**[a]	100 Hz	32 kHz	100 Hz	165 Hz
DC current cancelation	**Up to 10 μA**	NR	Up to 4 μA	53.6 μA (Ext. HPF)	100 μA
Integrated noise (RTI)	**486** $\mathbf{pA_{rms}}$[b]	NR	NR	2.2 nA_{rms}	600 pA_{rms}
Noise bandwidth	**10 Hz**	NR	NR	6 Hz	10 Hz
Integrated feature extraction	**Yes (HR/HRV)**	Yes (SpO_2[c])	No	No	No
Data compression	**Yes (8x, 10x, 30x)**	No	No	No	No
Power consumption (Readout)	**172 μW**[d]	400 μW	4 μW	600 μW	216 μW
Power consumption (LED driver)	**1200–43 μW**[e]	4400 μW	NA (ambient light)	NR	1125–120 μW

NA Not applicable, *NR* Not reported
[a]Average sampling frequencies corresponding to CRs 8x, 10x and 30x respectively
[b]TIA setting: $R_f = 50\,k\Omega$ and $C_f = 6\,pF$
[c]Blood oxygenation saturation measurement
[d]Includes AFE, ADC, DBE (while executing feature extraction) and bias power consumption, with power down mode disabled
[e]Off-chip LED driver. LED power consumption is subject to the SNR, skin tone of the subject and the efficiency of the LED used in the setup

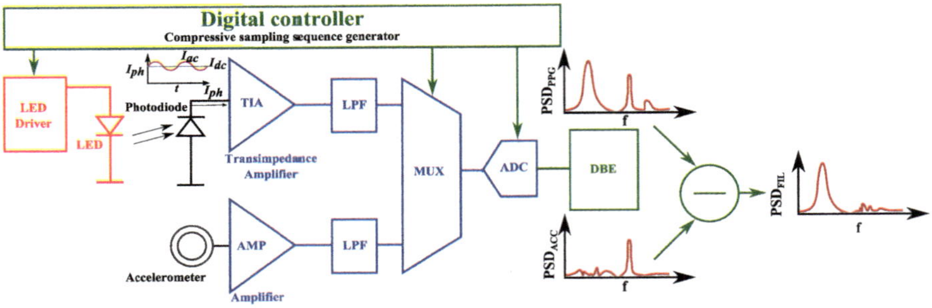

Fig. 8.11 Concept of motion artifact reduction using spectral subtraction

To demonstrate the efficacy of the proposed technique, PPG signals (Fig. 8.12a) are acquired using an internal PPG acquisition platform built from commercial off-the-shelf (COTS) components, from a subject under normal office working conditions. Simultaneously, accelerometer signals (Fig. 8.12c) are acquired using the same platform. The PPG and accelerometer signals are then randomly subsampled by a CR of 10x, and LSP is performed on both subsampled signals in MATLAB®. Spectral subtraction is finally performed on the normalized LSP of PPG and accelerometer signal and is rescaled. The rescaling process uses a scale factor that renormalizes the PSD of the *spectral subtracted* PPG signal, thereby restoring the amplitude of the peak in the PSD of the PPG signal. Figure 8.12e shows the spectral subtracted PSD of the PPG signal, and as can be seen, the spurious peak in the frequency range [2.7–3.2 Hz] that is correlated to the motion is significantly suppressed by spectral subtraction.

It must however be noted that while simulation results on a limited data set show promising results, extensive characterization of the technique is required under a variety of use case scenarios to arrive at a concrete conclusion regarding its efficacy under different motion artifact scenarios. Moreover, this technique is only a post-processing step and does not mitigate the requirement of a high channel DR. While an adaptive filter-based approach, presented in [25] for ECG, is promising to relax the DR requirements, it is challenging to design adaptive filters that work with randomly subsampled data.

8.5.2 Electrocardiogram (ECG)-Assisted Photoplethysmogram (PPG) Acquisition

Cuffless blood pressure (BP) monitoring using combination of ECG and PPG has been demonstrated in [23, 26]. The determination of BP is based on the relative timing between peaks in the ECG and PPG signals. Figure 8.13 shows the relevant timing information required for the BP estimation. Of interest is the pulse arrival time (PAT), which is the temporal difference between the peak in the ECG and

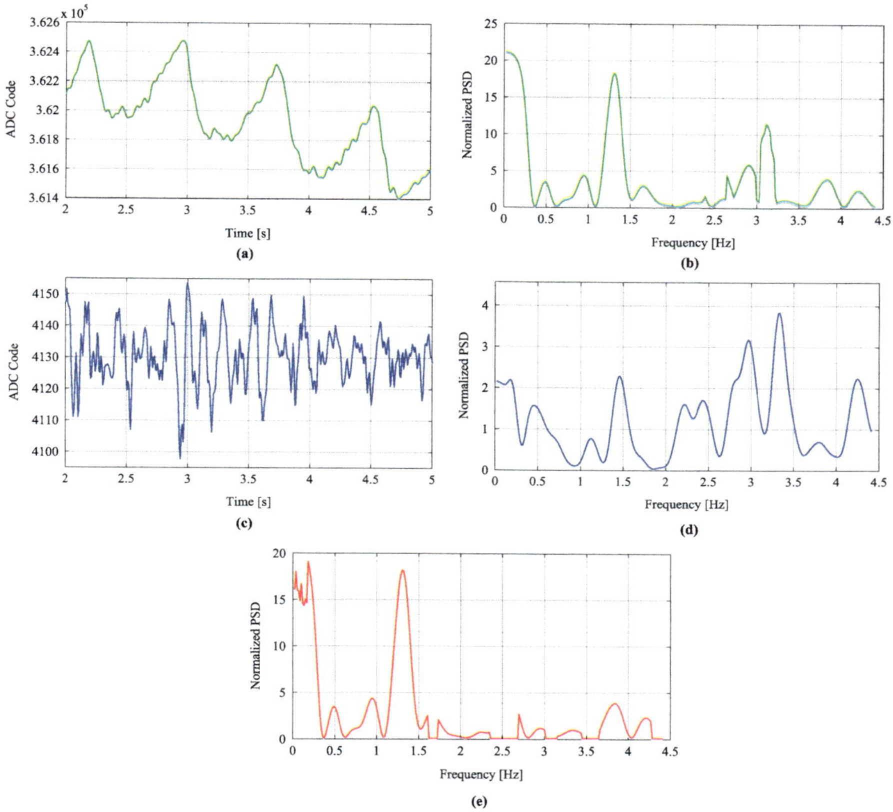

Fig. 8.12 (**a**) PPG signal acquired from a subject under normal office working conditions. (**b**) PSD of the PPG signal estimated using LSP after 10x random subsampling. (**c**) Accelerometer signal acquired simultaneous to PPG acquisition. (**d**) PSD of the accelerometer signal estimated using LSP after 10x random subsampling. (**e**) PSD of the PPG signal post spectral subtraction

the subsequent peak in the PPG signal. Once PAT is determined, BP is estimated using (8.5).

$$SBP = a_1 \cdot PAT + b_1 \cdot HR + c_1$$
$$DBP = a_2 \cdot PAT + b_2 \cdot HR + c_2 \tag{8.5}$$

where *SBP* and *DBP* are the systolic and diastolic blood pressure, respectively, while a_i, b_i and c_i, for $i = 1, 2$ are the calibration coefficients obtained through linear regression.

While the implementations in [23, 26] report achieving sufficient accuracy in determining BP for wearable applications, their power consumption is dominated by the PPG system, owing to the uniform stimulation and sampling. [12] demonstrated

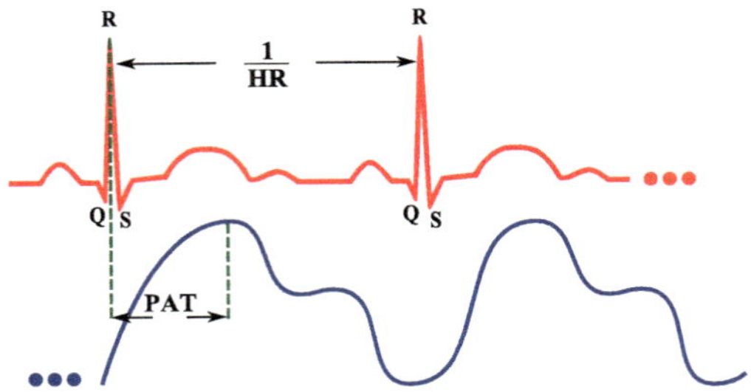

Fig. 8.13 Determination of PAT and HR for cuffless BP estimation

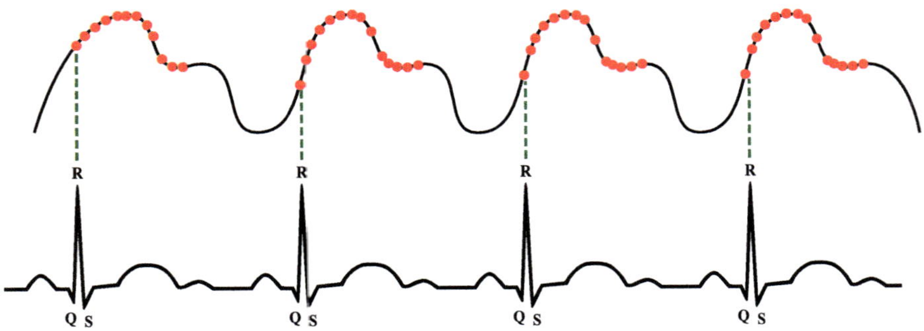

Fig. 8.14 ECG-assisted PPG acquisition for cuffless BP estimation

the use of CS-based PPG for cuffless BP estimation. However, [12] employs a full signal reconstruction process to perform BP determination from the reconstructed PPG signal, with the assumption of the availability of a powerful base station. As discussed in Sect. 8.3.2, the overhead in the reconstruction process can potentially cancel all power savings obtained from CS acquisition of PPG.

Alternatively, an event-driven approach that relies on the *assistance* from ECG to acquire PPG can be explored. Realizing that the peak in PPG signal is the aftereffect of the pumping action of blood through vessels by the heart, one can utilize the occurrence of the QRS complex to trigger the capture of the PPG signal. The acquisition can be stopped, when sufficient number of samples are acquired around the peak of the PPG signal. This approach is shown in Fig. 8.14. The presence of QRS complexes in the ECG can easily be detected using the activity detection process outlined in [27]. While a wide range of stopping criteria can be used for the PPG sampling, ranging from simple thresholding to more complex approaches based on learning, in this work a sum of slopes followed by thresholding is employed. Figure 8.15 shows a 10 s simultaneous ECG and PPG recording obtained through the COTS platform. The PPG signal is then adaptively resampled in ECG-*assisted* sampling mode, with the ECG signal acting as the trigger for PPG acquisition. For the recordings shown in Fig. 8.15, only 446 samples of PPG signal are acquired in the ECG-*assisted* acquisition mode as against 1280 in the uniform sampling mode, leading to an average stimulation and sampling frequency reduction

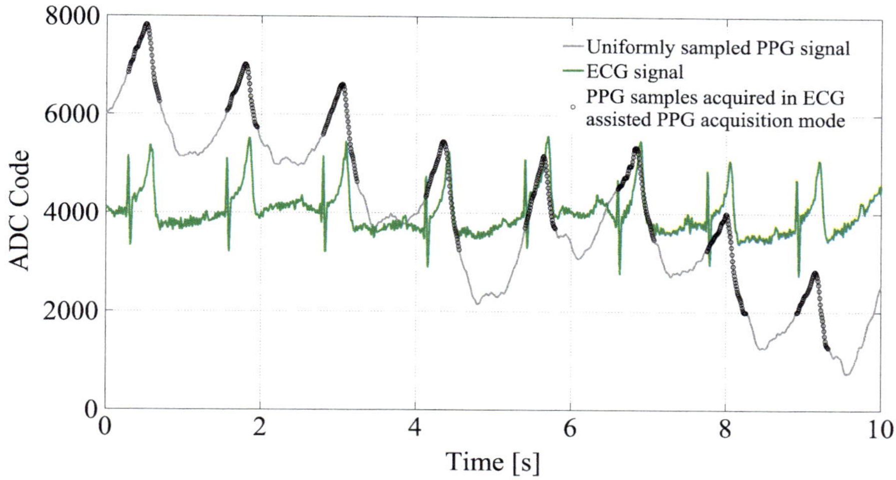

Fig. 8.15 A 10 s simultaneous ECG and PPG recording obtained through the COTS platform (average values are equalized for better representation). Both signals are sampled at 128 Hz

by a factor of 2.9. Since the relative timing information of interest is completely preserved in the ECG-*assisted* acquisition mode, both SBP and DBP are estimated with the same degree of accuracy as in the case of uniform sampling mode. It must however be noted that the amount of stimulation and sampling rate reduction depends on several factors including the relative placement of the ECG and the PPG sensors. To further validate this approach, MATLAB$^{\delta}$ simulations are run for ten different records, also obtained using the COTS platform. Simulation results demonstrate the possibility of reduction of the average stimulation and sampling frequency of PPG by a factor of 1.8 (averaged across the ten records). While this factor is small compared to the CS-based approach reported in [12], the proposed approach neither requires a base station nor involves a complex reconstruction process.

8.6 Conclusions

A compressive sampling (CS) photoplethysmographic (PPG) readout with embedded feature extraction is described. The feature extraction process exploits the Lomb-Scargle periodogram (LSP) to extract heart rate (HR) directly from CS PPG signal, thereby mitigating the need for complex reconstruction techniques. The implemented ASIC advances the state of the art by reducing the relative LED driver consumption by up to 30x while estimating the average HR with an accuracy conforming to the ANSI-AAMI standard for heart rate meters. The ASIC, implemented in a standard 0.18 μm CMOS process, consumes 172 μW of power from a 1.2 V supply, with the digital back-end (DBE) consuming only 7.2 μW, thus avoiding the energy penalties of wireless/wire line transmission and/or embedded signal reconstruction. In addition, a digital domain motion artifact suppression technique leveraging on multisensor fusion is presented. The proposed technique leverages on the *spectral subtraction* of the power spectral density (PSD) estimated for PPG signals and accelerometer signals. The efficacy of the technique is demonstrated through simulations. Finally, an electrocardiogram (ECG)-assisted PPG acquisition system is described for cuffless blood pressure (BP) monitoring. The proposed approach retains the relevant relative timing information between the ECG and the PPG signals, yet facilitating accurate BP estimation at a reduced average stimulation and sampling rate by a factor of 1.8 across ten records obtained using a COTS platform.

References

1. Wijsman, J., Grundlehner, B., Liu, H., Hermens, H., Penders, J.: Towards mental stress detection using wearable physiological sensors. In: 2011 Annual International Conference of the IEEE Engineering in Medicine and Biology Society, pp. 1798–1801, Aug 2011
2. Allen, J.: Photoplethysmography and its application in clinical physiological measurement. Physiol. Meas. **28**(3), R1 (2007)
3. Webster, J.G.: Design of Pulse Oximeters. Taylor & Francis Group, New York (1997)
4. Rhee, S., Yang, B.-H., Asada, H.: Artifact-resistant power-efficient design of finger-ring plethysmographic sensors. IEEE Trans Biomed. Eng. **48**(7), 795–805 (2001)
5. Alhawari, M., Albelooshi, N., Perrott, M.H.: A 0.5 V $<$ 4 μW CMOS photoplethysmographic heart-rate sensor IC based on a non-uniform quantizer. In: 2013 IEEE International Solid-State Circuits Conference Digest of Technical Papers, pp. 384–385 (2013)
6. Candès, E.J., Wakin, M.B.: An introduction to compressive sampling. IEEE Signal Process. Mag. **25**(2), 21–30 (2008)
7. Pamula, V.R., Verhelst, M., Van Hoof, C., Yazicioglu, R.F.: Computationally-efficient compressive sampling for low-power pulse oximeter system. In: 2014 IEEE Biomedical Circuits and Systems Conference (BioCAS) Proceedings, pp. 69–72 (2014)
8. Dixon, A.M., Allstot, E.G., Gangopadhyay, D., Allstot, D.J.: Compressed sensing system considerations for ECG and EMG wireless biosensors. IEEE Trans. Biomed. Circuits Syst. **6**(2), 156–166 (2012)
9. Ren, F., Marković, D.: A configurable 12–237 kS/s 12.8 mW sparse-approximation engine for mobile data aggregation of compressively sampled physiological signals. IEEE J. Solid-State Circuits **51**(1) 68–78 (2016)
10. Maechler, P., Studer, C., Bellasi, D.E., Maleki, A., Burg, A., Felber, N., Kaeslin, H., Baraniuk, R.G.: VLSI design of approximate message passing for signal restoration and compressive sensing. IEEE J. Emerging Sel. Top. Circuits Syst. **2**(3), 579–590 (2012)
11. Maechler, P., Greisen, P., Sporrer, B., Steiner, S., Felber, N., Burg, A.: Implementation of greedy algorithms for LTE sparse channel estimation. In: 2010 Conference Record of the Forty Fourth Asilomar Conference on Signals, Systems and Computers, Nov 2010
12. Baheti, P.K., Garudadri, H.: An ultra low power pulse oximeter sensor based on compressed sensing. In: 2009 Sixth International Workshop on Wearable and Implantable Body Sensor Networks, Jun 2009
13. Csavoy, A., Molnar, G., Denison, T.: Creating support circuits for the nervous system: Considerations for brain-machine interfacing. In 2009 Symposium on VLSI Circuits, Jun 2009

14. Pamula, V.R., Verhelst, M., Van Hoof, C., Yazicioglu, R.F.: A novel feature extraction algorithm for on the sensor node processing of compressive sampled photoplethysmography signals. In: 2015 IEEE SENSORS, pp. 1–4. IEEE (2015)
15. Yoo, J., Turnes, C., Nakamura, E.B., Le, C.K., Becker, S., Sovero, E.A., Wakin, M.B., Grant, M.C., Romberg, J., Emami-Neyestanak, A., Candes, E.: A compressed sensing parameter extraction platform for radar pulse signal acquisition. IEEE IEEE J. Emerging Sel. Top. Circuits Syst. **2**(3), 626–638 (2012)
16. Rajesh, P.V., Valero-Sarmiento, J.M., Yan, L., Bozkurt, A., Van Hoof, C., Van Helleputte, N., Yazicioglu, R.F., Verhelst, M.: A 172 μW compressive sampling photoplethysmographic readout with embedded direct heart-rate and variability extraction from compressively sampled data. In: 2016 IEEE International Solid-State Circuits Conference (ISSCC), pp. 386–387. IEEE, Piscataway (2016)
17. Glaros, K.N., Drakakis, E.M.: A sub-mW fully-integrated pulse oximeter front-end. IEEE Trans. Biomed. Circuits Syst. **7**(3), 363–375 (2013)
18. Pamula, V.R., Valero-Sarmiento, J.M., Yan, L., Bozkurt, A., Van Hoof, C., Van Helleputte, N., Yazicioglu, R.F., Verhelst, M.: A 172_W compressively sampled photoplethysmographic (PPG) readout ASIC with heart rate estimation directly from compressively sampled data. IEEE Trans. Biomed. Circuits Syst. **11**(3), 487–496 (2017). Available online at IEEE Xplore
19. Goldberger, A.L., Amaral, L.A., Glass, L., Hausdorff, J.M., Ivanov, P.C., Mark, R.G., Mietus, J.E., Moody, G.B., Peng, C.-K., Stanley, H.E.: Physiobank, physiotoolkit, and physionet components of a new research resource for complex physiologic signals. Circulation **101**(23), 215–220 (2000)
20. ANSI/AAMI-EC13: American National Standards for cardiac monitors, hearth rate meters and alarms (2002)
21. Tavakoli, M., Turicchia, L., Sarpeshkar, R.: An ultra-low-power pulse oximeter implemented with an energy-efficient transimpedance amplifier. IEEE Trans. Biomed. Circuits Syst. **4**(1), 27–38 (2010)
22. Wong, A.K., Pun, K.-P., Zhang, Y.-T., Leung, K.N.: A low-power CMOS front-end for photoplethysmographic signal acquisition with robust DC photocurrent rejection. IEEE Trans. Biomed. Circuits Syst. **2**(4), 280–288 (2008)
23. Winokur, E.S., O'Dwyer, T., Sodini, C.G.: A low-power, dual-wavelength photoplethysmogram (PPG) SoC with static and time-varying interferer removal. IEEE Trans. Biomed. Circuits Syst. **9**(4), 581–589 (2015)
24. Boll, S.: Suppression of acoustic noise in speech using spectral subtraction. IEEE Trans. Acoust. Speech Signal Process. **27**(2), 113–120 (1979)
25. Helleputte, N.V., Kim, S., Kim, H., Kim, J.P., Hoof, C.V., Yazicioglu, R.F.: A 160 μA biopotential acquisition IC with fully integrated IA and motion artifact suppression. IEEE Trans. Biomed. Circuits Syst. **6**(6), 552–561 (2012)
26. Poon, C., Zhang, Y.: Cuff-less and noninvasive measurements of arterial blood pressure by pulse transit time. In: 2005 IEEE Engineering in Medicine and Biology 27th Annual Conference (2005)
27. Pamula, V.R., Verhelst, M., Hoof, C.V., Yazicioglu, R.F.: A 17 nA, 47.2 dB dynamic range, adaptive sampling controller for online data rate reduction in low power ECG systems. In: 2016 IEEE Biomedical Circuits and Systems Conference (BioCAS), pp. 272–275, Oct 2016

Chapter 9
A 32 kHz DTCXO RTC Module with an Overall Accuracy of ±1 ppm and an All-Digital 0.1 ppm Compensation-Resolution Scheme

David Ruffieux, Nicola Scolari, Frédéric Giroud, Franz Pengg, Daniel Severac, Thanh Le, Silvio Dalla Piazza, and Olivier Aubry

9.1 Introduction

Timekeeping based on 32 kHz XTAL still remains the most popular, cost-effective, low-power, accurate solution for low-power portable applications. Simplest solutions with overall accuracies of a few 100 ppm are based on the combination of a through-hole or SMD XTAL together with an oscillator implemented as part of the application SoC (microcontroller, cell phone). As a single ppm error represents a deviation of 30s/year (nearly an hour/year at 100 ppm!), temperature-compensated XTAL or MEMS oscillators (TCXO, TCMO) used for timekeeping applications have received significant research attention over the last decade driven by further miniaturization, tighter accuracy and lower power consumption needs [1–4]. Combining both resonator and oscillator intimately or even better, in a single package, leads to superior stability, improved robustness and lower consumption by minimizing environmental effects (moisture, temperature gradients) and stray capacitance. Real-time clock (RTC) modules integrating further time, timer, calendar, timestamping and alarm functions become a key power management block capable of scheduling precise wake-up at user- or predefined intervals so that a more complex, energy-constrained application can be heavily duty cycled and left mostly hibernating (e.g. wireless sensor node). They are found in a variety of consumer, metering, medical, wearable, automotive, communication, outdoor, safety and automation applications and are a key component of the upcoming IoT revolution.

D. Ruffieux (✉) • N. Scolari • F. Giroud • F. Pengg • D. Severac • T. Le
CSEM, Neuchâtel, Switzerland
e-mail: David.RUFFIEUX@csem.ch

S.D. Piazza • O. Aubry
Micro Crystal, Grenchen, Switzerland

P. Harpe et al. (eds.), *Hybrid ADCs, Smart Sensors for the IoT, and Sub-1V & Advanced Node Analog Circuit Design*, DOI 10.1007/978-3-319-61285-0_9

9.2 Temperature Compensation

Tuning fork XTALs have a parabolic temperature coefficient of frequency (TCF), β, close to –35 ppb/°C^2 ±10%. Furthermore, their turnover temperature (ToT), the temperature at which the temperature sensitivity is naught, varies around 23 °C by ±5 °C depending on the quartz substrate cut and the device fabrication tolerances. The latter is further responsible for a 0.1% initial frequency accuracy that can be lowered to ±20 ppm after an individual laser trimming step vaporizing part of the electrode is applied to reduce the mass of the tines. Consequently, the resonance frequency of such a XTAL can vary by over 200 ppm over the −40 to 85 °C temperature range. Analog and digital temperature effect compensation schemes have been proposed and will be discussed hereafter before an improved, patented digital scheme is introduced.

9.2.1 Analog Temperature Compensation Scheme

The temperature effect might be compensated directly by varying the XTAL loading capacitance, C_L, exploiting the $f_{XO}(T, C_L)$ dependency. The principle is graphically illustrated in Fig. 9.1 together with the governing equation where C_m and C_O are the motional and dielectric capacitances of the XTAL, respectively, ΔT is the temperature offset equal to T-ToT and f_O is the XTAL series resonant frequency at the ToT. The XO frequency pulling range and hence the achievable temperature compensation one are determined by how much $C_{Lm} < C_L < C_{LM}$ can be varied as illustrated by the different $f_{XO}(T)$ and $f_{XO}(C_L)$ curves. Besides the limited compensation range, there are however several further drawbacks associated with this scheme. Due to a negative β, the biggest C_L is required near ambient temperature severely impacting the power consumption which scales at best quadratically with the loading capacitance when the gain transistor is operated in the

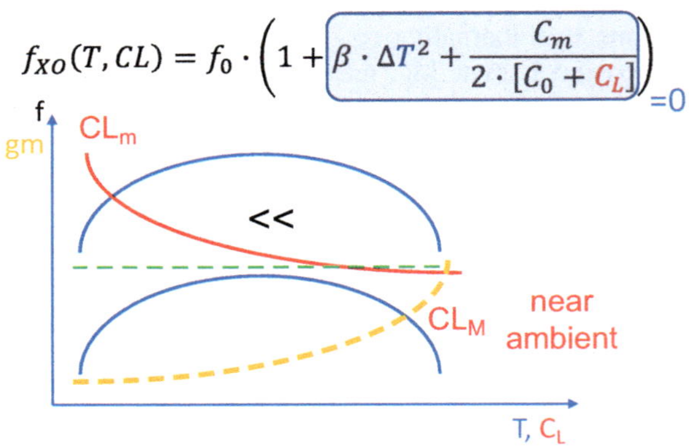

Fig. 9.1 Frequency vs temperature and loading capacitance (C_L) curves of 32 kHz tuning fork XO and required transconductance vs C_L

subthreshold regime ($gm \propto C_L{}^2$ when $C_L \gg C_O$). Secondly, the $C_L(\Delta T)$ polynomial that should be implemented for temperature compensation is highly non-linear further necessitating a temperature-to-digital converter, a high-order polynomial multiplier and eventually a DAC controlling a varactor or a capacitor bank leading to significant complexity, extra calibration burden and area penalty.

An advantage of that approach however is that it yields a truly accurate 32′768 Hz clock should it be of interest at the system level.

9.2.2 Digital Temperature Compensation Scheme

Temperature effect compensation may be implemented in a purely digital, indirect way by nulling the $f_{XO}(T)/N(T)$ dependency along the division chain. The simplicity of that scheme, called inhibition, has made it a popular choice for (temperature compensated) accurate wristwatches where only the accumulated time error matters. Figure 9.2 illustrates the concept graphically. The division ratio, N, is gradually reduced as the temperature deviates from ambient in a mostly quadratic dependent way resulting in a low-order polynomial compensating function.

Figure 9.3 illustrates the principle in the time domain. The top plot shows the 32 kHz signal and its divided-by-2 version at 16 kHz for a slightly too fast and too slow clock source, respectively. Assuming a divider by 32′768 generates the pulse-per-second (PPS) output, the two signals, respectively, the $2^{15\text{th}}$ and $2^{14\text{th}}$ edges, of the two signals will hence arrive too early and too late in this example. In the first case, an integer number of periods can be swallowed by gating the 32 kHz clock at the counter input, virtually lowering the clock frequency. In the other case, additional clock cycles could be injected by bypassing the first divide-by-2 stage accelerating the counter. In order to get the correcting division factor dN, 2^{15} should be subtracted from the computed $f_{XO}(T)$ quantity using a polynomial fitting function

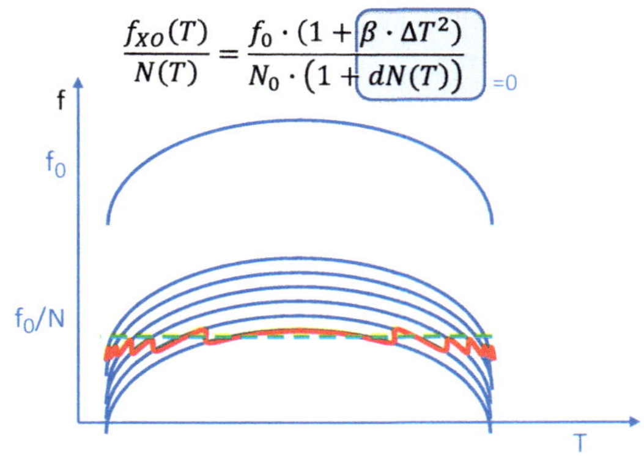

Fig. 9.2 Frequency vs temperature curves of 32 kHz XO illustrating a purely digital temperature compensation scheme obtained with variable divisors

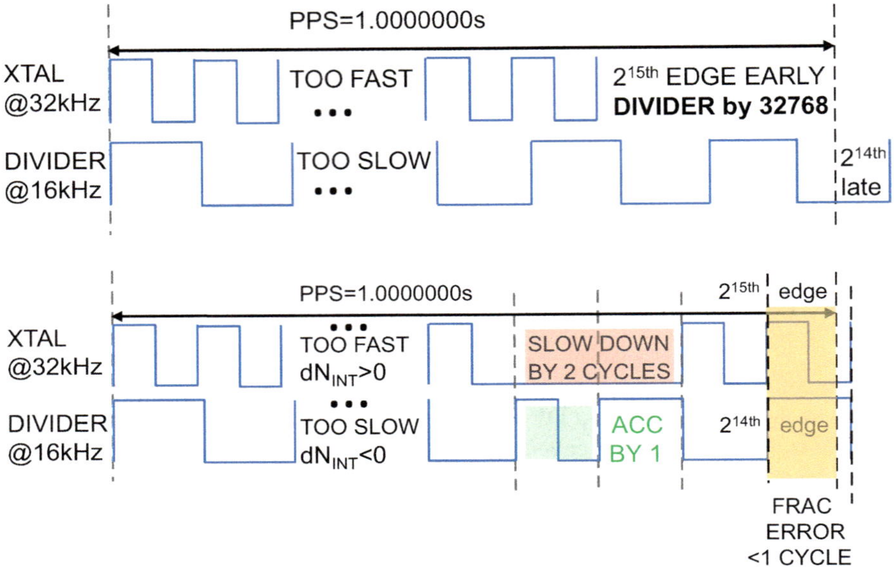

Fig. 9.3 Time domain illustration of the digital temperature compensation scheme known as inhibition

whose coefficients have been previously calibrated. The floor value of the integer part determines according to its sign how much clock periods should be swallowed (>0) or injected (<0). The fractional part, dN_{FRAC}, should be accumulated as in a first-order $\Delta\Sigma$ modulator. Upon overflow, an additional cycle is inhibited so as to guarantee that the maximum error is bounded to a single 32 kHz clock period. The coarse adjustment yields a resolution limited to 1/32 kHz or 31 μs that corresponds to 31 ppm over a 1 s averaging time. In order to reach a resolution measurement accuracy of 0.1 ppm, the associated integration time is 5 min making it almost impossible for such XO manufacturer or their customer to verify the individual module timing accuracy.

9.2.3 Improved Digital Temperature Compensation Scheme

To overcome that limitation, a novel patented scheme is proposed. It relies on an all-digital fractional or interpolated inhibition where the edge of the 1 Hz or PPS signal is delayed by the appropriate time using intermittently a faster oscillator. The novelty resides in reusing for that purpose the thermistor-based RC temperature sensor oscillator (RCO) proposed in [1] eliminating any additional analog hardware requirement. This dual oscillator measurement principle is used in most frequency metres and consists in counting a large bunch of periods, M, of an internal oscillator

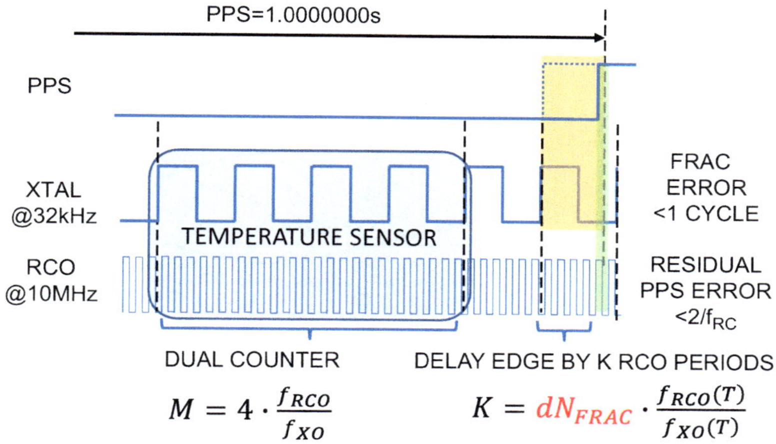

Fig. 9.4 Time domain illustration of the digital temperature compensation scheme using fractional interpolation

(~7 MHz RCO) over a gate time defined by N periods of the signal to be measured (XO). The frequency ratio (f_{RCO}/f_{XO} = M/N) needed for interpolation is obtained by a mere right shift if N is taken as a power of 2. N = 32 is chosen (1 ms conversion time) so as to duty cycle the RCO at 1 Hz with a 0.1% ratio. In addition to the conventional inhibition mechanism presented earlier, the interpolating edge is delayed by a number of RCO pulses corresponding to the accumulated fraction multiplied by the M/N ratio. The resulting error is now bounded to two clock periods of the RCO as the phases of the two clocks are uncorrelated. The concept is illustrated in Fig. 9.4 with N = 4. Calculating the compensation requires an extra four XTAL cycles, while the RC is maintained running. The power overhead is hence of 15% compared to that used for the temperature measurement. An additional benefit of this direct temperature-to-frequency conversion scheme is the elimination of an ADC resulting in low hardware complexity and area.

9.2.4 *Calibration Procedure*

Figure 9.5 shows an illustration of how the DTCXO should be calibrated. Here, a three-temperature trim is illustrated at which both the XO and RCO frequencies should be measured. The temperatures at which the calibration is performed should not be precisely known but should span the compensation range. Any thermal gradient between the XO and RCO should be avoided for accurate compensation. Knowing $f_{XO}(T)$ and $f_{RCO}(T)$ and substituting for T yield $f_{XO}(f_{RCO})$. With a three-point trim and provided the temperatures are spaced evenly, a second-order fit will lead to an exact compensation at the three calibration points and to the elimination

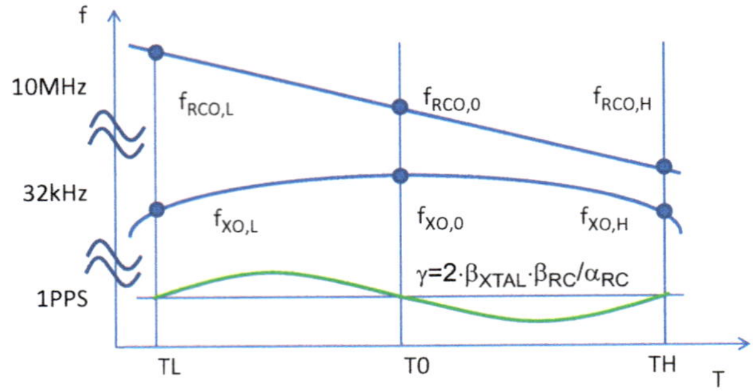

Fig. 9.5 Illustration of the calibration procedure and resulting compensated curve after a three-temperature trim

of the XO parabolic temperature dependency and ToT variations. A third-order term contributed by the XO third-order TCF that is usually negligible and by the second-order non-linearity of the RCO will be compensated partly by a linear term nulling the errors at the two extreme temperature points. The RCO-induced third-order TCF term is proportional to the XO quadratic one scaled by the ratio of the second- over first-order terms of the RCO. Typical material coefficients are usually scaled by three orders of magnitude each time the order is increased. Assuming such a non-linearity ratio, the temperature-compensated inaccuracy would be as high as $\pm$8 ppm over $-$40 to 85 °C, pretty much like the characteristic of an AT-cut XTAL. Higher-order compensation requiring calibration points at additional temperatures is hence required to get better accuracy. Performing an *n*th-order least mean square fit with more trimming data points than n + 1 yields a better overall stability, reducing the high-order coefficients that prove detrimental at the temperature extremes.

9.3 Circuit Architecture and Analog Block Schematics

The system architecture of the DTCXO RTC circuit that is depicted in Fig. 9.6 incorporates in addition to the two XO and RCO blocks mentioned in the previous section a fully integrated LDO so as to increase the power supply rejection and permit the circuit to operate over a wide 1.2–5.5 V range. A built-in POR and brownout detection (BOD) circuit together with an on-chip NVM used to store the calibration parameters complete the analog functions of the ASIC. In addition to the temperature compensation and PPS generation finite state machines, the circuit features all conventional RTC functions such as clock, calendar, timer and alarms as well as a timestamping unit triggered by an external event. All functions may generate interrupts that can be handled through an I^2C interface. Many different clocks are furthermore available on an output pad that can be gated externally so as

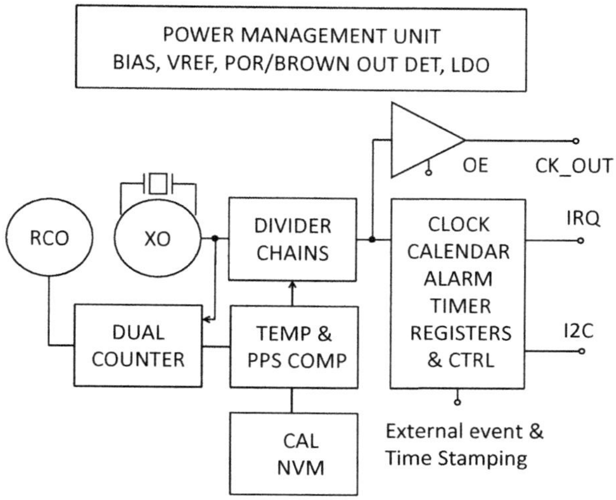

Fig. 9.6 System architecture

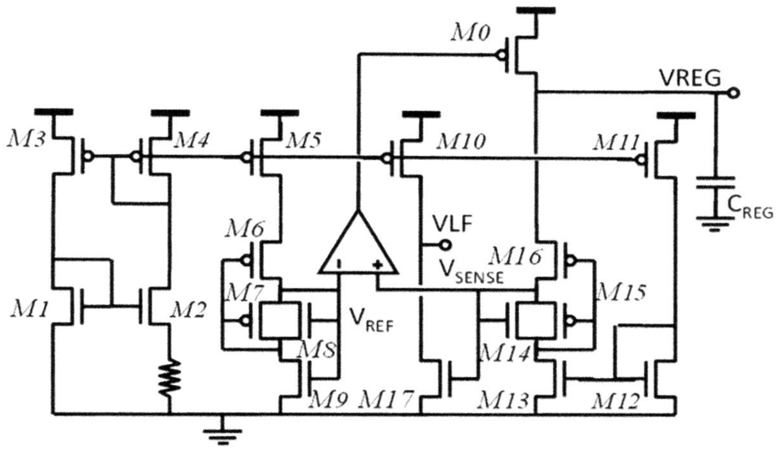

Fig. 9.7 System architecture

to provide 32 kHz, 1024 Hz and the temperature-compensated 1 Hz PPS signal or used to measure the RCO frequency.

9.3.1 LDO with POR and BOD Block Diagram

Figure 9.7 shows a transistor-level schematic of the LDO featuring a combined power-on reset and brownout (POR and BOD) detection circuits. A PTAT current reference (M_1–M_4) is used to bias the analog blocks and generate a 1.25 V voltage reference after passing through two pairs of P and NMOS transistors (M_6–M_9)

arranged in a series/parallel nested configuration. The drain voltage of M_6 is hence given by max($V_{TN} + V_{DSAT,P}$; $V_{TP} + V_{DSAT,N}$) yielding a robust low voltage PVT tracking reference. In order to minimize the power supply requirement, the OPAMP input V_{REF} is connected to the gate of M_8, whose voltage ensures through proper transistor sizing the saturation of the tail current source of the OPAMP NMOS differential pair without excessive overdrive. V_{SENSE} is generated almost similarly using a matched replica stack (M_{13}–M_{16}) whose current is controlled indirectly by the PTAT circuit via a current mirror formed with M_{11} and M_{12}. A current comparator implemented with M_{10} and M_{17} is used as a brownout detector, raising a voltage low flag (V_{LF}) when VDD goes too low forcing M_5 into the triode regime. The POR circuit combines this flag together with a fast power supply voltage ramp-up detector relying on a capacitor discharging time constant circuit (not shown) ensuring a suitable abrupt power-up detection. The amplifier driving the ballast transistor (M_0) is biased with a fraction of the load of the latter in an adaptive way to ensure minimum extra power dissipation whatever the circuit mode of operation, and the different poles are swept similarly. By proper compensation using a Miller capacitor, the LDO can operate in a stable way over a large range of load currents spanning from 100 nA in RTC mode to some 50 μA when the RCO, decimator and I^2C interface are running while relying solely on an on-chip 200 pF decoupling capacitance connected to the regulator output.

9.3.2 *XO Block Diagram*

The XO, whose simplified block diagram is shown in Fig. 9.8, is based on a standard Pierce topology (M_1, C_1, C_2) but features a non-linear amplitude limitation

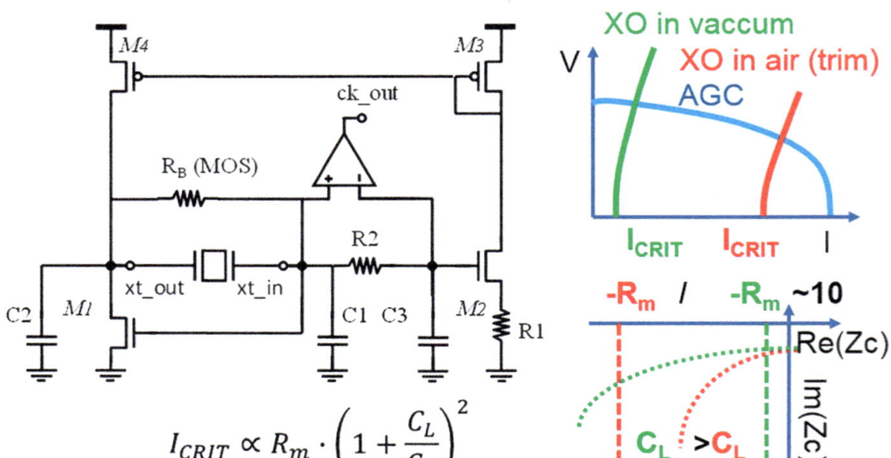

$$I_{CRIT} \propto R_m \cdot \left(1 + \frac{C_L}{C_O}\right)^2$$

Fig. 9.8 Simplified XO block and design explanation diagrams

mechanism (M_2–M_4, R_1) so that it can operate reliably over a very large range of XTAL ESR (50 kΩ < Rm < 1 MΩ) both in vacuum and at 1 bar during the frequency trimming step that is performed before the module is sealed under vacuum. This is illustrated in the top right part of Fig. 9.8 that shows the XO V_{OSC} (I_{MEAN}) behaviour in both vacuum and at 1 bar as well as the AGC regulator characteristic. At power-up ($V_{OSC} = 0$), a large current determined by R_1 and M_1–M_4 mirror ratios ensures a rapid and robust start-up whatever Rm. Steady state is reached once the AGC and XO curves intersect at $I_{MEAN} = {\sim}2I_{CRIT}$ minimizing the power consumption while guarantying a sufficient amplitude to drive the comparator generating a rail-to-rail clock signal. The critical current, I_{CRIT}, leading to oscillation build-up depends non-linearly on the loading capacitance, C_L, as shown in the equation on Fig. 9.8 valid in subthreshold regime. The capacitance should however be made big enough to ensure a sufficient negative resistance can be generated at start-up to overcome the XTAL losses. This is ruled by the circuit impedance, Z_C, defined as the impedance seen from the quartz terminals while excluding the motional elements (Rm, Lm, Cm). As a function of the transconductance, gm, it describes a circle whose radius is dependent on C_L (given by $C_1//C_2$ and any foot stray capacitance) and C_O, the XTAL dielectric capacitance (including any parasitic capacitance in parallel to the XTAL) that should, respectively, be maximized and minimized to obtain the biggest circle radius. This is illustrated at the bottom right of Fig. 9.8. With a too small C_L and too large C_O, the circuit would ultimately fail starting in air (or even in vacuum) whatever the current flowing through the gain transistor. For a small C_L, the frequency pulling (proportional to the distance from the Z_C, −Rm intersect point to the x-axis) becomes Rm dependent leading to a degraded frequency stability. The choice of C_L hence dictates a very careful optimization and modelling of the package parasitic. It should also be noted that the sensitivity of the frequency to a variation of the loading capacitance as determined by the equation shown in Fig. 9.1 is increased for small such values.

The differential comparator generating the rail-to-rail signal is connected between the gate of the main gain transistor and its DC-filtered version used to set the AGC time constant (R_2, C_3). Avoiding the combined use of the differential drain signal which sounds attractive at a first glance ensures superior PSRR, close to 50% clock duty cycle and more freedom to bias smartly the gain transistor using a low-loss non-linear MOS conductance instead of an area-consuming resistor. Most 32 kHz XOs necessitate large DC-blocking series capacitors to prevent the leakage current of ESD clamps or any moisture-induced board conductance from altering the DC operating point of the XO eventually stopping it. The most sensitive pin is the one connected to the gate of the gain transistor which can only be fed with some current through a high-impedance path (R_B) for proper biasing. In this design, such series capacitors could be avoided owing to a proper assisting circuit that turns on a lower-impedance path across the XTAL should the gate voltage of the gain transistor drop below its set point. The XO features additional non-linear current comparators that assist its initial biasing and detect when sufficient amplitude has been reached to turn on the comparator that could prevent proper XO start-up due to its Miller capacitance.

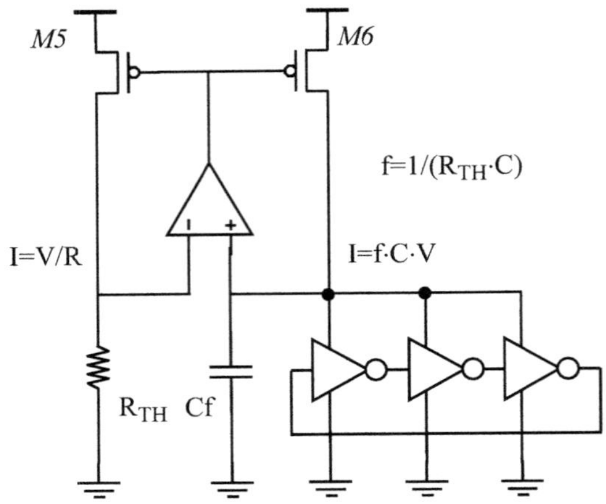

Fig. 9.9 Simplified RCO schematic

9.3.3 RCO Block Diagram

The RCO topology, pictured in Fig. 9.9, is such that both similar currents and voltages are biasing a ring oscillator of total gate capacitance C and a thermistor R_{TH}. With $I = f{\cdot}C{\cdot}V$ and $V = R_{TH}{\cdot}I$, one gets $f = 1/R_{TH}{\cdot}C$. An HR poly resistor is used as the thermistor yielding a RCO with a large, mostly linear, positive TCF. A $\sim$ 1 MHz seven-stage ring oscillator formed with large area inverters is used, and the signals from the different stages are combined to generate a 7$\times$ frequency reducing the quantization noise that scales inversely proportionally with the integration time, τ, by a similar factor. The temperature sensor resolution is equal to $1/\alpha_{RC}$, the inverse of the RCO first-order TCF, multiplied by the RCO Allan deviation, $\sigma_Y(\tau)$, which is a statistical measurement of $\Delta f/f$ over a time τ. The latter is linked to the accumulated jitter $\sigma_{ACC}(\tau) \propto \sigma_Y(\tau){\cdot}\tau$, which is itself related to the phase noise of the RCO. Both low thermal noise and 1/*f* noise corner frequency are desired so that the accumulated jitter remains white noise dominated for longer time intervals, improving the resolution of the sensor as $\tau^{-0.5}$. It should however be mentioned that while the proposed circuit offers a good PSRR, the thermal noise contribution of M_5–M_6 largely exceeds that of the thermistor, especially in low voltage designs where limited overdrive voltage can be applied to the gates of the current source transistors. This degrades the sensitivity significantly. 1/*f* noise minimization was achieved by increasing the transistor area and optimizing their biasing and geometry using standard simulation tools.

The importance of the temperature sensor linearity was evidenced in a previous section as it introduces higher-order polynomial compensation terms. This is a particularly sensitive design point as models of the different components, such as for the resistors, might lack sufficient accuracy to obtain predictable performances (higher-order TCR terms and their spreads). One should mention that the core ring oscillator contribution to the overall temperature sensor behaviour is far from negligible.

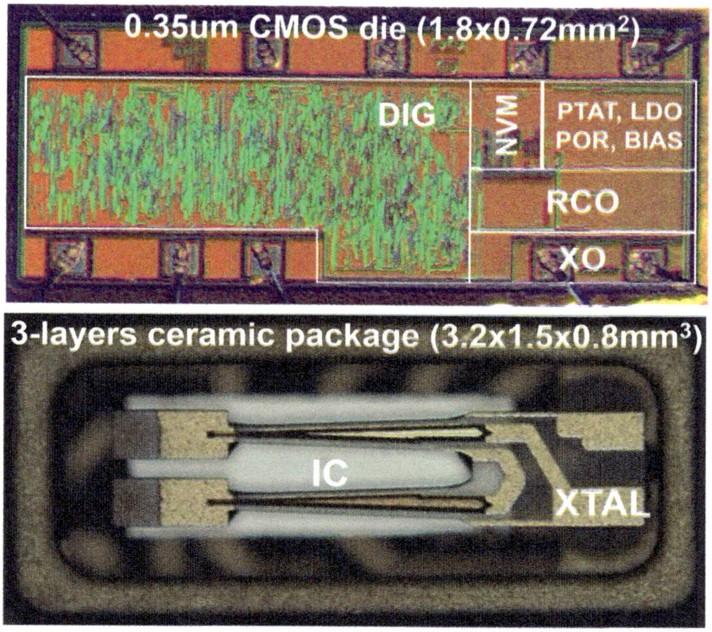

Fig. 9.10 Circuit micrograph with detailed floor plan and picture of the DTCXO module after assembly in the ceramic package

9.4 Circuit Floor Plan and TCXO Module Details

The circuit die photo highlighting the floor plan is shown in Fig. 9.10 together with a package assembly view. It is integrated in a 0.35 μm CMOS process and measures $1.8 \times 0.72\text{mm}^2$. It is flip chip bonded using Au studs on the bottom layer of a miniature $3.2 \times 1.5\text{mm}^2$ ceramic package containing eight IOs in addition to the two internal XTAL leads. The latter is assembled overhanging attached on its side on the second level of the package. Eventually, after laser trimming the XTAL, a metal lid is reflow soldered atop the third level to seal the package at low pressure (<0.1 mbar) to maximize the resonator Q factor. Parts are then individually calibrated and verified over the industrial temperature range in large batches before the compensation parameters are stored in the on-chip NVM.

9.5 Measurement Results

Figure 9.11 shows measurement data of the LDO output voltage and PTAT current over the complete supply range from 1.5 to 5.5 V obtained over ~50 samples. A DC PSSR of 38 dB can be extracted. The PTAT current varies by ±7.5% over the

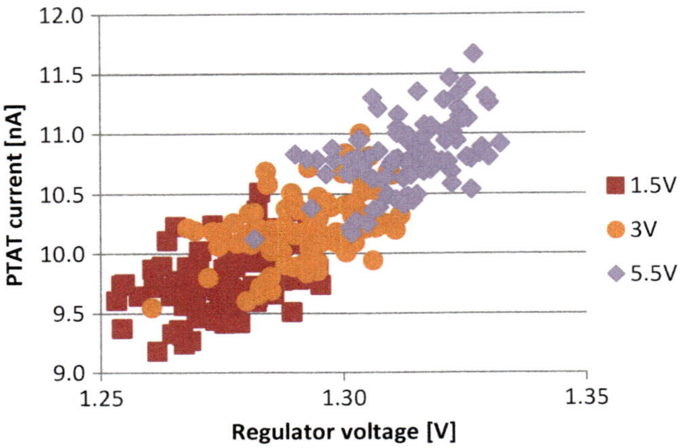

Fig. 9.11 PTAT current and LDO voltage spread at several supply voltages

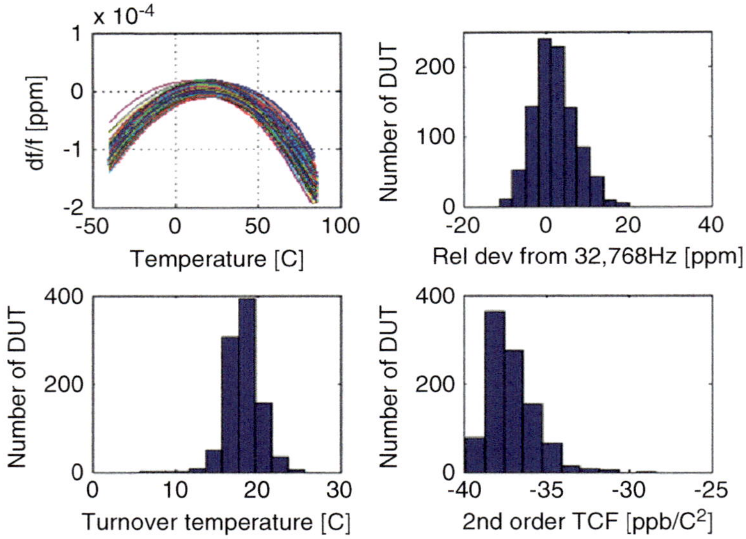

Fig. 9.12 Measured XO statistics on ~1000 samples

supply range corresponding to a conductance of 0.37 nS. Regarding the spread, σ/μ reaches 2.8% and 0.8% for the PTAT current and LDO output voltage, respectively.

Figure 9.12 shows XO calibration data obtained over ~1000 samples during the calibration procedure. Temperature gradients among parts in the oven are not compensated affecting the reported figures pessimistically. Over −40 to 85 °C, the XO's temperature stability varies by some 200 ppm. After individual trimming by laser, the frequency at ambient temperature lies well within the ±20 ppm claimed

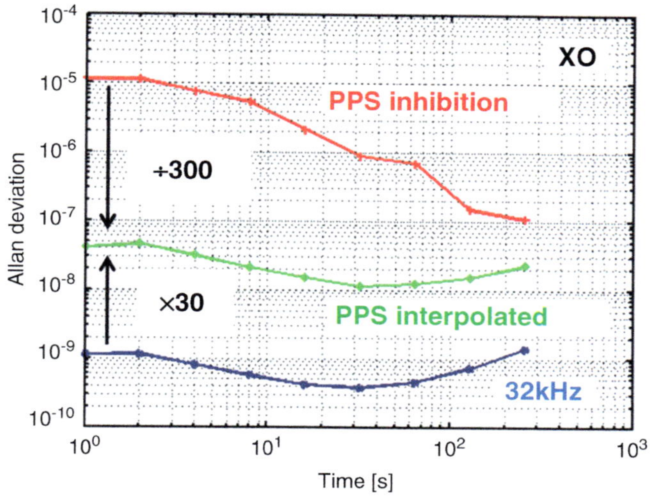

Fig. 9.13 Allan deviation of different clock sources derived from the XO

previously. The ToT (or similarly the first-order TCF) of the XO is reduced to a mean value of 18 °C—compared to 23 °C for the XTAL—most likely due to the effect of the temperature dependency of the XO circuit capacitance. The ToT variance is 2 °C mostly affected by variations in the tines etching. The parabolic coefficient reaches -37 ppb/°C^2, with a standard deviation over mean value of 3.8%.

Figure 9.13 shows the Allan deviation measurement for the 32 kHz XO, the generated 1PPS signal with usual integer and newly proposed fractional inhibition using higher-frequency interpolation. The measurement is performed at ambient temperature with the compensation disabled. The intrinsic 1 s short-term stability of the XO reaching 1 ppb is degraded by a factor 30 at the PPS due to interpolation quantization at f_{RCO}. Regarding the newly proposed scheme, a 300× improvement can be seen compared to that using classical inhibition corresponding to the frequency ratio. The Allan variance of the latter improves as τ^{-1} over time, as expected.

Figure 9.14 shows RCO calibration data obtained over ~1000 samples during the calibration procedure. Temperature gradients among parts in the oven are not compensated affecting the reported figures pessimistically. The mean RCO frequency is 7 MHz with a relative deviation of 3.6%. The first TCF is positive and rather large at 0.46%/°C with a relative variation of 2.1%. The temperature error without trim would span ±25 °C at 3σ at ambient, but after a single trim, it reduces to ±5 °C. The second-order TCF is a factor 1000 below the first one but has a larger relative deviation of some 30%. A three-point trim improves the temperature sensor inaccuracy down to a few degrees, but again, the fact that measurements are obtained over production data in a large oven should be taken into account.

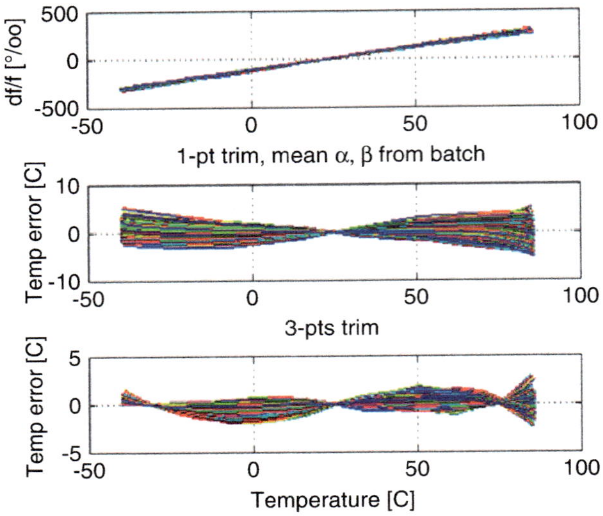

Fig. 9.14 Measured RCO statistics on ~1000 samples

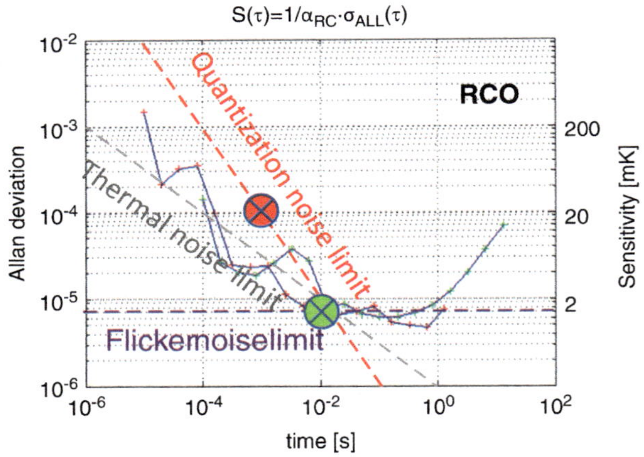

Fig. 9.15 Allan deviation and sensitivity of the RCO temperature sensor

Figure 9.15 shows the Allan deviation of the RCO at ambient temperature after having taken care of eliminating any airflow by placing the sample in a closed box. After division by the RCO first TCF, the temperature sensor resolution is obtained. It is shown on the right axis and reaches a floor of 1mK limited by 1/*f* noise. The thermal noise limit improving as expected as $\tau^{-0.5}$ over time reaches the level of the 1/*f* noise for an integration time of 10 ms. Quantization noise decaying as τ^{-1} is superimposed on the same plot. It also reaches the noise floor limit at ~10 ms. Operating the temperature sensor at a duty cycle of 1% would however represent as much

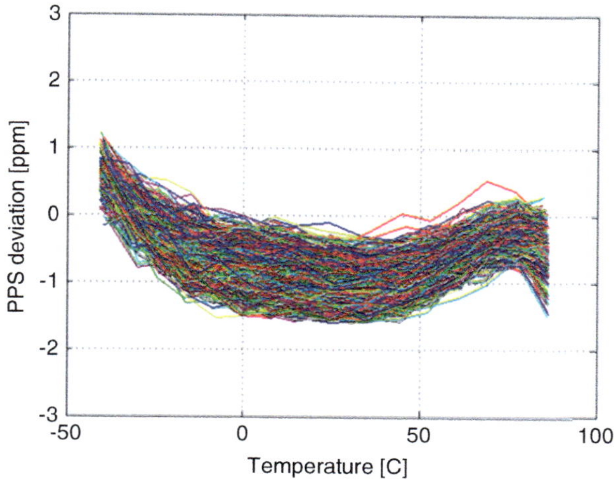

Fig. 9.16 Stability of ∼1000 DTCXO modules over temperature

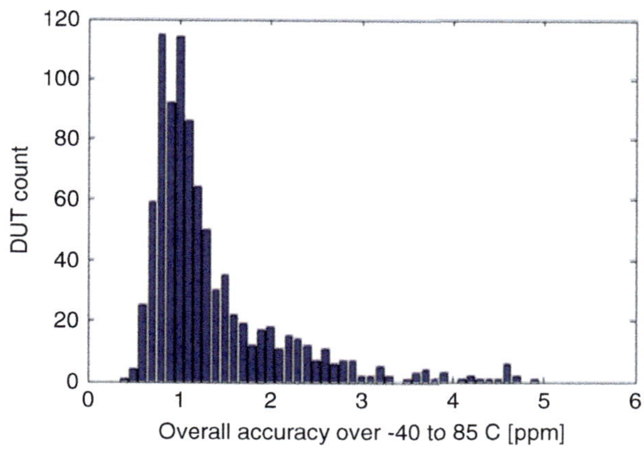

Fig. 9.17 Histogram of the maximum absolute frequency error of ∼1000 DTCXO modules over the industrial temperature range

power as the overall RTC including the LDO. At the cost of a lower resolution, the sensor is operated for only 1 ms reaching a ∼ 20mK quantization noise limited resolution that should be compared to 5mK when considering the thermal noise limit.

Figure 9.16 shows the temperature stability of ∼1000 DTCXOs over −40 to 85 °C. Most parts reach ±1 ppm stability over the complete industrial temperature range as evidenced by the histogram shown in Fig. 9.17 plotting the maximum error. Owing to the LDO and avoidance of PMOS ESD clamps, there is no noticeable influence of the supply voltage on the stability across the full 1.25–5.5 V range.

Figure 9.18 shows the RTC module current consumption breakdown. The XO consumes 16 and 30 nA for the core and buffer, respectively, (note that the buffer has been oversized for production testing where a 300 kHz clock is fed from one of

Block	I (nA)
PTAT	19
BIAS	14
XO core/buffer	16+30
RCO core/dig	7+28
DIG	75
POR	12
$V_{REF/SENSE}$	28
LDO	11
Complete RTC	240

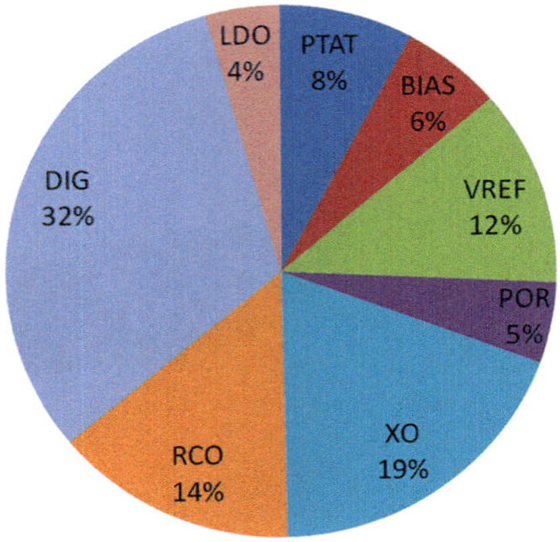

Fig. 9.18 Current consumption breakdown per function

Part number	RX8900CE	SiT1552	RV-8803-C7 (this work)
Type	DTCXO/RTC	DTCXO (4 pins)	DTCXO/RTC
VDD range [V]	2-5.5	1.5-3.63	1.5-5.5
Typ. current [uA]	0.70	0.99	0.24
Size [mm^3]	3.2x2.5x1	1.55x0.85x0.6	3.2x1.5x0.8 (DIE:1.8X0.72)
Stability [ppm]	±3.4	±5	±3

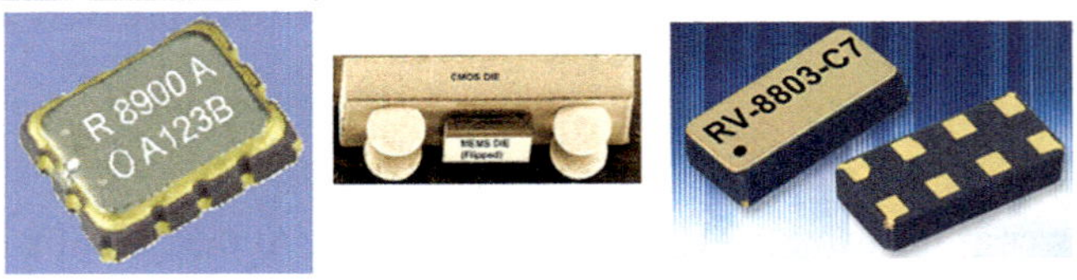

Fig. 9.19 Comparison table with commercially available DTCXO modules

the XTAL pin). The consumption of the RCO that is duty cycled with a 1/700 ratio reaches 7 and 28 nA for the core and decimation part, respectively. The digital part consumes about 32% of the overall budget at 75 nA, while the LDO including the current and voltage references a similar fraction.

The presented solution offers the widest voltage supply range, the best temperature stability and at 240 nA, a 4× power reduction compared to any other COTS TCXO/RTC state-of-the-art products as evidenced in Fig. 9.19.

9.6 Conclusions

This paper has presented a novel high-performance DTCXO module that achieves at 240 nA, a 4× power dissipation reduction compared to other COTS while being the most accurate on the market with a typical inaccuracy of ±1 ppm (±3 ppm max). It introduces a novel fractional inhibition all-digital temperature compensation scheme achieving a resolution of ~0.1 ppm at 1 s. It reuses for that purpose, to perform fine-grained timing interpolation, edges from a heavily duty-cycled 7 MHz RCO that operates mainly as a 20 mK resolution temperature sensor. Consequently, a very compact circuit eliminating the need for an ADC is obtained. The DTCXO RTC module is available commercially in large volumes.

References

1. Ruffieux, D., et al.: Silicon resonator based 3.2μ W real time clock with ±10 ppm frequency accuracy. J. Solid-State Circuits. **45**(1), 224–234 (2010)
2. Asl, S.Z., et al.: A 3 ppm 1.5 × 0.8 mm^2 1.0 μA 32.768 kHz MEMS-based oscillator. J. Solid-State Circuits. **50**(1), 1–12 (2015)
3. Ruffieux, D., et al.: A versatile timing microsystem based on wafer-level packaged XTAL/BAW resonators with sub-μW RTC mode and programmable HF clocks. J. Solid-State Circuits. **49**(1), 212–222 (Jan. 2014)
4. Park, P., et al.: A thermistor-based temperature sensor for a real-time clock with ±2 ppm frequency stability. J. Solid-State Circuits. **50**(7), 1571–1580 (2015)

Chapter 10
Energy-Efficient High-Resolution Resistor-Based Temperature Sensors

Sining Pan and Kofi A. A. Makinwa

10.1 Introduction

Integrated temperature sensors are often used for the temperature compensation of frequency references [1–3]. This is a demanding application, as it requires sensors that can achieve both high resolution and high energy efficiency. High resolution is essential for minimizing jitter in the compensated output frequency, while high energy efficiency is necessary to minimize the sensor's contribution to the reference's total energy budget. Furthermore, the sensor should be CMOS compatible, i.e., based on the generic devices or features of baseline CMOS technologies, so that it can be co-integrated in a low-cost manner with the rest of the electronics of the frequency reference.

The temperature dependencies of various CMOS-compatible devices, such as BJTs [5–7], MOSFETs [8–9], thermistors (temperature-dependent resistors) [1–3, 13–17], and electrothermal filters [10–11], have been used to realize temperature sensors. MEMS resonators have also been used to realize temperature sensors with excellent resolution and energy efficiency [12]. However, the resonators are fabricated in a dedicated process on separate die and so are not CMOS compatible.

A survey of smart temperature sensors [4] shows that resistor-based temperature sensors are the most energy-efficient class of CMOS temperature sensors. Their resolution figure of merit (FoM), defined as energy/conversion × resolution^2 [4], is about an order of magnitude less than that of conventional BJT-based sensors, and they can also achieve much higher (sub-mK) resolution [1].

Several high-resolution resistor-based temperature sensors have been reported [1–3, 15]. In [2, 3], temperature sensors based on Wien bridge RC filters are

---

S. Pan (✉) • K.A.A. Makinwa
Electronic Instrumentation Laboratory, Delft University of Technology, Delft, The Netherlands
e-mail: S.Pan@tudelft.nl

P. Harpe et al. (eds.), *Hybrid ADCs, Smart Sensors for the IoT, and Sub-1V & Advanced Node Analog Circuit Design*, DOI 10.1007/978-3-319-61285-0_10

presented, in which on-chip thermistors are used to generate a temperature-dependent phase shift. In [1, 15], half and full Wheatstone bridges are used to generate a temperature-dependent DC output. Both types of bridges achieve high resolution and energy efficiency. The Wien bridge in [3] achieves 2.8 mK resolution and an 8pJ·K^2 resolution FoM, while the Wheatstone bridge in [15] achieves 10 mK and 0.65 pJ·K^2, which represents the state-of-the-art for CMOS temperature sensors.

The resolution and energy efficiency of a resistor-based temperature sensor are mainly determined by the temperature coefficient (TC) of the chosen sensing resistor and the noise of its readout electronics. In [3], poly resistors were used, while in [15], both poly and diffusion resistors were used. The TCs of those resistors are low (<0.15%/°C). Moreover, their resolution was limited by the thermal noise [3], and 1/*f* noise [15], of their readout circuits. In [17], silicided resistors, which have a large TC (~0.3%/°C), were used. However, the sensor's resolution was limited by the quantization noise of the readout electronics.

In this paper, we propose two resistor-based sensors, a Wien bridge (WB) sensor and a Wheatstone bridge (WhB) sensor, both of which adopt silicide resistors as sensing elements, due to their superior TC and low 1/*f* noise. Since the readout electronics' efficiency is also important, energy-efficient continuous-time delta-sigma modulators (CTDSMs) are used to digitize their output signals: a phase-domain DSM for the WB sensor and a current-domain DSM for the WhB sensor. The WB sensor achieves a 410 μK resolution in a 5 ms conversion time and a resolution FoM of 0.13 pJ·K^2. The WhB sensor is even better, achieving a 160 μK resolution in 10 ms and a resolution FoM of 0.049 pJ·K^2.

This paper is organized as follows. First, the characteristics of CMOS resistors are discussed, and the choice of silicided resistors is motivated (Sect. 10.2). The analysis and design of the WB sensor and its readout circuitry are then presented (Sect. 10.3), together with its measurement results. Next, the analysis, design, and measurement results of the WhB sensor are presented (Sect. 10.4). Finally, the two sensors are compared with the state of the art (Sect. 10.5), followed by some conclusions (Sect. 10.6).

10.2 Resistor-Based Sensors

Compared to BJTs, on-chip resistors can operate over a wider temperature and voltage range. However, compared to the base-emitter voltages of BJTs, whose spread is typically in the order of a few millivolts, and so much less than 1% [6], on-chip resistors exhibit much larger spread and nonlinearity [2–3, 15]. These two disadvantages, however, can be compensated by trimming and systematic nonlinearity removal.

In CMOS processes, many resistors are available: metal resistors, diffusion resistors, poly-silicon (poly) resistors, N-well resistors, and silicided resistors. Their characteristics are summarized in Table 10.1.

Table 10.1 Characteristics of the resistors available in CMOS processes

Resistor type	Metal	Diffusion	N-well	Poly	Silicided
Temperature coefficient	Large	Medium	Large	Medium or small	Large
2nd-order temperature coefficient	Medium	Medium	Large	Medium	Small
Sheet resistance	Very small	Large	Large	Large	Medium
Supply dependency	Small	Medium	Large	Small	Small
$1/f$ noise	No	No	No	Large	Small?
Stress sensitivity	Small	Large	Large	Medium	Small?

To achieve both high resolution and energy efficiency, a temperature-sensing resistor should have a large TC, a low $1/f$ noise, and a stable resistance, i.e., small supply dependency and stress sensitivity.

Metal resistors can be quite stable and have large TCs, typically ranging from 0.3 to 0.4%/°C. However, their sheet resistance is extremely low (<100 mΩ/), resulting in either huge chip area or high power consumption. In comparison, poly-silicon resistors have smaller TCs (<0.15%/°C in the target 0.18 μm technology) and exhibit more $1/f$ noise [18]. The resistance of diffusion resistors is determined by their doping level, which is sensitive to the reverse bias between the resistor and the substrate (or well diffusion) and thus is voltage dependent. Although their TCs are higher, about the same as metal resistors, N-well resistors suffer even more from the voltage dependency, as their lower doping levels make them more sensitive to voltage variations.

Silicided resistors, however, have a large TC (> 0.3%/°C) and a relatively linear temperature dependency (i.e., small second-order TC). Since a silicide is essentially a silicon-metal alloy, a silicided resistor behaves more like a metal resistor than a silicon resistor. Compared to normal poly-silicon resistors, they have a lower sheet resistance (but not as low as that of metal resistors), and its $1/f$ noise should be less than that of a normal poly resistor with the same area and resistance [18]. They also have low supply dependency and stress sensitivity. In the adopted process (TSMC 0.18 μm), two types of silicided resistors are available, silicided poly resistors and silicided diffusion resistors. The main difference between them is that the former spread less than the latter. Thus, in this work, silicided poly resistors are employed as the main temperature-sensing elements.

10.3 Wien Bridge Sensor

10.3.1 Sensor Analysis

The structure of a Wien bridge (WB) sensor [2] is shown in Fig. 10.1a. It is a second-order band-pass RC filter, and its frequency domain voltage amplitude and phase transfer functions can be written as

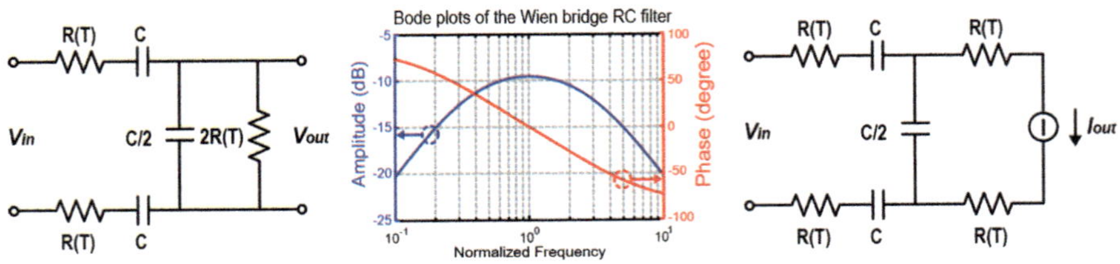

Fig. 10.1 (**a**) Wien bridge sensor, voltage readout scheme (**b**) Bode plots (**c**) Wien bridge sensor, current readout scheme

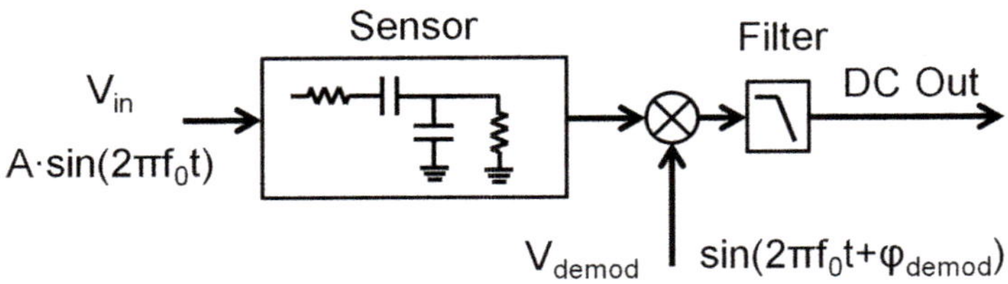

Fig. 10.2 Phase detection of a Wien bridge sensor

$$\mathrm{H}\,(j\omega) = \frac{RCj\omega}{1 - R^2C^2\omega^2 + 3RCj\omega} \tag{10.1}$$

$$\phi_{WB}\,(\omega) = -\tan^{-1}\left(\frac{R^2C^2\omega^2 - 1}{3RC\omega}\right) \tag{10.2}$$

Its Bode plot is shown in Fig. 10.1b, where the frequency is normalized to $f_0 = 1/(2\pi RC)$.

Given a fixed driving frequency f_{drive}, the phase output of the WB is determined by its resistors and capacitors. If capacitors with a negligibly low TC are used, such as metal-insulator-metal (MIM) capacitors, the bridge's phase shift is mainly a function of the resistor's temperature dependency. This phase shift can be determined either by measuring the voltage across the output resistor $2R(\mathrm{T})$ or the current flowing through it, as shown in Fig. 10.1c. In this paper, the former method is referred to as the voltage readout scheme and the latter as the current readout scheme.

An ideal phase detection model based on synchronous demodulation, shown in Fig. 10.2, can be used to estimate the achievable rms temperature resolution and the resolution FoM of the WB sensor. To simplify the analysis, both the driving signal and the demodulating signal are modeled as sine waves with the same frequency, but with a phase difference: $V_{\mathrm{in}} = \mathrm{A}{\cdot}\sin(2\pi f_0\mathrm{t})$ and $V_{\mathrm{demod}} = \sin(2\pi f_0\mathrm{t} + \varphi_{\mathrm{demod}})$, where $f_0 = 1/(2\pi RC)$, $\varphi_{\mathrm{demod}} = 90°$ and A is the amplitude of the driving signal. The signal of interest is the DC output after the low-pass filter.

Assuming that the resistors in the WB are the only noise contributors, the rms temperature resolution of a differential WB sensor for a voltage readout scheme can be expressed as

$$\Delta T_{v,WB} = \frac{3}{A \cdot 2\alpha} \cdot \sqrt{\frac{3kTR}{t_{\text{conv}}}}, \tag{10.3}$$

where α is the temperature coefficient of the resistors in the WB and t_{conv} is the conversion time.

In the current readout scheme, however, the noise contribution of the resistor $2R(T)$ is less attenuated than that in the voltage readout scheme. This results in a lower temperature-sensing resolution, which can be expressed as

$$\Delta T_{i,WB} = \frac{3}{A \cdot \alpha} \cdot \sqrt{\frac{3kTR}{2t_{\text{conv}}}}. \tag{10.4}$$

The FoM of a WB sensor can then be determined by combining its resolution with its energy consumption when it is driven at its center frequency. In the voltage readout scheme, $\text{FoM}_{\text{v, WB}} = 9kT/(4\alpha^2)$; while in the current readout scheme, $\text{FoM}_{i,\text{ WB}} = 9kT/(2\alpha^2)$. Interestingly, these expressions are independent of other design parameters such as resistance, capacitance, supply voltage, or conversion time. However, these results do not include the noise and energy consumption of the readout circuitry.

10.3.2 Sensor Design

As stated in Sect. 10.2, the silicided poly resistor is the chosen resistor type. In the actual design, the WB sensor is implemented with $R = 32$ kΩ, $C = 10$ pF, $f_{\text{drive}} = 500$ kHz, and $A = 0.9$ V. A current readout scheme is chosen for its low swing, thus relaxed readout linearity [2]. With $t_{\text{conv}} = 5$ ms, this results in a temperature-sensing resolution of 230 μK (rms) when driven by a sine wave. Within the target industrial temperature range of −40–85 °C, the output phase of the WB will then vary from about −7° to 10°. For comparison purposes, a WB sensor based on a non-silicided n-poly resistor (largest negative TC) has also been realized. With the same component values, this results in a resolution of 380 μK (rms). The corresponding $\text{FoM}_{\text{i,WB}}$ values are 2.3 fJ·K^2 and 8.1 fJ·K^2 for the silicided p-poly sensor and the non-silicided n-poly sensor, respectively.

To achieve high energy efficiency, the phase output of the WB sensors is digitized by a 1-bit second-order phase-domain CTDSM (PD-CTDSM). A simplified block diagram of the modulator is shown in Fig. 10.3. The input phase (at $f_0 = 500$ kHz) is firstly down-converted to DC by multiplying it by a phase reference at the same frequency ($f_{\text{demod}} = f_0$). Depending on the chosen phase references φ_1 or φ_2,

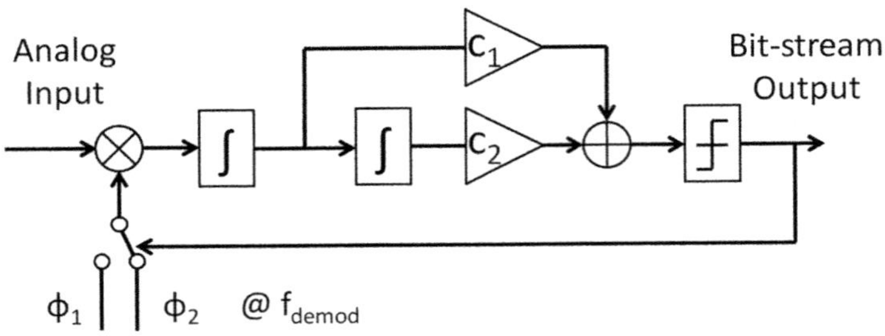

Fig. 10.3 Block diagram of the phase-domain CTDSM

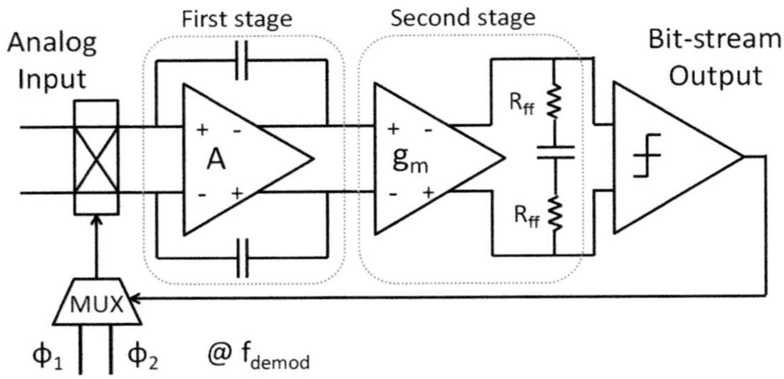

Fig. 10.4 Circuit diagram of the phase-domain CTDSM

the multiplier's DC output will either be positive or negative [11]. The result is integrated by the loop filter and then quantized. Due to a negative feedback, the quantizer will toggle the reference phases in such a way that the loop filter's average DC input will be zero. The output bitstream will then be a digital representation of the input phase. To leave enough margin for the spread of the resistors and capacitors of the WB sensor, the phase difference between the two references was designed to be 45° (+22.5° and −22.5°). These references are generated digitally from an 8 MHz master clock, which is 16× higher than the WB's driving frequency.

The circuit diagram of the phase-domain CTDSM is shown in Fig. 10.4. In the current domain, the analog input of the PD-CTDSM is the current output signal in Fig. 10.1c. The phase DAC is realized by controlling the direction of the current flow, i.e., by chopping the current input with a reference signal with different phases φ_1 and φ_2. The feedforward coefficient c_1 (Fig. 10.3) is achieved by R_{ff} at the output of the second stage.

To suppress the 1/*f* noise from the first integrating stage, the opamp is chopped. By choosing the chopping frequency f_{chop} the same as f_{demod}, the input chopper of the opamp can be merged with the input demodulator and becomes a single chopper in front of the integration capacitors, as shown in Fig. 10.5.

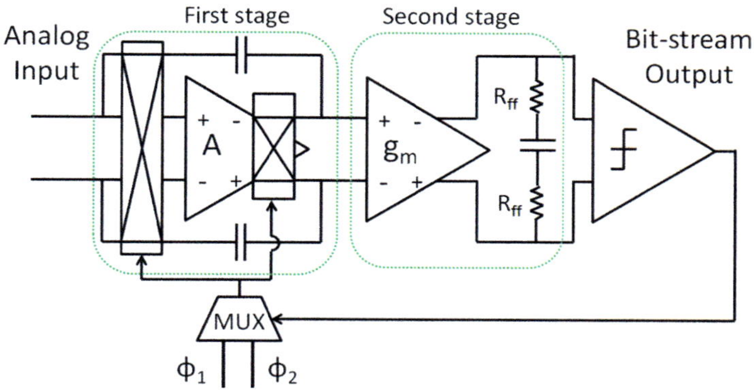

Fig. 10.5 Circuit diagram of the phase-domain CTDSM, with merged chopper

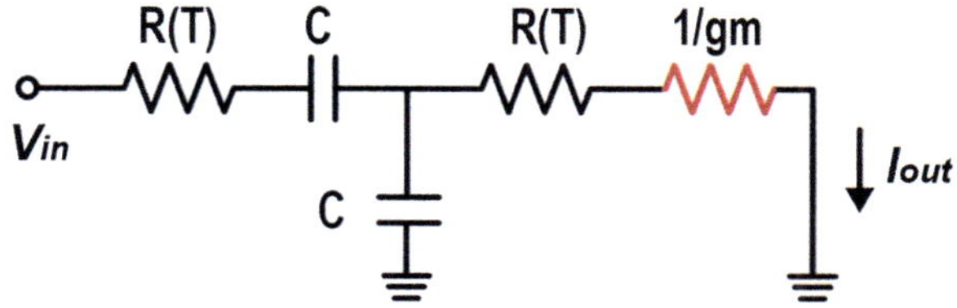

Fig. 10.6 Additive resistance on the output resistor of Wien bridge sensor when using a OTA-based first stage

In principle, the first stage amplifier can be an operational transconductance amplifier (OTA) which is often more energy efficient. However, the input resistance of the resulting integrator will be quite large ($1/gm$). Any variation in this resistance, which can be regarded as part of the output resistor of the WB shown in Fig. 10.6, will result in a temperature-sensing error. Assuming a nominal gm value of 1mS (I_d of about 50 μA, 4× larger than the maximum current output of the WB) and a variation of 10%, the variation of gm translates to more than 0.5 °C temperature error for the silicided p-poly sensor. For the non-silicided n-poly sensor, the error is even larger due to a smaller TC of the sensing resistor. The use of an opamp-based first stage, however, can suppress this error by significantly reducing the input impedance of the readout circuit.

In this work, the opamp in the first stage is made of a telescopic gm stage followed by two PMOS source followers, as shown in Fig. 10.7. The common-mode feedback of the gm stage is achieved by two PMOS transistors in their triode region. The tail current of the telescopic gm, which is 16 μA at room temperature, is optimized for low noise and power consumption. The source followers' DC current, which is 20 μA each at room temperature, is chosen to handle the WB's peak current (11 μA at room temperature, 16 μA at −40 °C). The integration capacitor of the first stage is 180 pF.

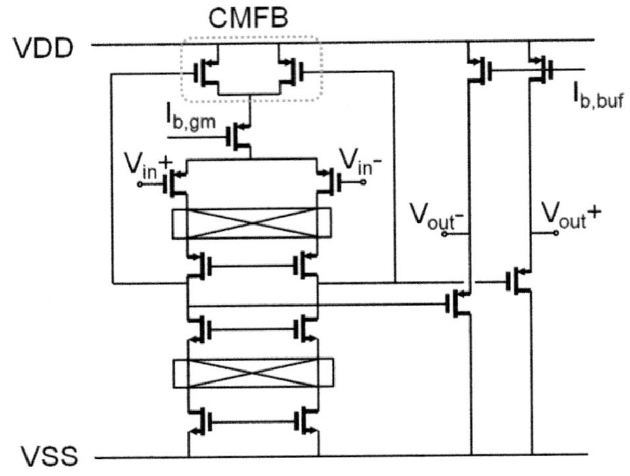

Fig. 10.7 Schematic of the opamp in the first stage

The g_m-C second stage uses an efficient telescopic OTA with source degenerated NMOS input pairs to improve its linearity, and it draws 4 μA, which is less than 10% of that of the first stage.

10.3.3 Measurement Results

The WB sensor chip [19] is fabricated in a 0.18 μm technology, and the chip micrograph is shown in Fig. 10.8. There are two sensors fabricated side by side: a silicided p-poly (s-p-poly) sensor and a non-silicided n-poly sensor. They share the same constant-gm biasing and phase generation circuits. Each sensor occupies about 0.72 mm^2, about 40% of which is occupied by the first integrator's capacitors (2 × 180 pF). Including the readout circuits, each WB sensor draws 87 μA from a 1.8 V power supply.

To determine the resolution of the proposed WB sensor design, the sensors are driven by a low-jitter (1 ps rms) frequency reference, which contributes less than 0.5% to the readout circuit's total noise power. The sensors are mounted in good thermal contact with a large aluminum block to minimize ambient temperature drift. Spectra of the bitstream outputs of both sensors are shown in Fig. 10.9a. The sensor's noise floor is dominated by the bridges' thermal noise. Figure 10.9b shows the temperature resolution vs. conversion time plot of both sensors obtained after a sinc^2 decimation of their bitstream outputs. In a 5 ms conversion time (2500 samples), the n-poly resistor achieves 880 μK rms resolution, while the s-p-poly achieves 410 μK rms resolution, due to its higher TC. The n-poly resistor exhibits a 1/*f* corner of about 10 Hz, while that of the p-poly sensor is below 1 Hz. These results demonstrate the effectiveness of the 1/*f* noise cancelation techniques of the readout circuit. The remaining 1/*f* noise, however, comes from the sensing resistors.

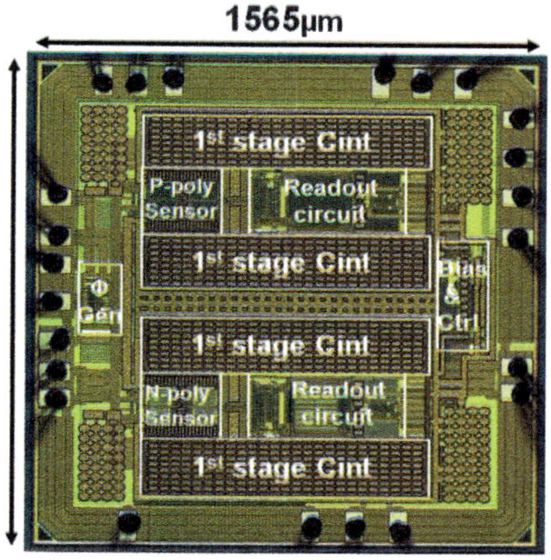

Fig. 10.8 Micrographs of the Wien bridge sensor chip

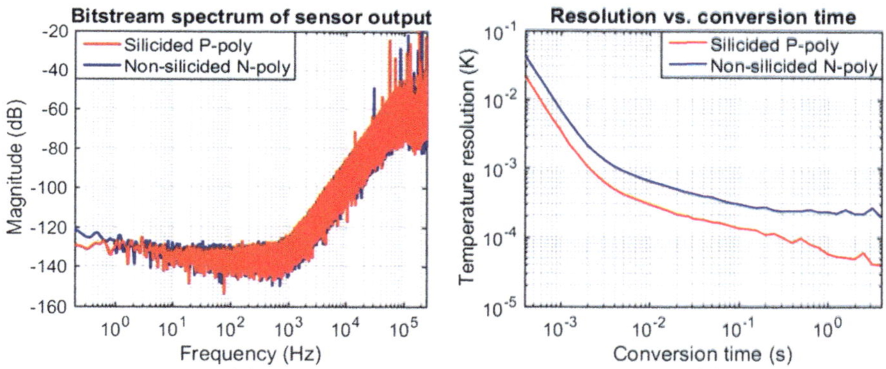

(a) (b)

Fig. 10.9 Bitstream FFT (**a**) and temperature resolution versus conversion time (**b**) of both Wien bridge sensors

Twenty samples from one wafer were characterized in ceramic DIL packages. At room temperature, supply sensitivities of −0.17 °C/V (s-p-poly bridge) and 0.34 °C/V (n-poly bridge) were observed for supply voltages ranging from 1.6 to 2 V. To improve the inherent nonlinearity of the phase domain CTDSM, a cosine nonlinearity compensation [10] is applied to the decimated bitstream before trimming, resulting in the phase output vs. temperature plots shown in Fig. 10.10.

After a first-order polynomial fitting followed by a fixed correction of the systematic nonlinearity on the phase shift vs. temperature plot, the silicided p-poly sensor achieves a 3σ inaccuracy of 0.07 °C over the industrial temperature range of −45–85 °C. The non-silicided n-poly sensor is less accurate. Its inaccuracy is 0.25 °C over the same range, as shown in Fig. 10.11.

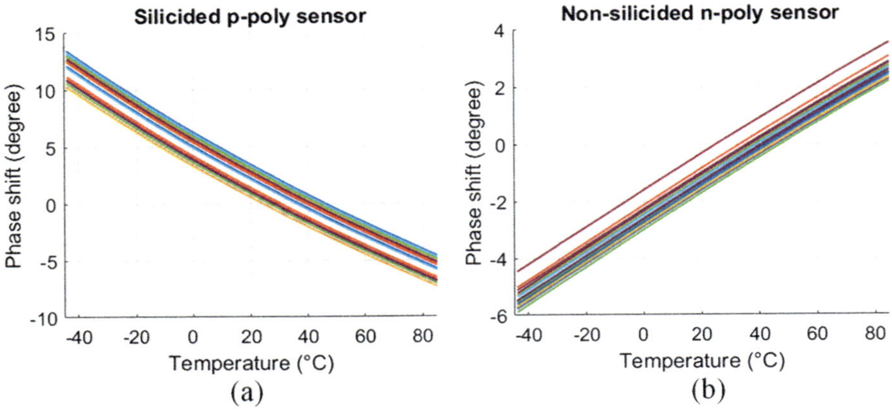

Fig. 10.10 Phase output vs. temperature plots for (**a**) silicided p-poly sensor (**b**) non-silicided n-poly sensor

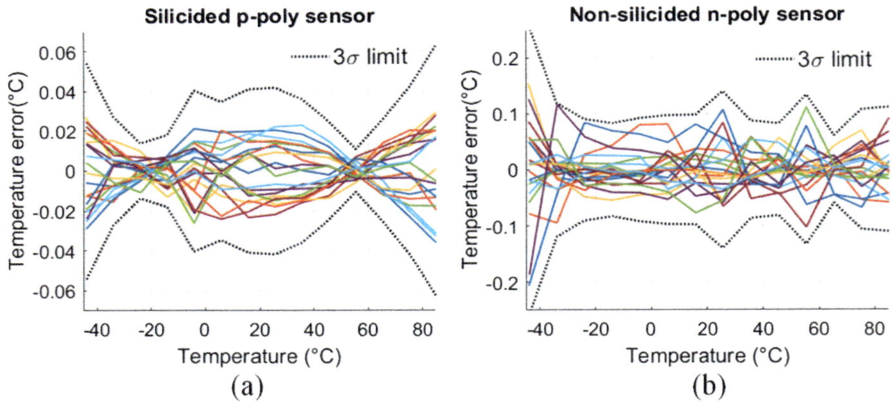

Fig. 10.11 Temperature error after first-order fit and systematic error removal for (**a**) silicided p-poly sensor (**b**) non-silicided n-poly sensor

10.4 Wheatstone Bridge Sensor

10.4.1 Sensor Analysis

Wheatstone bridges (WhBs) are often used to accurately measure differential resistance changes. By using two different resistor types (R_P and R_N) with different TCs, a temperature sensor can be made with either a voltage or current output as shown in Fig. 10.12a and b, respectively.

When the bridge is perfectly balanced ($R_P = R_N = R$), the rms temperature resolution of a WhB is

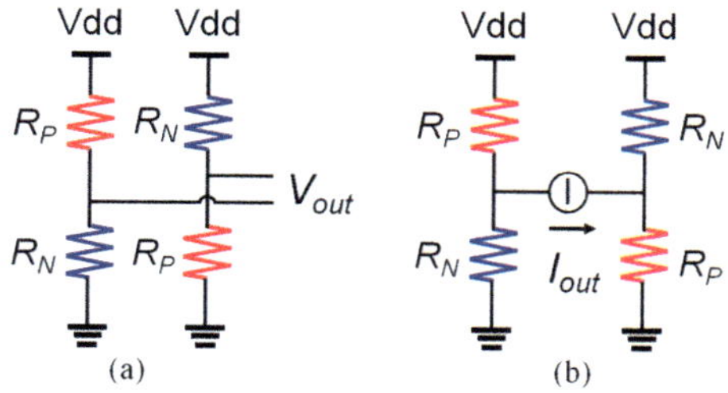

Fig. 10.12 Wheatstone bridge sensor: (**a**) voltage readout (**b**) current readout

$$\Delta T_{WhB} = \frac{1}{A \cdot (\alpha_P - \alpha_N)} \cdot \sqrt{\frac{2kTR}{t_{\text{conv}}}}, \tag{10.5}$$

where α_P and α_N are the TCs of R_P and R_N, respectively, and $A = \text{Vdd}/2$ is the voltage across one single resistor. Note that the resolution is independent of the nature of its output, i.e., whether it is a voltage or a current. For $\alpha = \alpha_{P-}\alpha_N$, the FoM of a WhB temperature sensor can be expressed as $\text{FoM}_{\text{WhB}} = 8k\text{T}/\alpha^2$, excluding the readout electronics' power consumption and noise.

10.4.2 Sensor Design

To maximize the resolution and the energy efficiency of WhB sensors, the TCs of R_P and R_N should have opposite polarities. In the target process, however, only the non-silicided poly resistors have negative TCs. To compare their performance, two bridges, based on non-silicided n- and p-poly resistors as the negative TC resistors (R_N), respectively, were realized. Both bridges used silicided p-poly resistors as the positive TC resistors (R_P). The two Wheatstone bridges, i.e., s-p/n-poly and s-p/p-poly bridge, achieve FoM values of 1.7 fJ·K^2 and 3.4 fJ·K^2, respectively (excluding the readout circuits).

The s-p/n-poly WhB sensor was designed to be slightly unbalanced at room temperature to leave a similar margin at the two extremes of the targeted temperature range (−40 to 85 °C). With $R_P = 105$ kΩ, $R_N = 95$ kΩ, and a supply voltage Vdd = 1.8 V, the sensor's resolution is 74 μK within 10 ms conversion time. The current output range of the WhB sensor is −2.0–3.0 μA. To achieve a similar current output range over the targeted temperature range, the resistors in the s-p/p-poly sensor were chosen to be $R_P = 67.5$ kΩ and $R_N = 64$ kΩ. These values result in a slightly worse resolution of 85 μK.

As for the WB sensor, a second-order 1-bit CTDSM is employed to read out the WhB sensor. Since the output of the WhB sensor is a DC current signal, a current-domain CTDSM is used as depicted in Fig. 10.13. The simplified circuit diagram of the designed CTDSM is shown in Fig. 10.14. The DC performance of the CTDSM is critical, so chopping is applied on the first stage opamp to suppress its offset

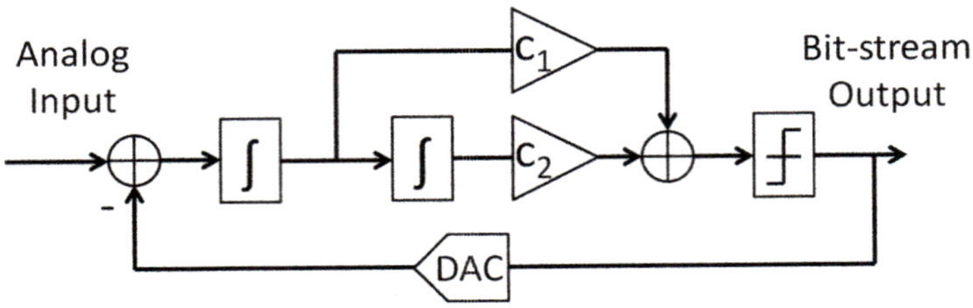

Fig. 10.13 Block diagram of the CTDSM

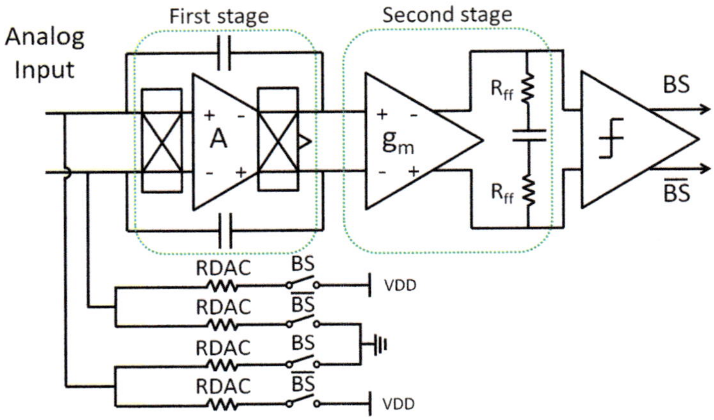

Fig. 10.14 Circuit diagram of the readout circuit of the Wheatstone bridge sensor

and 1/*f* noise. The feedforward coefficient is accomplished by the two feedforward resistors (R_{ff}) at the output of the gm stage. Because of the symmetry of the WhB, the common-mode input voltage of the readout circuit is Vdd/2 (0.9 V). A resistive DAC is chosen to implement the current feedback. To avoid introducing additional spread, the four DAC resistors (RDAC) are made of the same material as that of R_N, which has the smaller TC among the two types of sensing resistors. In this design, $R_{DAC} = 140\text{k}\Omega$, for both the s-p/n-poly sensor and the s-p/p-poly sensor.

To avoid aliasing high frequency quantization noise at the chopping transitions [20], the chopping frequency is set to be equal to the sampling frequency (500 kHz). Since no phase generation circuits are required, the readout electronics of the WhB sensor employs a lower reference clock frequency than that of the WB sensor, i.e., 4 MHz instead of 8 MHz, which lowers the power consumption of its digital circuits.

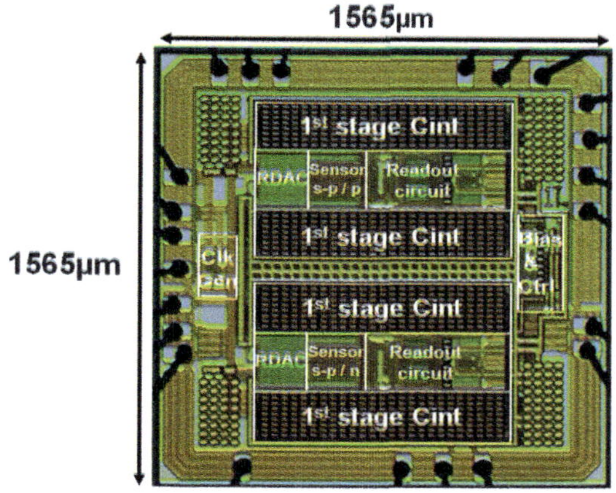

Fig. 10.15 Micrograph of the Wheatstone bridge sensor chip

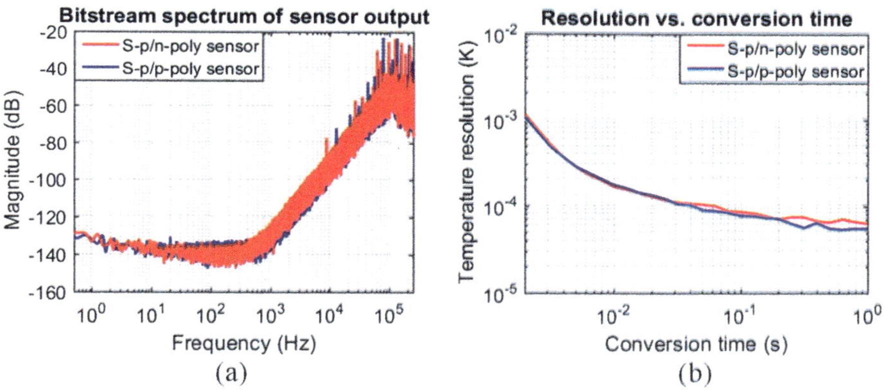

Fig. 10.16 Bitstream FFT (**a**) and temperature resolution over conversion time (**b**) of both Wheatstone bridge sensors

10.4.3 Measurement Results

The WhB sensor chip is fabricated in the same 0.18 μm CMOS technology, and the chip micrograph is shown in Fig. 10.15. There are two sensors fabricated side by side: a silicided p-poly/non-silicided n-poly (s-p/n-poly) sensor and a silicided p-poly/non-silicided p-poly (s-p/p-poly) sensor. They share the same constant-gm biasing and clock generation circuits. Under a 1.8 V power supply, the current consumption is 101 μA for each sensor, including the readout circuits. The chip area is the same as the WB sensor.

With similar measurement setup as that for the WB sensor (but without the low-noise frequency reference), the bitstream spectra and the conversion time vs. resolution plots of the two WhB bridge sensors are shown in Fig. 10.16. The $1/f$ corner of both sensors is around 10 Hz. After a sinc^2 decimation of their bitstream

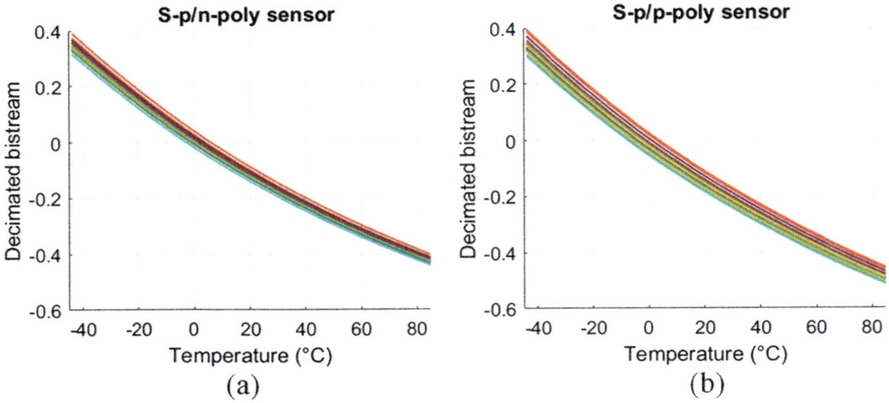

Fig. 10.17 Decimated bitstream vs. temperature plots for (**a**) s-p/n-poly sensor (**b**) s-p/p-poly sensor

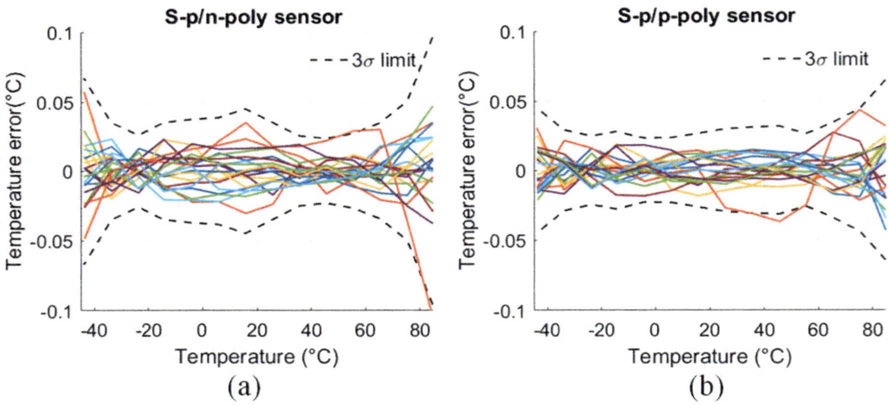

Fig. 10.18 Temperature error after first-order fit and systematic error removal for (**a**) s-p/n-poly sensor (**b**) s-p/p-poly sensor

outputs, the s-p/n-poly sensor achieves a temperature resolution of 164 μK in a 10 ms conversion time (5000 samples). In the same conversion time, the s-p/p-poly sensor's resolution is about 170 μK.

Twenty samples from one wafer in ceramic DIL packages were characterized. At room temperature, supply sensitivities of maximum 16 m°C/V (s-p/n-poly sensor) and 26 m°C/V (s-p/p-poly sensor) were observed for supply voltages ranging from 1.6 to 2 V. The decimated bitstream vs. temperature plots are shown in Fig. 10.17. After a first-order polynomial fitting followed by a fixed correction of the systematic nonlinearity, the s-p/n-poly sensor achieves a 3σ inaccuracy of 0.10 °C over the industrial temperature range. The s-p/p-poly sensor is more accurate with 0.07 °C inaccuracy over the same range, as shown in Fig. 10.18.

10.5 Comparison of the Sensors

10.5.1 Comparison Between the Two Sensors

The two sensors represent different ways of realizing the reference impedance required to digitize a temperature-sensing resistance. The WB sensor's reference is a capacitive reactance, i.e., the impedance of a capacitor driven by a fixed frequency. In contrast, the reference of the WhB sensor is simply another resistance.

The realization of a reference resistance is more straightforward, as it does not require a stable frequency reference (although this is not a problem in a frequency reference). Also, the associated readout electronics can be more efficient, because its input signals are at DC and do not need to be down-modulated. Moreover, the use of two different types of resistors in the WhB sensor results in a higher effective TC than in the WB sensor. As a result, the FoM of the WhB sensor is 2.7× better than the WB sensor.

However, apart from FoM, other important performance criteria for temperature sensors are their accuracy and stability. The temperature dependency of a WhB sensor is determined by two different types of resistors with different (non-linear) TCs and spread, while a WB employs only a single type. As a result, a WhB sensor can be expected to be somewhat less accurate than a WB sensor, especially over corners. Moreover, the observed $1/f$ noise of silicided poly resistors is much better than that of normal poly resistors, which also makes WB sensors better in this respect.

10.5.2 Comparison with the State of the Art

The performance of the two resistor-based temperature sensors is compared to the state of the art in Table 10.2. The energy efficiency of the two presented sensors is better than any other CMOS temperature sensor, and close to that of the MEMS-based sensor, which is not fully CMOS compatible [6]. The two sensors also achieve inaccuracies of better than 0.1 °C with only a first-order polynomial fitting followed by a systematic error removal, which is better than previous designs. As shown in the temperature resolution vs. energy/conversion plot (Fig. 10.19), the two sensors introduced in this paper are by far the most energy-efficient CMOS temperature sensors.

10.6 Conclusion

In this paper, two resistor-based temperature sensors are implemented in a 0.18 μm CMOS technology. One is based on a Wien bridge RC filter, while the other is based on a Wheatstone bridge. The silicided p-poly resistor, because of its high TC and

Table 10.2 Performance summary of the Wien bridge (WB) sensor and the Wheatstone bridge (WhB) sensor compared to previous high-resolution energy-efficient temperature sensors

	WB sensor	WhB sensor	JSSC13 [1]	JSSC15 [15]	JSSC17 [12]	TIE17 [5]
Sensor type	Resistor	Resistor	Resistor	Resistor	MEMS	BJT
Tech (μm)	0.18	0.18	0.18	0.18	0.18	0.7
Area (mm^2)	0.72	0.72	0.18	0.43	0.54	1.5
Power (mW)	0.16	0.18	13	0.065	19	0.16
Temp. range (°C)	−40–85	−40–85	−40–85	−40–125	−40–105	-40–130
Resolution (mK)	0.41	0.16	0.1	10	0.02	3
T_{conv} (ms)	5	10	100	0.1	5	1.8
Trim point	2[a]	2[a]	6	2[b]	–	1
Inaccuracy (3σ)	±70 mK	±100 mK	±15 mK[c]	±400 mK[c]	–	±300 mK
Res. FoM ($pJ{\cdot}K^2$)	0.13	0.049	13	0.65	0.04	3.2

[a]First-order fit
[b]One-point trim with first-order curve fitting
[c]Min or max

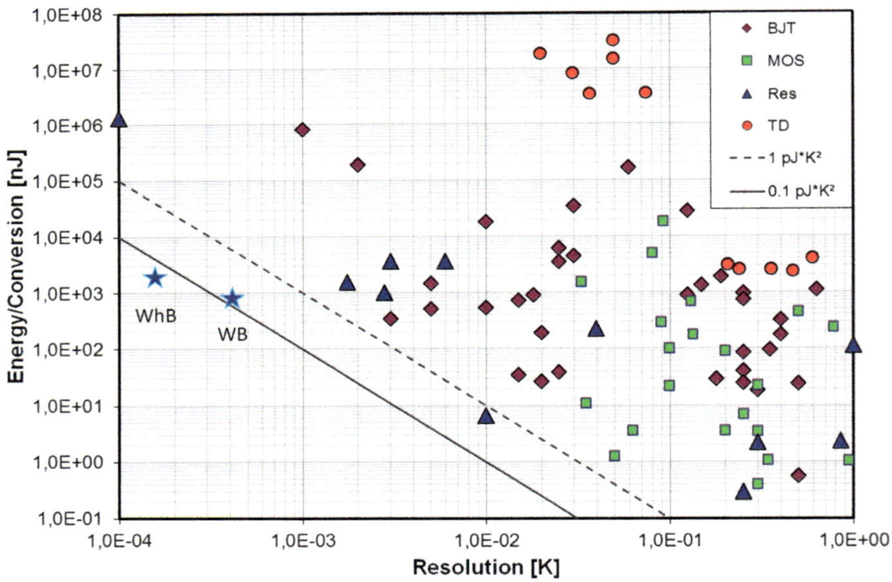

Fig. 10.19 Energy efficiency of the two measured sensors compared to other CMOS temperature sensors [4]

low 1/*f* noise, is employed in both bridges. Energy-efficient CTDSMs are adopted for the readout circuits for both sensors, and chopping is applied to suppress the offset and 1/f noise of the readout electronics. The Wien bridge sensor achieves a 410 μK resolution in a 5 ms measurement time and a resolution FoM of 0.13 $pJ{\cdot}K^2$. The Wheatstone bridge sensor, however, achieves a 160 μK resolution in a 10 ms measurement time and a resolution FoM of 0.049 pJ·K^2. These results clearly show that resistor-based temperature sensors can be used to realize the high-resolution and energy-efficient sensors for the temperature compensation of frequency references.

Acknowledgments The authors would like to thank Yanquan Luo and Saleh Heidary Shalmany for their contributions to the Wien bridge sensor design and Hui Jiang for his work on the Wheatstone bridge sensor. The authors would also like to thank Burak Gönen, Vincent van Hoek, and Said Hussaini for proofreading and suggestions.

References

1. Perrott, M.H., et al.: A temperature-to-digital converter for a MEMS-based programmable oscillator with < ±0.5-ppm frequency stability and < 1-ps integrated jitter. IEEE J. Solid-State Circuit. **48**(1), 276–291 (2013)
2. Shahmohammadi, M., Souri, K., Makinwa, K.A.A.: A resistor-based temperature sensor for MEMS frequency references. In: 2013 Proceedings of the ESSCIRC (ESSCIRC), Bucharest, pp. 225–228 (2013)
3. Park, P., Ruffieux, D., Makinwa, K.A.A.: A thermistor-based temperature sensor for a real-time clock with ±2 ppm frequency stability. IEEE J. Solid-State Circuit. **50**(7), 1571–1580 (July 2015)
4. Makinwa, K.A.A.: Smart Temperature Sensor Survey. [Online]. Available: http://ei.ewi.tudelft.nl/docs/TSensor_survey.xls
5. Wang, G., Heidari, A., Makinwa, K.A.A., Meijer, G.C.M.: An accurate BJT-based CMOS temperature sensor with duty-cycle-modulated output. IEEE Trans Ind Electron. **64**(2), 1572–1580 (2017)
6. Pertijs, M.A.P., Makinwa, K.A.A., Huijsing, J.H.: A CMOS smart temperature sensor with a 3σ inaccuracy of ±0.1°C from −55°C to 125°C. IEEE J. Solid-State Circuit. **40**(12), 2805–2815 (2005)
7. Yousefzadeh, B., Shalmany, S.H, Makinwa, K.: A BJT-Based Temperature-to-Digital Converter with ±60mK (3σ) Inaccuracy from −70°C to 125°C in 160nm CMOS. In: 2016 IEEE Symposium on VLSI Circuits (VLSI-Circuits), Honolulu, HI, pp. 1–2 (2016)
8. Souri, K., Chae, Y., Thus, F., Makinwa, K.: A 0.85V 600nW all-CMOS temperature sensor with an inaccuracy of ±0.4°C (3σ) from −40 to 125°C. In: 2014 IEEE International Solid-State Circuits Conference Digest of Technical Papers (ISSCC), San Francisco, CA, pp. 222–223 (2014)
9. Anand, T., Makinwa, K.A.A., Hanumolu, P.K.: A VCO based highly digital temperature sensor with 0.034 °C/mV supply sensitivity. IEEE J. Solid-State Circuit. **51**(11), 2651–2663 (2016)
10. Kashmiri, S.M., Xia, S., Makinwa, K.A.A.: A temperature-to-digital converter based on an optimized Electrothermal filter. IEEE J. Solid-State Circuit. **44**(7), 2026–2035 (July 2009)
11. van Vroonhoven, C.P.L., Makinwa, K.A.A.: A CMOS Temperature-to-Digital Converter with an Inaccuracy of ± 0.5° C (3σ) from −55 to 125°C. In: 2008 IEEE international Solid-state circuits Conference – Digest of technical papers, San Francisco, CA, pp. 576–637 (2008)
12. Heidarpour Roshan, M., et al.: A MEMS-assisted temperature sensor with 20-μK resolution, conversion rate of 200 S/s, and FOM of 0.04 pJK^2. IEEE J. Solid-State Circuit. **52**(1), 185–197 (2017)
13. Horng, J.J. et al.: A 0.7V resistive sensor with temperature/voltage detection function in 16 nm FinFET technologies. In: 2014 Symposium on VLSI Circuits Digest of Technical Papers, Honolulu, HI, pp. 1–2 (2014)

14. Wu, C.K., Chan, W.S., Lin, T.H.: A 80kS/s 36μW resistor-based temperature sensor using BGR-free SAR ADC with a unevenly-weighted resistor string in 0.18 μm CMOS. In: 2011 Symposium on VLSI Circuits – Digest of Technical Papers, Honolulu, HI, pp. 222–223 (2011)
15. Weng, C.H., Wu, C.K., Lin, T.H.: A CMOS thermistor-embedded continuous-Time Delta-sigma temperature sensor with a resolution FoM of 0.65 pJ°C^2. IEEE J. Solid-State Circuit. **50**(11), 2491–2500 (2015)
16. Sankaragomathi, K.A., Koo, J., Ruby, R., Otis, B.P.: A ±3ppm 1.1mW FBAR frequency reference with 750MHz output and 750mV supply. In: 2015 IEEE International Solid-State Circuits Conference – (ISSCC) Digest of Technical Papers, San Francisco, CA, pp. 1–3 (2015)
17. Tang, X., Pun, K.P., Ng, W.T.: A 0.9V 5kS/s resistor-based time-domain temperature sensor in 90 nm CMOS with calibrated inaccuracy of −0.6°C/0.8°C from −40°C to 125°C. In: 2013 IEEE Asian Solid-State Circuits Conference (A-SSCC), Singapore, pp. 169–172 (2013)
18. De Graaff, H.C., Huybers, M.T.M.: 1/f noise in polycrystalline silicon resistors. J. Appl. Phys. **54**(5), 2504–2507 (1983)
19. Pan, S., Luo, Y., Shalmany, S.H., Makinwa, K.A.A.: 9.1 A resistor-based temperature sensor with a 0.13pJ·K^2 resolution FOM. In: 2017 IEEE International Solid-State Circuits Conference (ISSCC), San Francisco, CA, pp. 158–159 (2017)
20. Billa, S., Sukumaran, A., Pavan, S.: A 280μW 24kHz-BW 98.5 dB-SNDR chopped single-bit CTΔΣM achieving < 10Hz 1/f noise corner without chopping artifacts. In: 2016 IEEE International Solid-State Circuits Conference (ISSCC), San Francisco, CA, pp. 276–277 (2016)

Chapter 11
A High-Resolution Self-Oscillating Integrating Dual-Slope CDC for MEMS Sensors

J. P. Sanjurjo, Enrique Prefasi, C. Buffa, C. Rogi, and R. Gaggl

11.1 Introduction

Integrated microelectromechanical systems (MEMS) capacitive sensors are widely used for measuring different physical quantities as they are quite small, dissipate little power, and can be readily integrated into miniaturized inertial measurement units (IMUs). Indeed the current trend is to integrate ten sensing axes in a single chip: accelerometers and gyros are mature, several solutions are available for the three magnetic axes, but research is ongoing for pressure sensors [1]. Other important advantages of capacitive sensors are their high level of stability and low drift over temperature. In addition, due to the demand for the low-power sensors required for the Internet of Everything, the use of such sensors is steadily increasing.

Nowadays, many different types of capacitive sensor interfaces are in use. To reduce development costs and time to market, however, a generic capacitance-to-digital converter (CDC) would seem to be the next logical step. This could then be digitally configured to meet the requirements of different applications (e.g., humidity, pressure, temperature, gas, etc.). To show the performance of the CDC proposed in this work, a MEMS pressure sensor has been chosen as a representative of a low bandwidth sensing application.

High-resolution CDCs typically use charge transfer between capacitors to convert a sampled capacitance into voltage. This is then digitized by a high-order multi-bit switched capacitor (SC) $\Sigma\Delta$ ADC with large area and power-demanding blocks [2–4]. Alternative approaches (like period modulation or delay-chain discharge)

J.P. Sanjurjo • E. Prefasi (✉) • C. Rogi
Electronic Technology Department, Carlos III University, Madrid, Spain
e-mail: eprefasi@ing.uc3m.es

C. Buffa • R. Gaggl
Infineon Technologies, RF & Sensors, Villach, Austria

P. Harpe et al. (eds.), *Hybrid ADCs, Smart Sensors for the IoT, and Sub-1V & Advanced Node Analog Circuit Design*, DOI 10.1007/978-3-319-61285-0_11

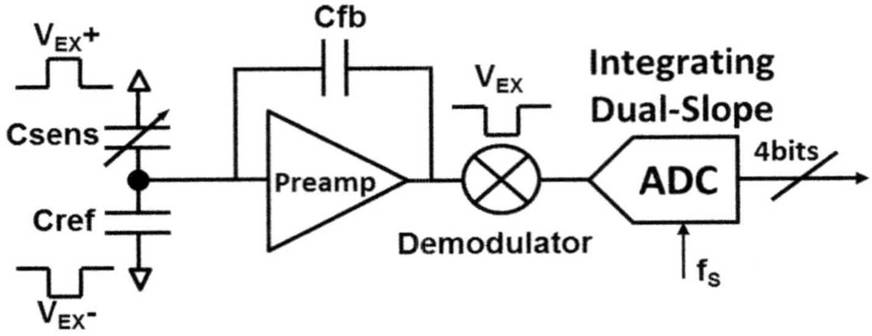

Fig. 11.1 Block diagram of the proposed architecture. Sensors are connected in a Wheatstone bridge to a preamplifier, whose output is digitized by an ADC

have been proposed to reduce area and power consumption [5, 6]. However, it has proven quite challenging to match the high resolution of multi-bit SC $- \Delta\Sigma$ ADCs.

In order to save power and area without reducing the performance of the CDC, this paper presents a topology that is able to detect small capacitance changes while reducing ADC complexity, which is a key determinant of CDC area and power. In order to be area and energy efficient, this work is based on an integrating dual-slope topology [7–9]. The proposed converter maps amplitude information into the time domain. In addition, like traditional $\Sigma\Delta$ converters [10], it uses quantization-noise shaping to reduce measurement time. However, unlike these, it does not require flash quantizers or n-bit DACs to generate a multi-bit digital output. Instead, it achieves high resolution by using single-bit circuitry and operating in the time domain. The main strengths of the proposed CDC are (1) intrinsically small sensitivity to temperature and process variations, (2) ease of trimming offset and gain to correct the sensor parameter spread, and (3) area and energy-efficient implementation while maintaining the performance of traditional approaches. A block diagram of the proposed solution is shown in Fig. 11.1. Among the main blocks are an excitation circuit that drives a MEMS bridge and which also provides a reference signal for synchronous demodulation, an analog front-end and an integrating DS ADC.

11.2 MEMS Topology

MEMS are microscale systems made of both mechanical moving parts and electronics. Such structures can sense the variation of mechanical quantities or actuate. Most MEMS devices consist of a mass which is free to move in one or more direction in 3D space with respect to the substrate, to which it is anchored by springs. Different methods are commonly used to sense the displacement of the moving mass: capacitive, piezoresistive, optical, and resonant sensing. Capacitive readout MEMS are based on the measure of a capacitance variation due to the displacement of a suspended microscopic structure in presence of an external applied force. Moving electrodes (also called *rotors* borrowing mechanical terminology) are mechanically

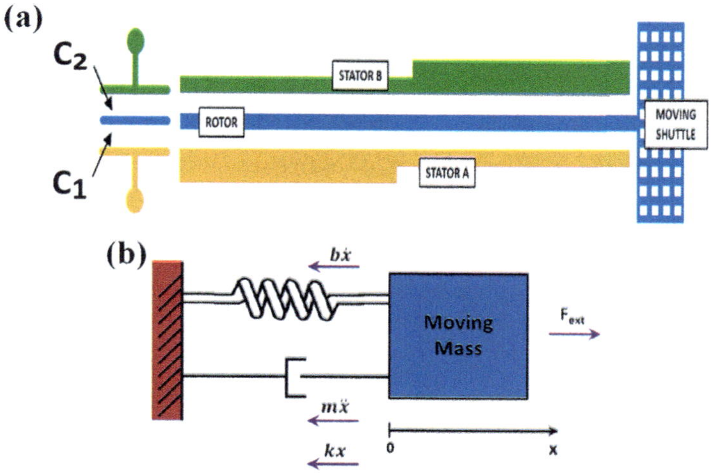

Fig. 11.2 (**a**) Sketch of a typical differential capacitive sensing cell for a MEMS structure. Stators A and B are anchored to chip substrate and they form a differential capacitor with rotor. (**b**) Diagram with forcing acting on a suspended mass

anchored to the moving structure and fixed electrodes (as a consequence called *stators*) are anchored to the substrate. Figure 11.2a shows a differential capacitive sensing cell with a moving electrode anchored to a suspended shuttle (on the right) and forming a couple of capacitors with stators A and B.

A microelectromechanical system can be modeled as a lumped parameter spring-mass-damper system [11], as shown in Fig. 11.2b; a mass is connected via a spring to a fixed support, being pulled by an external force F_{ext}. A dashpot is used to represent a mechanical damping element. All these three elements share the same displacement x with respect to a rest position. For the sake of simplicity, only a one-axis model is considered now, neglecting secondary vibrating modes; this analysis can be easily extended to a 3-DOF system in an inertial frame of reference. Applying Newton's second law of motion, stating that the net force on a body is equal to the product of acceleration and mass of the body, $F = m^*a$, the classical equation of motion describing the dynamics of a suspended micromachined structure can be derived:

$$m\cdot\ddot{x} + b\cdot\dot{x} + k\cdot x = F_{\text{ext}} \tag{11.1}$$

being the elastic force proportional to the displacement x, the viscous force to the velocity $\dot{x}$.

By applying Laplace transform to Eq. (11.1), the frequency behavior of MEMS can be studied with respect to frequency, and two main parameters can be highlighted to describe the behavior of the mechanical element.

The resonance frequency is defined as

$$f_r = \frac{1}{2\pi} \cdot \sqrt{\frac{k}{m}} \tag{11.2}$$

A heavier moving mass resonates at lower frequency; the stiffer is the spring, the higher is the resonance frequency.

The quality factor Q is a dimensionless parameter useful to characterize how over- or under-damped a MEMS resonator is. Equivalently, for large values, Q also characterizes resonator bandwidth Δf relative to its center frequency f_r, and it is related to MEMS parameters according to

$$Q = \frac{f_r}{\Delta f} = \frac{\omega_r \cdot m}{b} = \frac{\sqrt{k \cdot m}}{b} \tag{11.3}$$

Micromechanical devices are affected by thermal noise like all dissipative systems. In particular, dimensions scaling is attractive for a higher density integration, but small moving parts become more susceptible to mechanical noise due to molecular movement. Especially in sensors targeted for very low signals applications, mechanical noise may be a limiting factor. The power spectral density of the noise force can be written as [12]

$$S_{Fn} = 4k_{\mathrm{B}}Tb \tag{11.4}$$

where T being the absolute temperature, k_{B} the Boltzmann constant, and b damping coefficient previously introduced. It should not be surprising that the resulting expression for mechanical noise is very similar to Johnson noise in resistors, $SV_n = 4k_{\mathrm{B}}TR$, as they both have the same physical origin and dissipation.

In gas damped systems, like MEMS working either at ambient pressure or in a package at a lower pressure, mechanical noise is mainly due to random paths of molecules which hit the suspended structure. The result of this statistic process is an unwanted random displacement of the moving mass which is nevertheless detected by position sense interface.

11.3 Proposed Architecture

A possible configuration of the MEMS, in this case based on a differential Wheatstone bridge, is shown in the left part of Fig. 11.3. This bridge has two different types of capacitive MEMS sensors. C_{sen} is a micromachined sensing element whose capacitance has a major dependency with external pressure. This capacitor is made with metallic membranes that, due to the variation in pressure, change the distance between plates and thus its capacitance. It is designed to work in a range of pressure from 500 mbar to 1200 mbar, which corresponds to a capacitance change of a few pF

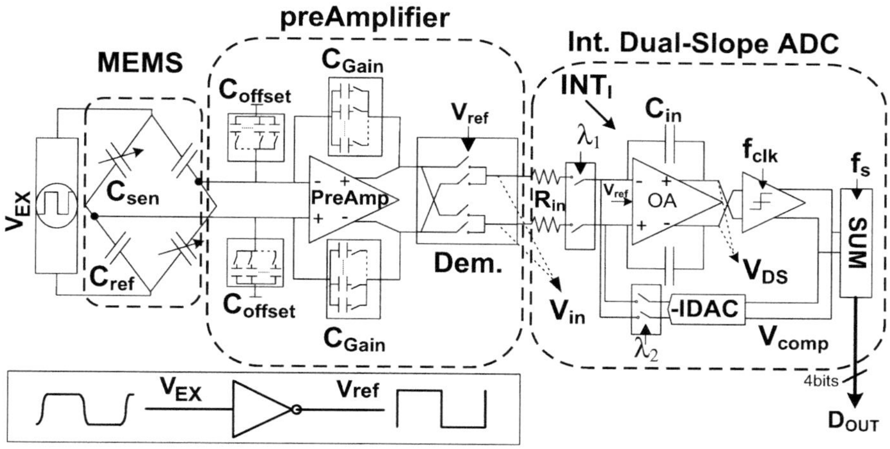

Fig. 11.3 Circuit schematic of the MEMS sensor and CDC

in C_{sen}. C_{ref} is a reference capacitor, whose capacitance does not depend on pressure and has the same temperature variation as C_{sen}. The bridge is differentially ($V_{EX}{}^{+}$ and $V_{EX}{}^{-}$) modulated by a smoothed (in order not to excite mechanical higher order resonances) square wave, which results in a signal at the output of the bridge with its amplitude proportional to bridge capacitance unbalancing.

11.3.1 Preamplifier and Demodulator

As mentioned before, the differential capacitor bridge is AC excited (i.e., modulated; see Fig. 11.3). Using this technique, the low frequency signal information is transposed to the odd harmonic frequencies of the modulation signal [13]. The modulated signal is then processed by a voltage amplifier (PreAmp in Fig. 11.3) to perform a capacitive to voltage (C–V) conversion. Depending on the sensor, the amplifier's gain can be adjusted to boost the bridge signal to the full-scale level of the DS ADC. C_{offset} and C_{Gain} are programmable capacitors which are used to amplify or attenuate the bridge signal and to compensate for bridge offset at rest (i.e., unbalancing of parasitics). In order to obtain the DC information of the modulated and amplified signal of the bridge, a demodulation stage is needed. This stage is driven by a square signal, V_{ref} (Fig.11.3), generated using the excitation signal of the bridge. The phase shift between the modulation and demodulation signal should be aligned to the phase shift of the amplifier stage in order to keep the DC gain of the amplifier stage at a maximum.

It is important to mention that offset and low frequency (i.e., flicker) noise introduced by the PreAmp is modulated only once by the demodulation stage and is therefore transposed to odd harmonics of the demodulation frequency, ideally

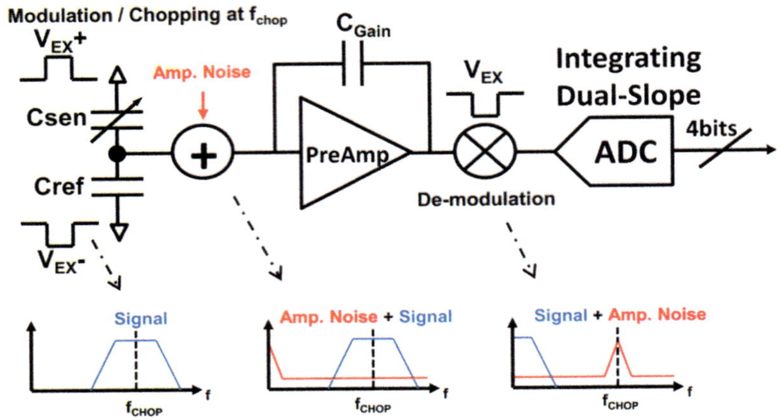

Fig. 11.4 Modulation scheme and chopping

leaving the demodulated signal without any offset and low frequency noise. This effect is illustrated in Fig. 11.4. This procedure is referred to as chopping [13].

Note, that the PreAmp is not chopped separately, but rather intrinsically chopped by the demodulation stage in the signal chain. In this way, the DS ADC only digitizes the voltage difference between C_{sen} and C_{ref}, as output by the PreAmp. The ideal relation between the bridge capacitors and the DS converter input voltage V_{in} (see Fig.11.3) is shown in (11.5).

$$V_{in} = -V_{EX} \cdot \frac{C_{sen} - C_{ref}}{C_{Gain}} \tag{11.5}$$

11.3.2 Integrating Dual-Slope ADC

11.3.2.1 Classical Dual-Slope Converter

As mentioned in the introduction, classical DS ADCs are good candidates for measuring low bandwidth signals with low power, simple architecture, and high resolution. The goal of this technique is to store an amount of charge proportional to the analog input in a capacitor and count the time necessary to discharge it with a fixed current. This technique works in two steps using one integrator.

The behavior can be seen in Fig. 11.5. To start with the conversion (*Phase I*), the DS ADC begins to integrate the input voltage (Vin in Fig. 11.3) through to the integrating capacitor. This integration ends after a fixed amount of time, T_1. In the second step of the conversion (*Phase II*), the input of the integrator is connected to a reference voltage using a 1-bit DAC (IDAC in Fig. 11.3). This voltage is fixed independently of the input. In this way, the integrator will discharge the voltage

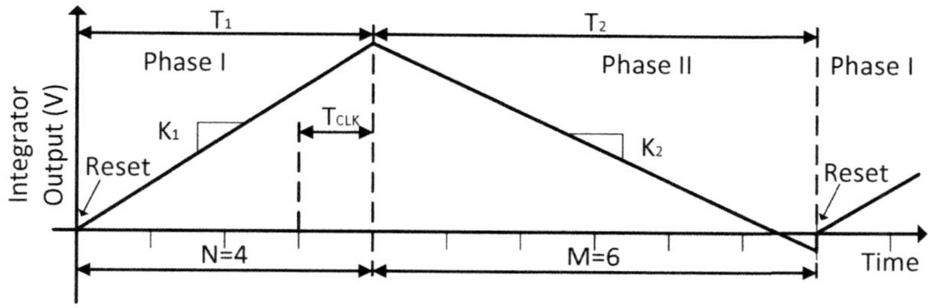

Fig. 11.5 Classical dual-slope time diagram

obtained in *Phase I* with a fixed slope until a zero crossing in the comparator. Then the conversion is over, and the integrator moves to the reset position to eliminate the charge in the integrating capacitor. In classical DS converters, T_1 is always fixed, i.e., in Fig. 11.5 this time is 4 times the clock period ($N = 4$). The time needed to discharge the voltage accumulated at the end of *Phase I* (T_2 in Fig. 11.5) then depends on the charge accumulated in the integrating capacitor during *Phase I* (in the example of Fig. 11.5, this time is equal to six clock cycles, $M = 6$). The resolution of this method is then proportional to the number of clock cycles needed to discharge the voltage integrated when input voltage is the full-scale value (M_{MAX}). Equations (11.6) and (11.7) show the equivalent number of bits:

$$\mathrm{LSB} = \frac{V_{FS}}{M_{\mathrm{MAX}}} \tag{11.6}$$

$$N_{\mathrm{bits}} = \log_2\left(M_{\mathrm{MAX}}\right). \tag{11.7}$$

However, in order to achieve high resolution and low power, classic DS converters [10] still are not efficient enough compared with other techniques. The main reason is the time cost of achieving high resolution. It can be seen in Eq. (11.7) that, in order to increase the resolution, $\mathrm{M_{MAX}}$ needs to increase exponentially. This would lead to a higher clock rate and power consumption. In addition, to keep high resolution, the comparator must resolve very small LSB voltages. This implies the need of high performance and power hungry comparators.

11.3.2.2 Self-Oscillating Integrating Dual-Slope ADC

To improve performance of classical DS ADC converters, a new topology called integrating DS was presented in [7, 9]. The basic principle behind this technique is its ability to map multi-bit amplitude information into time domain using single-bit circuitry (one comparator and 1-bit DAC). In addition, compared to traditional DS converters, it introduces quantization error noise shaping to reduce measurement time. To do so, the integrating DS converter stores the quantization error in the

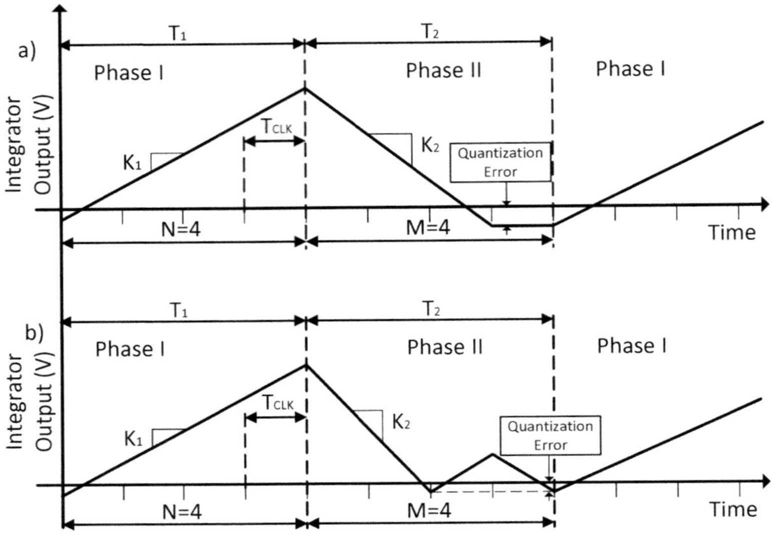

Fig. 11.6 (**a**) Standard integrating dual-slope and (**b**) proposed self-oscillation approach time diagrams

integrating capacitor (instead of resetting its voltage), at the end of Phase II, as it can be seen in Fig. 11.6a. If the integrating DS converter is compared with standard ΣΔ modulators, it does not require multi-bit circuits (i.e., flash quantizers or n-bit DACs) to keep the same resolution and performance. In [7, 9], it was demonstrated that the maximum signal-to-noise ratio (SNR) that can be achieved using this architecture is similar to the one of a first-order multi-bit ΣΔ modulator:

$$\mathrm{SNR}(dB) = 6.02 \cdot N_{\mathrm{bits}} + 30 \cdot \log(OSR) - 5.17 \tag{11.8}$$

where N_{bits} is equal to Eq. (11.7) and OSR represents the oversampling ratio between the sampling frequency of the converter (fs) and the signal bandwidth (f_{BW}) divided by 2: OSR $= fs/(2{\cdot}f_{\mathrm{BW}})$. According to Eq. (11.8) the resolution of the integrating DS converter can be modified by these two factors, similar to a ΣΔ modulator. However, to do so, it is needed to have a fixed sampling time ($Ts = T_1 + T_2$). For this, a slight modification of the DS ADC equations is needed. In this case, the variables of the converter must be selected in a way that the system is able to discharge completely the integrating capacitor C_{in} (see Fig. 11.6) in Phase II when the input signal is at its maximum value. The control of the discharge in *Phase II* depends on the following two equations:

$$K_1 \cdot N \ \ cdotp T_{\mathrm{clk}} + V_{\mathrm{LSB}} = K_2 \cdot M \cdot T_{\mathrm{clk}} \tag{11.9}$$

$$K_2 = \frac{I_{\text{feedback}}}{C_{\text{int}}} \tag{11.10}$$

where two main clocks must be defined: $f_{\text{clk}} = 1/T_{\text{clk}}$ and $fs = 1/Ts$. According to the rate $R = f_{\text{clk}}/fs$, and assuming $N = M$, then $R = 2^{\text{Nbits-1}}$. In this way, Eq. (11.8) can be rewritten as follow:

$$\text{SNR}(dB) = 6.02 \cdot \left(\frac{\text{Log}(R)}{\text{Log}(2)} + 1\right) + 30 \cdot \log\left(\frac{f_s}{2 \cdot f_{\text{BW}}}\right) - 5.17 \tag{11.11}$$

Using Eq. (11.11) it is easy to scale the integrating DS converter in the same way a multi-bit ΣΔ modulator is done. However, still some circuit design-related issues must be solved.

First, the integrating capacitor must keep its charge constant after the zero crossing until the next sampling period. But parasitics of the circuit play an important role in this case and take a significant unwanted amount of charge from the capacitor (leakage). This leads to a value of the quantization error that, at the end of the sampling period, is different from the estimated one after the zero crossing. Second, the transfer function is not linear. The possible output digital values are in the range from -Dn to Dn. However, there are two different "0", "0+" and "0−." To compensate this, an extra digital logic is needed (two different approaches has been already shown in [7, 9]), increasing the area and power consumption of the converter.

In this work, the integrating DS of [7, 9] has been modified in order to solve some of the issues mentioned before. The main difference is shown in Fig. 11.6b. In this case, the output of the DS integrator is kept toggling after the zero crossing in the discharging phase of the integrator. By adopting this modification, the quantization error is still kept for the next sampling period as it is shown in [8]. Since now, the zero crossing is resolved intrinsically by the operation of the DS converter, and the digital control complexity and comparator precision can be extremely relaxed, saving power and area. Also, this modification introduces a zero self-compensation, removing also the digital logic necessary to avoid the nonlinear effect of the transfer function. In addition, the proposed solution relaxes significantly the leakage issue mentioned before. The integrating capacitor does not need now to "freeze" the quantization error information during a long period of time, reducing the associated error.

11.3.2.3 Capacitive-to-Digital Converter Using Self-Oscillating Integrating Dual-Slope Converter

This section describes the solution adopted in this work to connect the proposed self-oscillating integrating DS converter with the MEMS and C–V stage. The DS converter needs to be connected with the voltage amplifier and the demodulation block in a way that the information is not disturbed. As it is mentioned before,

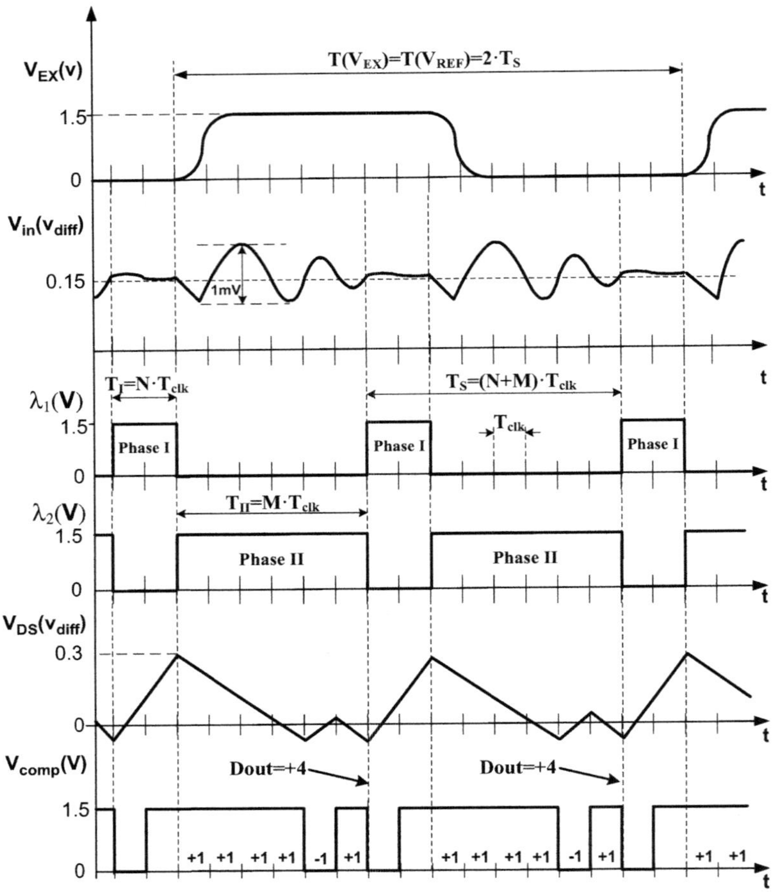

Fig. 11.7 Time diagram of the CDC

the readout circuit is modulating the input signal to high frequencies to remove offset and flicker noise in first part of the chain. Figure 11.7 shows a timing diagram with the different signals along the full readout chain. V_{EX} represents the excitation signal of the bridge. It is shaped in a pseudo-trapezoidal wave in order to reduce stimulation of high frequency of MEMS sensor and therefore reduce ringing in the output voltage of the bridge. The excitation of the MEMS together with the demodulation block makes the chopping of the first part of the CDC. The signals that drive these blocks (V_{EX} and V_{ref}) are synchronized and have the same frequency ($f_{chopping} = f_s/2$). The chopping frequency has to be carefully chosen: it has to be high enough to modulate flicker above bandwidth of interest, but low enough, to allow the signal stabilize after the ringing. The output of the demodulator is the input for the integrating DS ADC (V_{in} in Figs. 11.1 and 11.3). As it was explained, the signal is affected by the ringing of the MEMS but with a stable period of time (due the frequency of chopping and the pseudo-trapezoidal waveform). This period of time will be assigned for the *Phase I* of the DS ADC (as a track of a track and hold). The signals λ_1 and λ_2 represent the two phases of the DS ADC sampling period. In order to achieve high resolution with the lowest clock frequencies (to reduce power consumption), N and M are selected unequal. N is equal to two clock periods, which is equal to the stable time of V_{in}. M is equal to six.

Now V_{DS} (Fig. 11.3) is the output of INT_I (Fig. 11.1). During *Phase I*, V_{DS} is charged proportional to $(V_{in} \cdot T_I)/(R_{IN} \cdot C_{IN})$. Then, at the beginning of *Phase II* (signal λ_2 high), the input of the DS converter is connected to the feedback DAC (IDAC in Fig. 11.2), and V_{DS} is discharged with a constant slope proportional to IDAC$\cdot T_{II}/C_{IN}$. Due to the self-oscillation behavior, during *Phase II*, when the loop sees a change of polarity in the output of the comparator, the system starts to toggle until the end of the phase, keeping the quantization error for the next sampling period (λ_1 high). Signal Vcomp represents the output of the clocked comparator. This signal will be used to drive IDAC and to generate the multi-bit digital output of the CDC (D_{OUT} in Fig. 11.3). A simple counter (SUM in Fig. 11.3) can be used to generate this signal.

This block makes the logic addition of the output digital data only during *Phase II* (high level adds a 1 and low level subtracts a 1) every clock period Ts. D_{OUT} is then proportional to the input amplitude of the DS converter (V_{in}) and therefore to the input pressure of the CDC. In this system, the sampling period is defined as $Ts = T_I + T_{II}$, where $T_I = N \cdot \text{Tclk}$, $T_{II} = M \cdot T\text{clk}$, and Tclk is the clock period of the comparator.

11.4 Circuit Design

The analog fronted is based on a closed loop capacitive voltage amplifier (preAmp in Fig. 11.3) built with a telescopic gain-boosted OTA in order to be power efficient. This amplifier does not need any specific technique to reduce low frequency noise as it is already chopped by the modulated/demodulated scheme proposed in this architecture.

The power consumption of the DS ADC is given by the RC integrator (INT_I in Fig. 11.3). The OTA used in the integrator is a two-stage class A/AB pull-up-down topology, and it is Miller-compensated. A simplified schematic is shown in Fig. 11.8. This architecture allows a better power trade-off with respect to folded cascode and two-stage class A topologies. However, in the class A/AB topology, each gain stage requires a separate common mode feedback (CMFB) circuit. This is the consequence of using current mirrors in the second stage: the common mode output voltage of the first stage affects the bias condition of the second stage but does not affect the second stage output voltage. Moreover, to cope with the $1/f$ noise introduced by the OTA input differential pair, a chopping modulation technique has been adopted. The signal demodulation is applied on low impedance nodes (cascode nodes; see Fig. 11.8). This chopping configuration does not suffer from the opamp

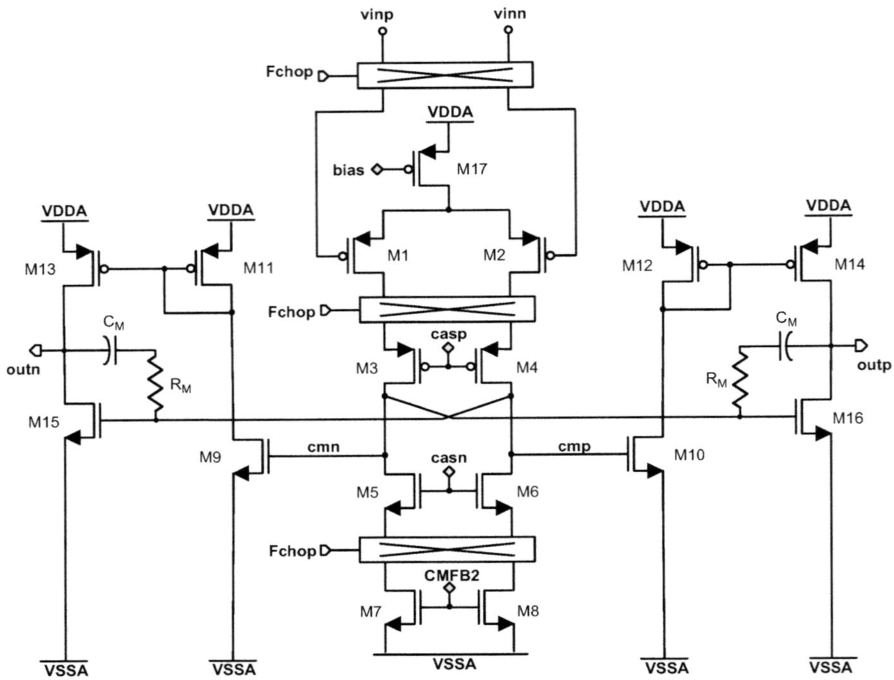

Fig. 11.8 Simplified schematic of OTA used in the dual-slope integrator $\mathrm{INT_I}$

limited bandwidth and from the distortion introduced by the chopping switches. This is because the demodulation of the chopping stage is performed before the dominant pole and not at the output of the OTA. Signal Vref in Fig. 11.3 is also used to make the chopping of this OTA, this way aligning it with the timing of the system. The gain bandwidth product of this OTA has been set to four times the clock frequency $G \times \mathrm{BW} = 4{\cdot}f\mathrm{clk}$ ($f\mathrm{clk} = 1/T\mathrm{clk}$) in order to deal with the DAC pulses.

The switches that control the different phases in the DS ADC do not need to be done by special circuit because their distortion is shaped by the noise shaping behavior of the converter. The current DAC uses non-return-to-zero topology. In order to deal with the $1/f$ noise at the output of the IDAC, the current mirrors that drive the current cells are designed with large size PMOS ($W = 9\ \mu\mathrm{m}/L = 94\ \mu\mathrm{m}$) and NMOS ($W = 9\ \mu\mathrm{m}/L = 150\ \mu\mathrm{m}$) transistors. A two-stage regenerative low-power clocked comparator is used for the single-bit conversion. Its output is a PWM waveform that can be directly connected to the $\mathrm{I_{DAC}}$ and to the counter SUM to generate D_{OUT}.

In order to compute the final CDC output measurement, the generated multi-bit output D_{OUT} must be interpreted appropriately. Since the signal information is contained in the DC component of the D_{OUT} data stream, an efficient digital averaging filter is applied. To improve performance further, this digital filter additionally implies a special windowing function. After scaling the averaged and filtered sum of D_{OUT}, the measurement of the CDC is ready to be processed further.

11.5 Measurements

The designed CDC was fabricated in a standard digital 0.13 μm CMOS technology. It was bonded together with a pressure sensor MEMS in the same carrier to minimize interconnection or parasitic capacitances between them. Figure 11.9 shows a die photograph of both the MEMS sensor and the CDC. The CDC core, with an area of 0.317 mm^2 and clocked at $f$clk = 1.28 MHz (equivalent to a sampling frequency of fs = 160 kHz), consumes 146 μA from a single 1.5 V power supply. This current includes the analog, digital, and excitation signal generator blocks. The bridge of the capacitive MEMS is excited with a signal of 80 kHz, which is also used to demodulate the output voltage of the preAmp and to chop the integrator INT_I. The spectrum of the digital output (D_{OUT}) of the CDC is shown in Fig. 11.10. It represents an input pressure of 1050 mbar, equivalent to a voltage of −16dBFS at the input of the DS converter, where the full scale is VFS = 1 V.

The measured equivalent integrated noise over a bandwidth of 10 Hz is 2.9 μVrms, leading to an ENOB of 17bits. Also, in Fig.11.10, the first-order noise shaping from the integrating DS converter as well as the DC tone representing the input pressure can be observed. In addition, the modulation between the DC signal and the clock is present. This behavior is well known in first-order noise shaping ADCs, but these tones at high frequencies do not corrupt the in-band noise floor. In order to measure the pressure resolution of the CDC in a more precise way, another experiment has been set. A special high precision pressure generator was used in the lab. The prototype (MEMS + CDC) was inserted in this special equipment where the air pressure can be controlled with a precision of 0.1 Pa.

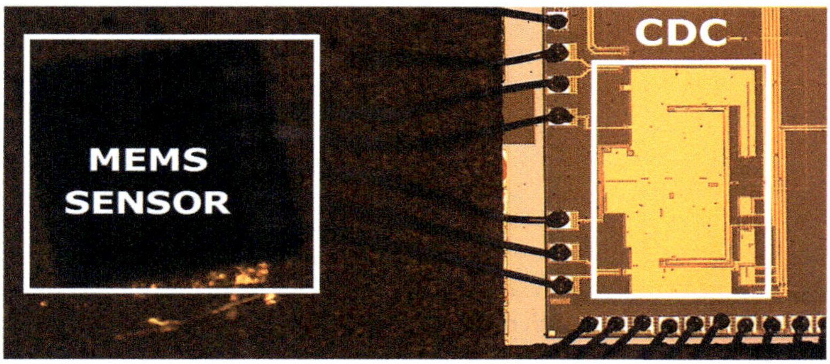

Fig. 11.9 Die photo of the MEMS sensor and the CDC in the same package

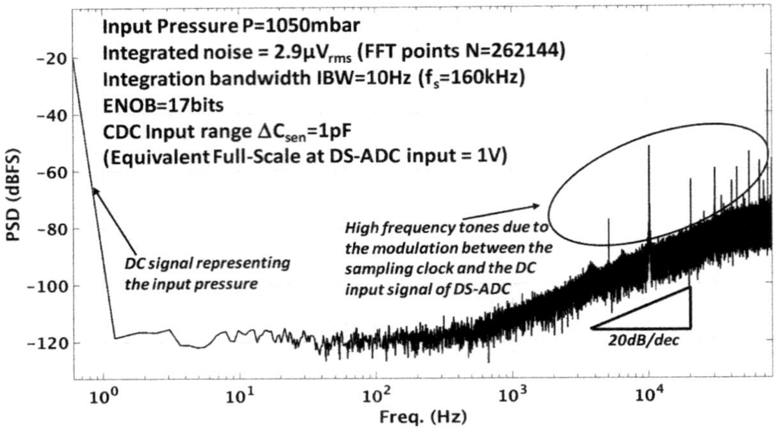

Fig. 11.10 FFT of the CDC D_{OUT} for an input pressure of 1050 mbar

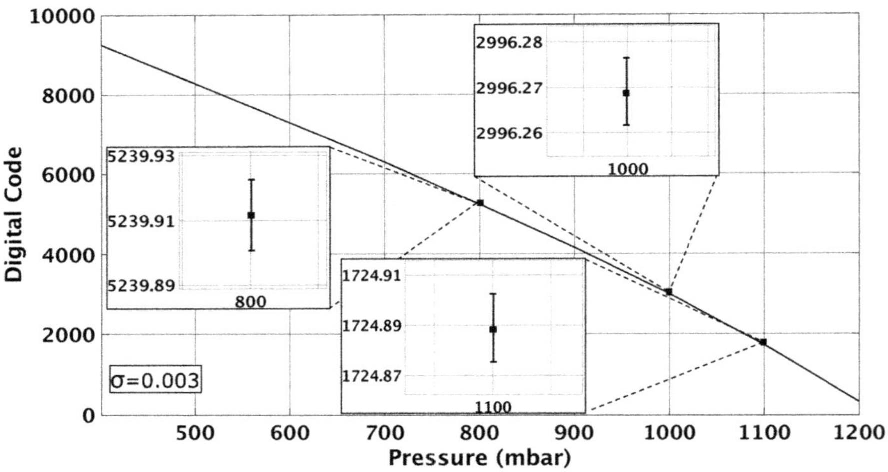

Fig. 11.11 Digital output code of the CDC vs input pressure

The output of the CDC D_{OUT} integrated and averaged over 20 ms is plotted versus the input pressure in Fig. 11.11. Three different values of pressure in the range of specifications were used to calculate the resolution of the CDC. For each of these three points, 100 different measurements were done. The standard deviation is calculated for every group of data related to a certain input pressure. The equivalent resolution of each measurement is equal to a digital standard deviation of $\sigma_D = 0.003_{\text{digital-code}}$ or equivalent voltage standard deviation of $\sigma_V = 3.12\ \mu V_{rms}$ (assuming a full scale at the input of the DS converter of VFS = 1 V). In order to calculate the CDC resolution using the standard deviation of each measurement, the following definition can be used:

$$\mathrm{ENOB} = \log_2\left(\frac{2\cdot \mathrm{VFS}}{\sqrt{2}\cdot \sigma_v}\right) \tag{11.12}$$

Using Eq. (11.12) the CDC achieves an ENOB of 17.5 bits. This result is in agreement with the spectral density shown in the FFT of Fig. 11.10. Using these results, the CDC is able to resolve difference in capacitance of $\Delta C\mathrm{sen} = 5.4\mathrm{aF}$ (for a maximum capacitive range in the sensor of $C\mathrm{sen} = 1$ pF), which means a pressure resolution equal to 0.8 Pa.

11.6 Comparison with State of the Art

The demand for low power and reduced area CDCs is significantly increasing in the last years due to the fast development of technology demanded in the Internet of Things (IoT). To optimize the trade-off between performance and power consumption, different approaches are being used: ΣΔ modulation, SAR, and incremental and DS converters are the main topologies. ΣΔ ADCs can achieve higher resolution than other topologies, but this is often at the expense of area and power. On the other hand, SAR converters are not able to achieve as much resolution as ΣΔ ADCs, but in terms of power and area, these converters are more efficient. On the other hand, Incremental converters are becoming popular for these applications. They are able to achieve moderate to high resolutions with some advantages: a simpler decimation filter, ease of multiplexing, low latency, and the absence of idle tones. But the circuit complexity of these converters is still high. A new family of converters (integrating DS), presented in this work, is able to achieve the same performance of an incremental converter (in the same range of power consumption) with a simpler configuration. As it was said before, this configuration includes an analog front end and uses a simple decimation filter. Also (due to the digital control) it is able to multiplex different inputs and change the resolution of the converter in an efficient way. In Fig. 11.12, a plot with the state-of-the-art converters and this work is presented using the data from Table 11.1.

$$\mathrm{SNRcap}(dB) = 20\log\cdot\left(\frac{\mathrm{In.Range}/2\sqrt{2}}{\mathrm{Re\,solution}}\right) \qquad \mathrm{FoM} = \frac{\mathrm{Power}\times \mathrm{Meas.Time}}{2^{(\mathrm{SNRcap}-1.76)/6.02}}$$

As shown in Table 11.1, the topology presented in this work has similar SNR and FoM as the state-of-the art CDCs for the same capacitance resolution. As mentioned before, and shown in Table 11.1, ΣΔ modulators are able to process the sensor information with higher resolution at the expense of higher power consumption. On the other hand, incremental and SAR converters are power efficient, but their resolution is moderated. Finally, classical DS topologies are not efficient enough for these applications. As a summary, it can be seen in Fig. 11.12 that there is a clear trade-off between measurement time, power, and resolution in all the converters.

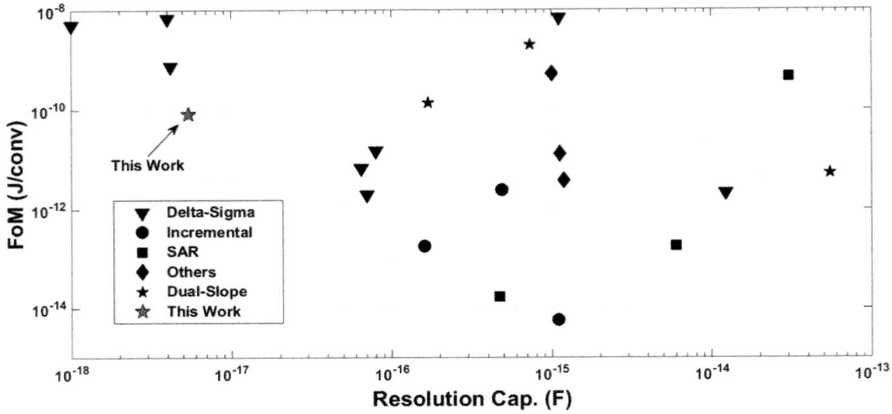

Fig. 11.12 State of the art. FoM vs resolution

Table 11.1 CDCs state of the art

Refs.	Type	Meas. time (ms)	Power (μW)	Sensor range (pF)	Resolution cap. (aF)	SNR_{cap} (dB)	FoM (pJ/conv)
[5]	ΔΣ	0.019	1.84	0.7	12,300	26.1	**2.13**
[6]	DS	7.6	211	6.8	170	83	**139**
[10]	DS	6.4	0.1	25.4	55,300	44.2	**5.31**
[2]	ΔΣ	0.020	15,000	10	65	94.71	**6.75**
[14]	ΔΣ	1090	3750	8	4.2	116.6	**742**
[15]	ΔΣ	100	60,000	4	1	123	**5190**
[16]	ΔΣ	13.3	6000	0.16	4	83	**6900**
[4]	ΔΣ	0.8	10	1	70	74	**2**
[17]	ΔΣ	10.2	10	2	80	78,9	**14.9**
[18]	ΔΣ	100	7	0.4	1110	42.1	**6730**
[19]	Inc.	0.23	33.7	24	160	94.5	**0.179**
[20]	Inc.	0.001	1440	1	490	57.2	**2.44**
[21]	Inc.	0.001	7.5	5	1100	64.1	**0.006**
[22]	SAR	4	0.16	72.8	60	72.6	**0.183**
[23]	SAR	0.005	6.7	3.2	470	67.6	**0.017**
[24]	SAR	100	0.8	18.5	30,400	46.7	**455**
[25]	Digital	1	0.27	0.3	1200	38.9	**3.74**
[26]	Current	0.004	725	0.75	1130	47.4	**13.2**
[27]	DS	0.02	15,800	0.4	733	45.7	**2010**
This work	Int DS	20	220	1	5.4	96.3	**82.2**

In this scenario, the proposed work offers a power-efficient CDC using a very simple and robust implementation. In addition, as it interchanges amplitude by time resolution, the proposed converter can have a more efficient trade-off in terms of circuit implementation for low-voltage technologies between measurement time, power, and resolution compared with other CDCs.

11.7 Conclusions

This paper shows experimental results of a high-resolution CDC based on a self-oscillating integrating DS converter that can work for different types of MEMS sensors. It achieves a capacitive resolution of 5.4aF that is equivalent to a resolution of 0.8 Pa. Compared with the state of the art, this CDC does not require multi-bit circuits to keep high resolution; instead, it uses time-domain circuitry to exchange amplitude by time resolution. In addition, the proposed ADC can have a more efficient trade-off between measurement time, power, and resolution compared with other CDCs. The main strengths of the proposed CDC are (1) intrinsically small sensitivity to temperature and process variations, (2) simplicity of trimming offset and gain to correct the sensor parameter spread, and (3) energy-efficient implementation. The CDC presented in this work seems a good candidate for sensor readout circuits in the Internet of Everything applications.

Acknowledgments This work has been funded by Marie Curie project SIMIC, Grant Agreement No. 610484, and funded by grants from the European Union (Research Executive Agency) and TEC2014-56879-R of CICYT, Spain.

References

1. Langfelder, G., Buffa, C., Frangi, A., Tocchio, A., Lasalandra, E., Longoni, A.: Z-Axis magnetometers for MEMS inertial measurement units using an industrial process. IEEE Trans. Ind. Electron. **60**(9), 3983–3990 (2013)
2. Xia, S., Makinwa, K., Nihtianov, S.: A Capacitance-to-Digital Converter for Displacement Sensing with 17b Resolution and 20μs Conversion Time. In: International Solid-State Circuits Conference Digest of Technical Papers, San Francisco, CA, pp. 198–200 (2012)
3. Oh, S., Jung, W., Yang, K., Blaauw, D., Sylvester, D.: 15.4b incremental sigma-delta capacitance-to-digital converter with zoom-in 9b asynchronous SAR. In: Symposium on VLSI Circuits Digest of Technical Papers, Honolulu, HI, pp. 1–2 (2014)
4. Tan, Z., Chae, Y., Daamen, R., Humbert, A., Ponomarev, Y.V., Pertijs, M.A.P.: A 1.2V 8.3nJ energy-efficient CMOS humidity sensor for RFID applications. In: Symposium on VLSI Circuits Digest of Technical Papers, Honolulu, HI, pp. 24–25 (2012)
5. Jung, W., Jeong, S., Oh, S., Sylvester, D., Blaauw, D.: 27.6 A 0.7pF-to-10nF fully digital capacitance-to-digital converter using iterative delay-chain discharge. In: International Solid-State Circuits Conference Digest of Technical Papers, San Francisco, CA, pp. 1–3 (2015)
6. Tan, Z., Shalmany, S.H., Meijer, G.C.M., Pertijs, M.A.P.: An energy-efficient 15-bit capacitive-sensor Interface based on period modulation. J. Solid-State Circuits. **47**(7), 1703–1711 (2012)
7. Maghari, N., Temes, G.C., Moon, U.: Noise-shaped integrating quantisers in $\Delta\Sigma$ modulators. Electron. Lett. **45**(12), 612–613 (2009)
8. Hernández, L., Prefasi, E.: Multistage ADC based on integrating quantiser and gated ring oscillator. Electron. Lett. **49**(8), 526–527 (2013)
9. Cannillo, F., Prefasi, E., Hernández, L., Pun, E., Yazicioglu, F., Van Hoof, C.: 1.4V 13μW 83dB DR CT-$\Sigma\Delta$ modulator with Dual-Slope quantizer and PWM DAC for biopotential signal acquisition. In: Proceedings of ESSCIRC, Helsinki, pp. 267–270 (2011)
10. Oh, S., et al.: A Dual-slope capacitance-to-digital converter integrated in an implantable pressure-sensing system. J. Solid-State Circuits. **50**(7), 1581–1591 (2015)

11. Senturia, S.D.: Microsystem Design. Springer, Dordrecht (2001)
12. Gabrielson, T.B.: Mechanical-thermal noise in micromachined acoustic and vibration sensors. IEEE Trans. Electron Devices. **40**(5), 903–909 (1993)
13. Enz, C.C., Temes, G.C.: Circuit techniques for reducing the effects of op-amp imperfections: autozeroing, correlated double sampling, and chopper stabilization. Proc. IEEE. **84**(11), 1584–1614 (1996)
14. Analog devices AD 7746
15. Gozzini, F., Ferrari, G., Sampietro, M.: An Instrument-on-Chip for Impedance Measurements on Nanobiosensors with attoFarad Resoution. In: International Solid-State Circuits Conference Digest of Technical Papers, San Francisco, CA, pp. 346–347 (2009)
16. Amini, B.V., Pourkamali, S., Ayazi, F.: A 2.5V 14-bit ΣΔ CMOS-SOI capacitive accelerometer. In: International Solid-State Circuits Conference Digest of Technical Papers, San Francisco, CA, pp. 314–323 (2004)
17. Tan, Z. et al.: A 1.8V 11μW CMOS smart humidity sensor for RFID sensing applications. In: Asian Solid-State Circuits Conference Digest of Technical Papers, Jeju, pp. 105–108 (2011)
18. Bracke, W., Merken, P., Puers, R., Van Hoof, C.: Ultra-low-power Interface Chip for autonomous capacitive sensor systems. IEEE Trans. Circuits Syst I: Reg. Papers. **54**(1), 130–140 (2007)
19. Oh, S., Jung, W., Yang, K., Blaauw, D., Sylvester, D.: 15.4b incremental sigma-delta capacitance-to-digital converter with zoom-in 9b asynchronous SAR. In: VLSI Circuits Digest of Technical Papers, Honolulu, HI, pp. 1–2 (2014)
20. Shin, D.Y., Lee, H., Kim, S.: A Delta–sigma Interface circuit for capacitive sensors with an automatically calibrated zero point. IEEE Trans. Circuits Syst II Express Brief. **58**(2), 90–94 (2011)
21. Sanyal, A., Sun, N.: A 55fJ/Conv-Step Hybrid SAR-VCO ΔΣ Capacitance-to-Digital Converter in 40nm CMOS. In: Proceedings of ESSCIRC, Lausanne, pp. 385–388 (2016).
22. Ha, H., Sylvester, D., Blaauw, D., Sim, J.Y.: 12.6 A 160nW 63.9fJ/conversion-step capacitance-to-digital converter for ultra-low-power wireless sensor nodes. In: International Solid-State Circuits Conference Digest of Technical Papers, San Francisco, CA, pp. 220–221 (2014)
23. Jiang, H., Wang, Z., Liu, L., Zhang, C., Wang, Z.: A combined low power sar capacitance-to-digital / analog-to-digital converter for multisensory system. In: Proceedings MWSCAS, Boise, ID, pp. 1000–1003 (2012)
24. Xiao, Y., Zhang, T., Mak, P.I., Law, M.K., Martins, R.P.: A 0.8 μW 8-bit 1.5∼20-pF-input-range capacitance-to-digital converter for lab-on-chip digital microfluidics systems. In: Proceedings of BioCAS, Hsinchu, pp. 384–387 (2012)
25. Danneels, H., Coddens, K., Gielen, G.: A fully-digital, 0.3V, 270 nW Capacitive Sensor Interface without External References. In: Proceedings of ESSCIRC, Helsinki, pp. 287–290 (2011)
26. Singh, T., Saether, T., Ytterdal, T.: Current-mode capacitive sensor Interface circuit with single-ended to differential output capability. IEEE Trans. Instrum. Meas. **58**(11), 3914–3920 (2009)
27. Bruschi, P., Nizza, N., Piotto, M.: A current-mode, dual slope, integrated capacitance-to-pulse duration converter. IEEE J. Solid State Circuits. **42**(9), 1884–1891 (2007)

Chapter 12
Ultra-low Power Charge-Pump-Based Bandgap References

Shikhar Tewari and Aatmesh Shrivastava

12.1 Introduction

An ideal voltage reference is independent of variation of power supply or temperature or manufacturing process. The term voltage reference means that output voltage coming out of the reference circuit is very precise and can be used to measure and compare other voltages in an integrated circuit (IC). Voltage reference circuit is needed in a voltage regulator circuit, in RF blocks, analog to digital converter (ADC), and many other circuits inside an IC. In a voltage regulator circuit, output V_{DD} is controlled through the voltage reference. A precise reference voltage gives a precise V_{DD}. In an ADC circuit, the analog voltage from external world is converted into digital bits. The external signal is compared with the reference voltage for conversion. The accuracy of the conversion depends on the accuracy of the reference voltage. A voltage reference circuit is an essential component in a system on chip (SoC) and greatly determines the performance of various blocks inside it. In this paper, we present a voltage reference circuit suitable for ultra-low power (ULP) applications.

A voltage reference circuit is also needed for energy harvesting and power management systems in an Internet of things (IoT) device for maintaining control over the power supply. The design of voltage reference is very critical for an IoT device, and it needs to meet several design requirements. Apart from the typical requirement that a voltage reference should meet which includes a very small variation with process, temperature, and voltage, additional requirements should also be met for ULP systems running on harvested energy [1]. One of the key requirements for reference voltage for IoT devices is the ultra-low power

S. Tewari • A. Shrivastava (✉)
Northeastern University, Boston, MA, USA
e-mail: tewari.s@husky.neu.edu; aatmesh@ece.neu.edu

P. Harpe et al. (eds.), *Hybrid ADCs, Smart Sensors for the IoT, and Sub-1V & Advanced Node Analog Circuit Design*, DOI 10.1007/978-3-319-61285-0_12

consumption in the nano-watt regime. The area of reference voltage also needs to be very small to keep the size of the IoT device small. Further, the reference voltage circuit should become operational at lower voltage. This is very critical for ULP IoT devices because of the following reasons. The voltage at which reference voltage becomes operational is typically the system start-up voltage. The start-up voltage determines the voltage at which all the power supplies are active. The value of start-up voltage determines the lifetime of an IoT device. A lower start-up voltage means a higher lifetime. In this paper, we present a charge-pump-based voltage reference circuit which has a start-up voltage below 400 mV and a power consumption below 20 nW in simulation. We measure the operation of the bandgap reference from 500 mV supply at a power consumption of 32 nW.

Temperature compensated zener diodes were used for generating reference voltage in the very beginning where the breakdown of the zener would determine the reference level. However, zeners did not meet many requirements of a typical reference as they were higher power and noisy apart from being a discrete component. The bandgap voltage reference circuit meets all of the above requirements and is primarily used as reference voltage in most of the ICs that are sold today. The bandgap reference circuit produces a temperature-independent reference voltage which is equal to the silicon bandgap. The temperature independence is achieved by adding a voltage that is proportional to absolute temperature (PTAT) with a voltage that is complimentary to absolute temperature (CTAT). This concept was first proposed by Widlar [2] providing a reference at 5 V. Other bandgap reference circuits utilized the similar concept to provide voltage reference ICs producing 10 V [3] and 2.5 V [4].

While the architecture of the bandgap reference has undergone changes, the concept however has remained the same. Various voltage reference circuits have been proposed as an alternative to bandgap reference. Some of the recent publications in this area are listed in bibliography [5–7]. These circuits are MOS transistor based and achieve good temperature stability. However, bandgap reference provides better performance in one or more critical parameters. The alternative circuits have found limited use in some application space. Further, the alternative circuits can provide lower-voltage operation at ULP levels. There is a need to achieve ultra-low power consumption with the bandgap reference. In this paper, we present a bandgap voltage reference circuit with a new architecture suitable for IoT devices. The proposed circuit is designed to meet the power, area, and voltage requirements of a ULP system.

12.1.1 Conventional Bandgap Reference

Figure 12.1 shows the concept of the operation of the bandgap reference circuit. The voltage V_{BE} which is obtained from a BJT in diode configuration is complementary to absolute temperature (CTAT) with a slope of -2.2 mV/°C. The thermal voltage, V_t, on the other hand is proportional to absolute temperature (PTAT) with a slope of

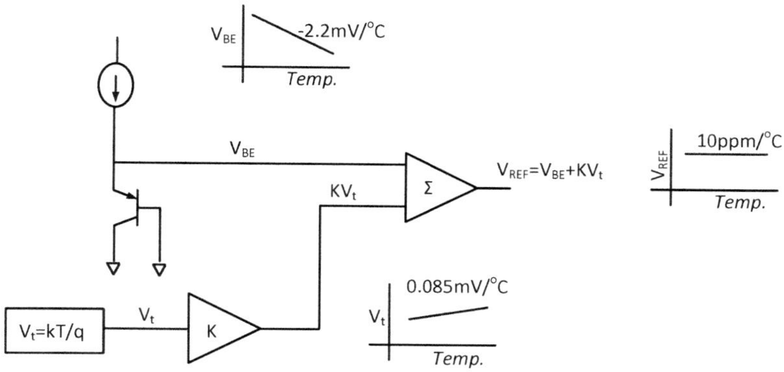

Fig. 12.1 Principle of bandgap voltage reference circuit

Fig. 12.2 PTAT and CTAT voltage generation circuit

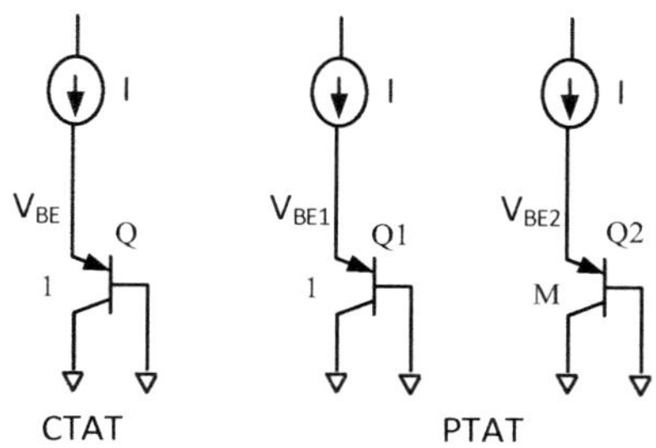

0.085 mV/°C. The PTAT voltage is scaled with a constant K and added to the CTAT voltage. The addition cancels the first-order temperature variation and achieves a constant voltage which is close to the silicon bandgap. The value of K is carefully chosen to cancel the temperature dependence of CTAT and PTAT, and the reference voltage V_{REF} becomes temperature-independent voltage reference. This circuit can achieve temperature stability in the order of 10 ppm/°C.

12.1.1.1 Generation of Bandgap Reference

Figure 12.2 shows the CTAT and PTAT voltage generation circuit used for bandgap. The bipolar junction transistors (BJTs) are used for the generation of PTAT and CTAT voltages. The CTAT voltage is simply generated by connecting the BJT transistor Q into a diode configuration. The CTAT voltage is given by the base-emitter voltage, V_{BE} of the transistor. As temperature increases, the voltage V_{BE} decreases because of the increase in the number of carriers. Since the number of carrier increases, the conductivity of the transistor increases which decreases the V_{BE} voltage. The expression of V_{BE} with temperature is given by [8, pp. 155],

$$V_{BE} = V_{G0}\left(1 - \frac{T}{T_0}\right) + V_{BE0}\left(\frac{T}{T_0}\right) + \frac{\gamma kT}{q} ln\left(\frac{T}{T0}\right) + \frac{kT}{q} ln\left(\frac{J_C}{J_{C0}}\right) \tag{12.1}$$

where V_{G0} is the bandgap voltage of silicon, T is the temperature, T_0 is the room temperature, V_{BE0} is the V_{BE} voltage at room temperature, J_C and J_{CO} are the current densities, and γ, k, q, and α are various physical constants. The equation shows that the value of V_{BE} will decrease with temperature. The difference in the voltage of two BJT transistors with different sizes but biased with the same current will provide the PTAT voltage as shown in Fig. 12.2. The transistor Q2 is bigger than Q1 by multiplicity factor M. The difference between V_{BE1} and V_{BE2} is called ΔVBE which is given by using Eq. (12.1),

$$\Delta V_{BE} = V_{BE1} - V_{BE2} = \frac{kT}{q} ln\left(\frac{J_{C1}}{J_{C2}}\right) = V_t ln\left(\frac{J_{C1}}{J_{C2}}\right) \tag{12.2}$$

The ΔVBE voltage increases linearly with the temperature making it a PTAT voltage. Thus, Eqs. (12.1) and (12.2) provide the expressions of CTAT and PTAT that are generated from the circuit shown in Fig. 12.2. The V_{REF} voltage which will be independent of temperature can be obtained by adding CTAT and PTAT voltages,

$$V_{REF} = V_{BE1} + K\Delta V_{BE} \tag{12.3}$$

The value of K is chosen in such a way that it cancels the temperature variation. The optimal value of K is given by

$$K = \frac{V_{G0} - V_{BE0} - (\gamma - \alpha) V_t}{V_{t0}} \tag{12.4}$$

The value of V_{REF} is determined as,

$$V_{REF} = V_{G0} + (\gamma - \alpha) V_{t0} \tag{12.5}$$

The reference voltage V_{REF} comes out very close to the silicon bandgap and hence is known as bandgap reference. The circuit implementation of bandgap reference is conventionally realized using BJTs and resistors in a feedback network using opamp [4]. The ratio, K, is typically obtained using resistors, although capacitors are also used to realize the ratio [9].

The bandgap reference circuit in [4] utilizing resistors and opamp and providing voltage given by Eq. (12.5) provides an almost ideal on-chip voltage reference circuit which performs very well with the variation of voltage, temperature, and process. However, the circuit has couple of limitations when used in ULP space. The output voltage of V_{REF} comes out to be close to 1.2 V being close to silicon bandgap, which requires the power supply to be higher than 1.2 V. The circuit cannot work

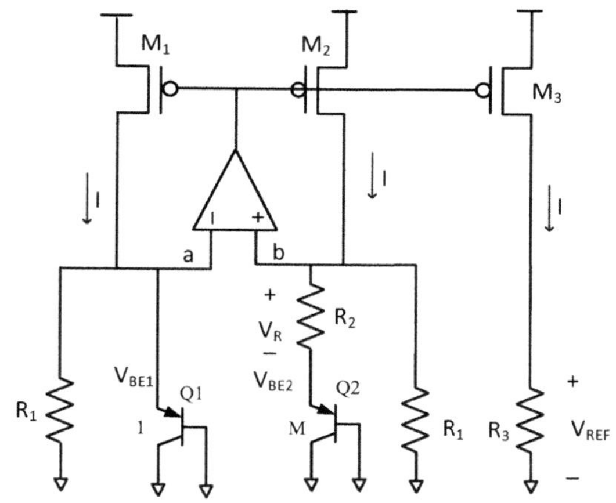

Fig. 12.3 Architecture of the low-voltage bandgap reference [10]

below a V_{DD} of less than 1.2 V. Further, the circuit provides a reference voltage of 1.2 V, which is higher than the power supply level for modern V_{DD}. Some of the recent architectures which can overcome these limitations were presented in [9–11]. In the next section, we will present circuit architectures which are used to provide reference voltage at lower V_{DD}.

12.1.2 Low-Voltage Bandgap Reference

A bandgap reference voltage architecture that can operate from lower power supply and provide lower reference level was proposed in [10] overcoming the limitation of a conventional bandgap. Figure 12.3 shows the architecture of the low-voltage bandgap reference. It works in the following manner. Nets *a* and *b* are set at the same voltage by the amplifier:

$$V_a = V_b = V_{BE} \tag{12.6}$$

The current I through the node *b* is given by

$$I = \frac{V_b}{R_1} + \frac{V_b - V_{BE2}}{R_2} \tag{12.7}$$

Using Eq. (12.6), *I* will be given as

$$I = \frac{V_{BE1}}{R_1} + \frac{V_{BE1} - V_{BE2}}{R_2} = \frac{V_{BE1}}{R_1} + \frac{\Delta V_{BE}}{R_2} \tag{12.8}$$

Now, transistors M_1, M_2, and M_3 form current mirrors. Their output resistance needs to be very high for the bandgap reference to work. V_{REF} is given by

$$V_{\mathrm{REF}} = \frac{R_3}{R_1}\left(V_{BE1} + \frac{R_1}{R_2}\Delta V_{BE}\right) \tag{12.9}$$

Equation (12.9) gives us an expression of bandgap reference which is similar to the expression of bandgap reference in Eq. (12.5) with a scaling factor given by R_3/R_1 and K given by R_1/R_2.

The minimum operating voltage, V_{MIN}, for this circuit is given by

$$V_{\mathrm{MIN}} = V_{BE1} + V_{DS} \tag{12.10}$$

While the minimum operating voltage using this circuit can be brought down to 750 mV, even lower V_{MIN} is needed for ultra-low power application. Further, a lower area solution is desired which is not possible with this architecture while achieving lower power consumption. In the next section, we present a bandgap voltage reference circuit that starts operating at 400 mV. Further, the power consumption of the proposed circuit is 20 nW in simulation. The proposed circuit is also smaller in area.

12.2 Charge-Pump-Based Bandgap Reference

One of the key limitations of bandgap reference circuit reported in [10] and other bandgap reference circuits is that they cannot operate below 750 mV because the BJT diode needs to be biased with a current source. A charge-pump-based bandgap reference [12] overcomes this limitation. The biasing of a BJT diode in charge-pump-based bandgap reference is achieved through the charge-pump circuit. The use of charge pump provides following advantages to the bandgap reference. It enables operation from lower input voltage which enables ultra-low power consumption. Further, the use of switched capacitor network also enables a lower power solution at a lower area cost.

Figure 12.4 shows the use of voltage doubler charge-pump circuit for our bandgap circuit. The output of the charge pump is connected to the transistor Q1.

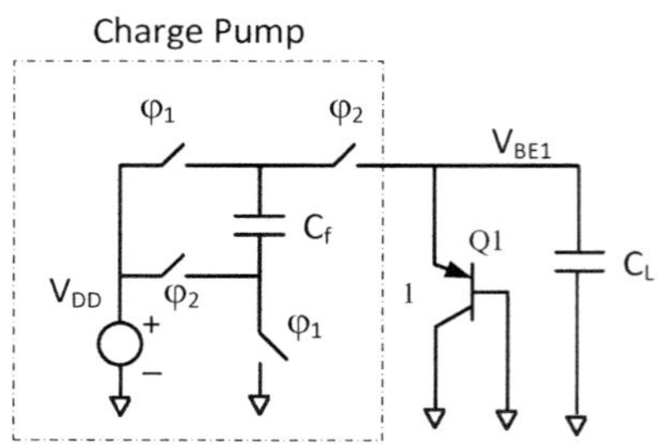

Fig. 12.4 A charge-pump circuit biasing a BJT

In the absence of Q1, the output will go to 2V_{DD}. However, connecting Q1 at the output will restrict the voltage to V_{BE}. The transistor Q1 is connected in a voltage clamp configuration. It sinks the additional current from the voltage doubler circuit and restricts the output at V_{BE}. A key advantage of this circuit is that the voltage V_{DD} needed for generating V_{BE} is smaller than V_{BE}. The minimum voltage for the bandgap to be operational is given by

$$V_{\text{MIN}} > \frac{V_{BE}}{2} \tag{12.11}$$

It is easy to see that if voltage trippler or higher-order charge pumps are used, then even lower V_{MIN} can be achieved using this technique. Figure 12.4 shows the circuit used to generate V_{BE1}. For generating V_{BE2}, same circuit is used with transistor Q2 which is M times bigger than Q1. Figure 12.5 shows the simulation result of V_{BE} and ΔV_{BE} generated from the proposed circuit. Figure 12.5 shows the CTAT behavior of V_{BE1} and V_{BE2} and the PTAT behavior of ΔV_{BE}. These voltages are generated using V_{DD} of 0.4 V. The weights of voltages of V_{BE1} and ΔV_{BE} are added to generate the bandgap voltage. We propose the generation of bandgap voltage in the form of

$$V_{\text{REF}} = a\left(V_{BE1} + b\Delta V_{BE}\right) \tag{12.12}$$

where constants a and b are needed to generate the weights for V_{BE} and ΔV_{BE} to generate V_{REF}. These constants are obtained by employing switched capacitor circuit techniques as opposed to the use of resistor in [10]. The use of resistor increases the area of the circuit for low power systems. The power consumption of the bandgap depends on the value of the resistors used in the design. The power consumption can be brought down by using large resistors in this design. For a 200 nW bandgap reference circuit, 14 MΩ resistors are needed which can be very large in area. The use of switched cap circuit on the other hand can generate these constants with lower area. One disadvantage of the switched capacitor scheme is that it increases the noise on the reference voltage which can be reduced by using bigger capacitors.

12.2.1 Theoretical Background

While the simulation results in Fig. 12.5 show PTAT and CTAT characteristics for ΔV_{BE} and V_{BE}, we also study the underlying theory of the proposed circuit. Figure 12.4 shows the charge-pump biasing circuit. The transistor Q1 maintains output voltage V_{BE} with a bias current which can be obtained through the charge transfer happening during switching cycle. The charge across the plates of capacitor C_f 2x-charge-pump circuit in phase ϕ_1 is given by

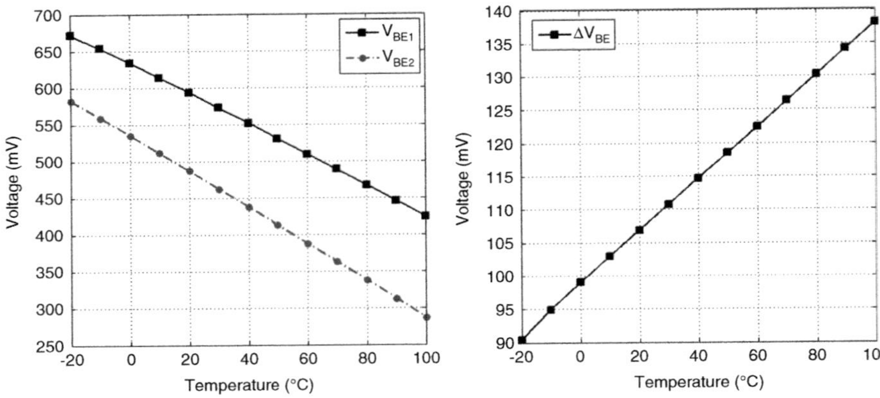

Fig. 12.5 Simulation result of V_{BE} and ΔV_{BE} generated from the proposed circuit

$$Q_{\phi_1} = C_f\,(2V_{DD} - V_{DD}) \tag{12.13}$$

However, the voltage gets clamped to V_{EB} by the diode. The charge that remains across the flying capacitor C_f is

$$Q_{\phi_2} = C_f\,(V_{EB} - V_{DD}) \tag{12.14}$$

Therefore, the current that biases the BJT is given by

$$\frac{\Delta Q}{\Delta T} = \frac{Q_{\phi 1} - Q_{\phi 2}}{\Delta T} => I_C = C_f f\,(2V_{DD} - V_{EB}) \tag{12.15}$$

where f $(\frac{1}{\Delta T})$ is the clock frequency.

Further, the current flowing through the BJT transistor is defined as,

$$I = I_S \exp^{\frac{V_{EB}}{V_t}} \tag{12.16}$$

where V_t is thermal voltage and I_S is the saturation current whose temperature dependency is given as

$$I_S = CT^n \exp^{\frac{-V_{G0}}{V_t}} \tag{12.17}$$

where C is a constant, n is the temperature dependency order, and V_{G0} is the silicon bandgap voltage. Comparing currents in Eqs. (12.17) and (12.16), the V_{EB} can be written as

$$V_{EB} = V_{EB0}\left(\frac{T}{T_0}\right) + V_{G0}\left(1 - \frac{T}{T_0}\right) - nV_t \ln\left(\frac{T}{T_0}\right) + V_t \ln\left(\frac{2V_{DD} - V_{EB}}{2V_{DD} - V_{EB0}}\right) \tag{12.18}$$

where V_{EB0} is emitter-base voltage at the normalized temperature of room temperature T_0. Equation (12.18) shows the characteristics of the V_{EB} voltage with temperature. This expression is very similar to the equation of V_{EB} given by the Eq. (12.1). The current density is determined by the power supply in this case. The expression in Eq. (12.18) shows that the characteristics of the charge-pump-based V_{EB} is indeed CTAT and can be used to generate the reference voltage.

12.2.2 Reference Generation Circuit

Figure 12.6 shows the circuit diagram of a charge-pump-based bandgap reference circuit [12]. It uses switched capacitor techniques to generate the bandgap voltage. The circuit consists of charge pumps, BJT transistors Q1 and Q2, capacitors C_f and C_L for charge pumps, and a summing circuit. The charge-pump circuits are used to generate voltages V_{BE} and ΔV_{BE} as shown in Fig. 12.6. The summing circuit generates the constants *a* and *b* for V_{BE} and ΔV_{BE} which are then added to generate the bandgap voltage. In Fig. 12.6, the outputs of the charge pumps are connected to Q1 and Q2, which clamps the outputs to V_{BE1} and V_{BE2}. The use of charge-pump circuit enables the low-voltage operation of the bandgap circuit. Further, the clock used for the charge pump can be made to operate at lower frequency to reduce the power. The lower V_{DD} and lower clock frequency for charge pumps enable lower power consumption compared to the prevalent bandgap circuits.

12.2.2.1 ΔV_{BE} Constant Generation Circuit

Figure 12.7 shows the circuit for generating constants for ΔV_{BE}. Figure 12.7(i) shows the circuit that generates ΔV_{BE}. The capacitor C_b is connected between V_{BE1} and V_{BE2} generated from the charge-pump circuit shown in Fig. 12.7. Therefore, the voltage across C_b is ΔV_{BE}. For generating different bandgap reference voltages, ΔV_{BE} needs to be multiplied by different constants. In this circuit, we present ways to generate three constants for ΔV_{BE}, namely, one, two, and three. Figure 12.7 shows the circuits that can generate $2\,\Delta V_{BE}$ and $3\,\Delta V_{BE}$. Figure 12.7(ii) shows the circuit to generate $2\,\Delta V_{BE}$. It uses the two nonoverlapping phases of clock ϕ_1 and ϕ_2. In phase ϕ_2, the voltages V_{BE1} and V_{BE2} are connected across the capacitors C_{b1} and C_{b2}. In phase ϕ_1, the connections for the capacitors are rearranged, and the top plate of C_{b1} is connected to the bottom plate of C_{b2} as shown in Fig. 12.7(ii). So, the voltage appearing at the top plate of C_{b2} is $2\,\Delta V_{BE}$. This essentially is the voltage doubler scheme. Similarly, a voltage trippler scheme is presented in Fig. 12.7(iii) to generate $3\,\Delta V_{BE}$. In design we can choose from ΔV_{BE}, $2\,\Delta V_{BE}$, and $3\,\Delta V_{BE}$, which can be used to generate different reference voltages.

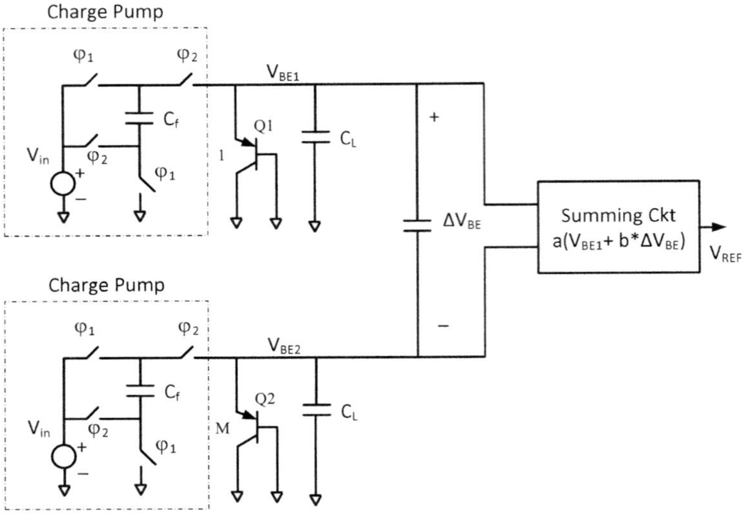

Fig. 12.6 Architecture of the charge-pump bandgap reference

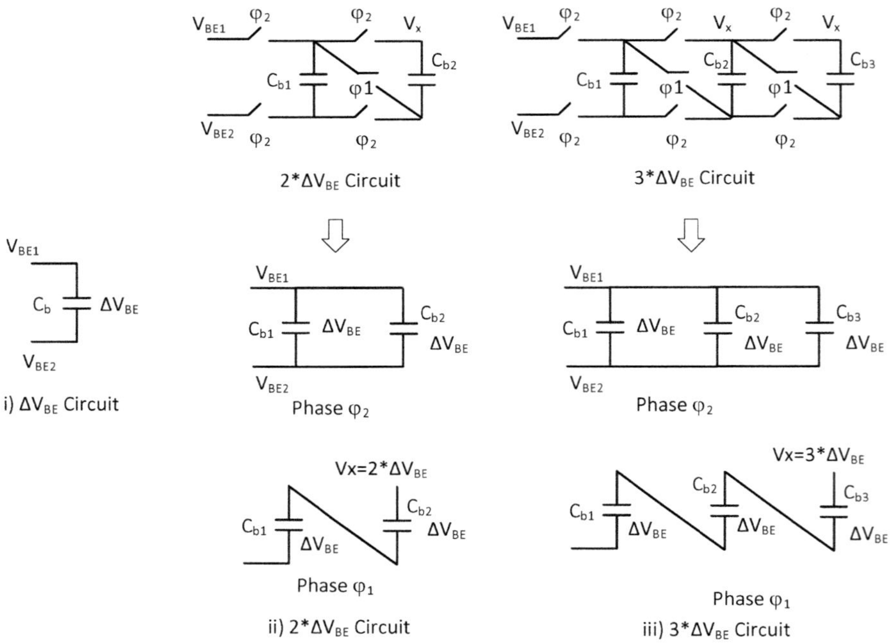

Fig. 12.7 ΔV_{BE} constant generation circuit

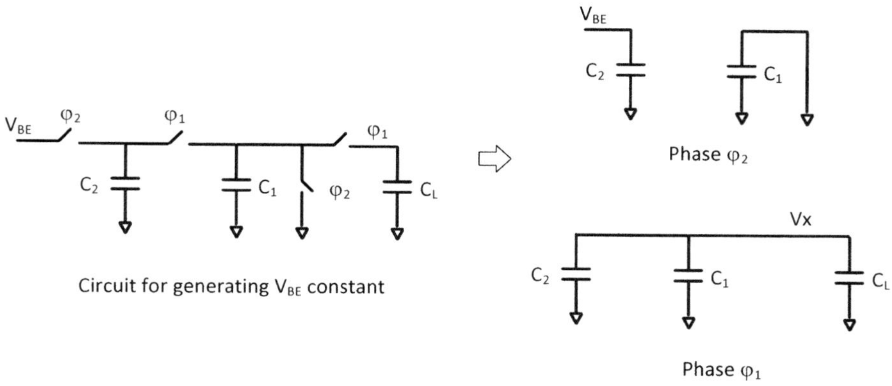

Fig. 12.8 V_{BE} constant generation circuit

12.2.2.2 V_{BE} Constant Generation Circuit

Figure 12.8 shows the circuit that is used to generate the rational number constant for V_{BE}. It also uses switched capacitor circuit with nonoverlapping clock phases ϕ_1 and ϕ_2. In phase ϕ_1, capacitor C_2 is connected to V_{BE}, while C_1 is connected to ground. In phase ϕ_1, the capacitors C_1 and C_2 are connected together as shown in Fig. 12.8. The total charge on the capacitors remains the same. Therefore, V_X is given by

$$V_X = V_{BE}\frac{C_2}{C_1 + C_2} \tag{12.19}$$

By selecting C_2 and C_1, a value of V_X is obtained which is a fraction of V_{BE} as given by Eq. (12.12). We have generated constants for both V_{BE} and ΔV_{BE}. Now we need to add them together to generate the bandgap voltage. Next section shows the summing circuit.

12.2.2.3 Summing Circuit

Figure 12.9 shows the summing circuit of the bandgap. It consists of circuits used for generating constants for V_{BE} and ΔV_{BE} and uses switched capacitor scheme to generate the sum. It also uses the nonoverlapping phases of the clock ϕ_1 and ϕ_2. Figure 12.9a shows the summing circuit with all the signals. In phase ϕ_2, the switches connected with signal ϕ_2 are closed, and the circuit is configured as shown in Fig. 12.9b. The capacitor C_{a1} is discharged to ground, while the top plate of C_{a2}, C_{b1}, C_{b2}, and C_{b2} is connected to V_{BE1}. The bottom plate of C_{a2} is connected to ground, while the bottom plate of C_{b1}, C_{b2}, and C_{b3} is connected to V_{BE2}. So the

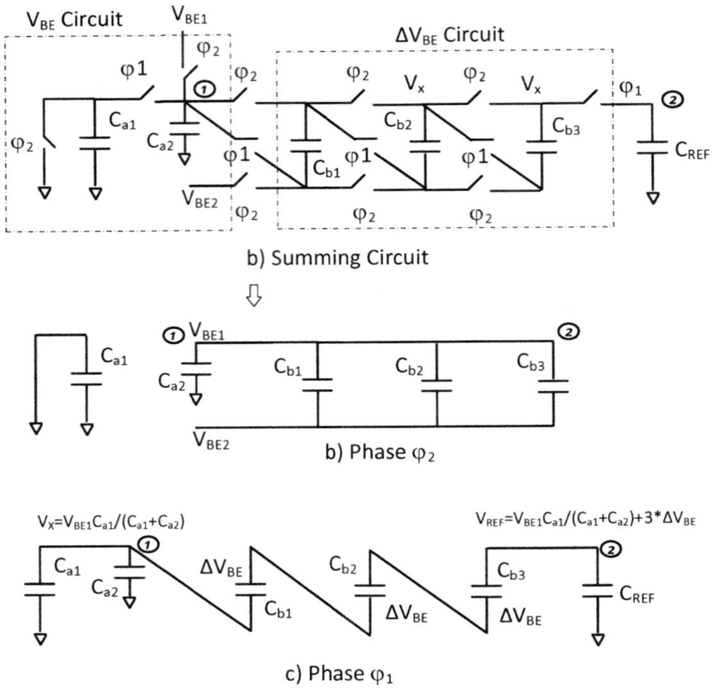

Fig. 12.9 Summing circuit to generate the reference voltage

voltage across C_{a2} is V_{BE1}, while the voltage across C_{b1}, C_{b2}, and C_{b3} is ΔV_{BE}. In phase ϕ_1, the switches are reconfigured and the circuit is arranged as shown in Fig. 12.9c. First, the capacitors C_{a1} and C_{a2} are connected and charge share to generate V_{BE} component of the bandgap. The reference voltage generated is given by

$$V_{\text{REF}} = V_{BE1} \frac{C_{a2}}{C_{a1} + C_{a2}} + 3\Delta V_{BE} \tag{12.20}$$

12.2.2.4 Top-Level Implementation

The bandgap reference uses switched capacitor circuit which uses two nonoverlapping phases of a clock. Therefore, a clock source is needed for the working of this circuit. Therefore, the power consumption of the clock source needed for the bandgap can be minimized by operating it at a very low frequency. The frequency of the clock source should be enough to maintain the bias voltage of transistors

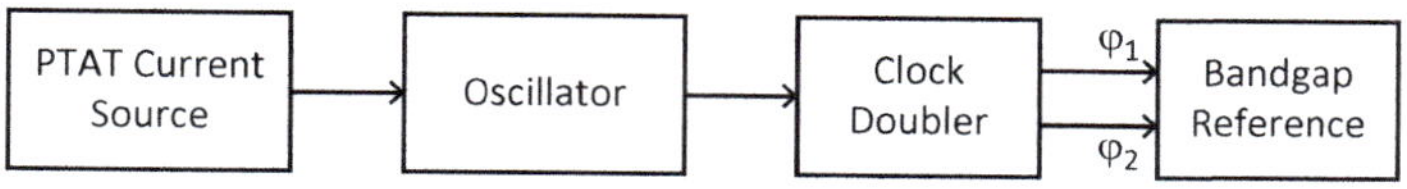

Fig. 12.10 Top-level implementation of bandgap reference

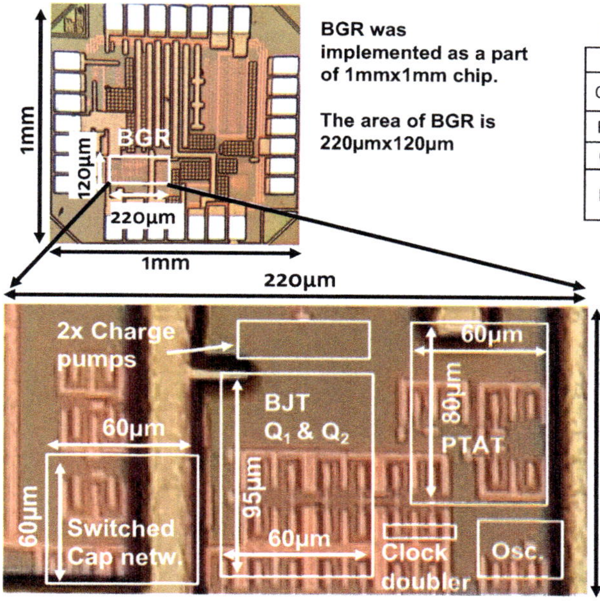

Power Breakdown at 0.5V

Block	Current (Power)
Current Source	17.4nA(8.7nW)
Ring Oscillator	3.6nA(1.8nW)
Clock Doubler	3.8nA(1.9nW)
Bandgap Core	39.4nA(19.7nW)

Area Breakdown

Block	Area(μm^2)
Current Source	80x60
BJT Q_1 and Q_2	95x90
SCN	60x60

Fig. 12.11 Die photo of the bandgap reference [12]

Q1 and Q2. A low-frequency, low-power clock source is needed for the bandgap circuit. Further, the switches used for bandgap need to pass V_{BE}, which is a voltage higher than V_{DD}. Therefore, the clock phases ϕ_1 and ϕ_2 need to swing from 0 to $2V_{DD}$. Figure 12.10 shows the block diagram for the clock generation scheme of the bandgap circuit. A PTAT current source is used for current controlled ring oscillator. The ring oscillator produces a clock of 30 kHz at 0.4 V V_{DD} and consumes 2 nW of power. Further, a clock doubler circuit is used to double the swing of output clock to enable the switches which can pass V_{BE} voltage level. The details of the circuit are covered in [12].

12.3 Implementation of the Charge-Pump Circuit

The proposed bandgap circuit was implemented in 130 nm CMOS process. Figure 12.11 shows the annotated chip. It has an area of 0.0264 mm^2. The capacitors are implemented using nMOS capacitors and MIM capacitors. The load capacitors for the V_{BE} generation circuit and the fraction generation switched capacitor circuit were implemented using nMOS capacitors, whereas the load capacitors for the bandgap output and the ΔV_{BE} trippler circuit were implemented using MIM capacitors to avoid bottom plate parasitic. The total area of the proposed circuit is much smaller than conventional low power bandgap circuit because it does not use big resistors. It consumes 19.2 nW power at 0.4 V V_{DD} in simulation. We were able to measure the performance of the circuit only after 0.5 V V_{DD} as the measurement below 0.5V was not achieved due to limitation of the measurement circuit.

The circuit was verified for temperature range of 0–80 °C. While this range is quite large for the intended ULP applications, the performance of the circuit in this range is crucial for it to compare with the state-of-the-art circuits. Figure 12.12 shows the variation of the bandgap output for a temperature change of 0–80 °C. The proposed bandgap circuit is designed to provide an output voltage of 500 mV and the output voltage achieving a performance of 75 ppm/°C. The performance of the bandgap circuit with temperature is in line with the reported work. A better performance can be achieved at higher output power.

Figure 12.13 shows the output of bandgap circuit with respect to process and mismatch variation and with input voltage variation. The circuit achieves a 3σ variation 2%. The variation of the output voltage can be reduced by trimming the

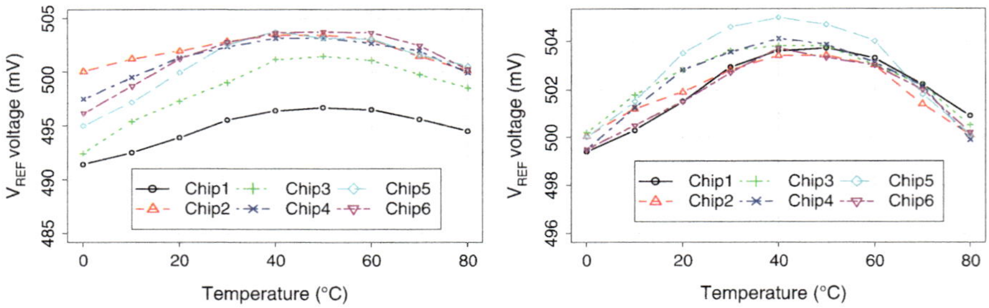

Measured (untrimmed) BGR output voltage across 6 chips from 0 to 80°C at 0.5V V_{in} varies from 492mV to 504mV

Measured (trimmed) BGR output voltage across 6 chips from 0 to 80°C at 0.5V V_{in} with best stability of 75ppm/°C

Fig. 12.12 Measurement of the bandgap reference with temperature variation [12]

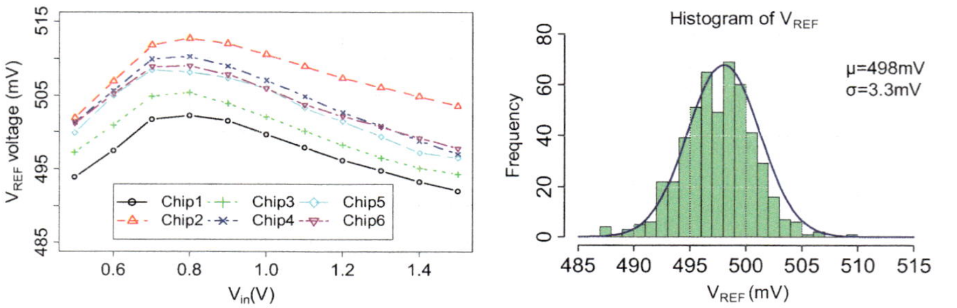

Measured (untrimmed) BGR output voltage across 6 chips from 0.5 to 1.5V V_{DD} at 20°C shows variation of 2%/V

Simulation result of 500 point Monte-Carlo at 0.5V V_{DD} and 20°C shows a 3σ process variation of 2% (untrimmed)

Fig. 12.13 Power supply and process variation of V_{REF} voltage [12]

bandgap output using the capacitors used in switched capacitor circuits to generate the constants for bandgap. Figure 12.13 also shows the variation of bandgap with V_{DD}. The output varies by 1% for 1 V variation of V_{DD}. The power supply variation of the bandgap reference is quite high which we address in the next version of the design.

12.4 Design Improvement in Charge-Pump Bandgap Reference

The charge-pump-based bandgap reference explained in previous sections provides high stability, low power, and lower area solution. However, the circuit suffers from high variation with power supply variation. The output of the bandgap reference varies by 5 mV for a variation of 1 V in power supply. The high variation of the bandgap reference circuit manifests because of the large variation of the power supply which causes a large variation in the bias current for BJT transistors. Figure 12.14 presents the circuit that overcomes the power supply variation by utilizing a current source-based design. The circuit takes advantage of the PSRR of the current source. As V_{DD} changes, the current through the transistor M_{PC} will change only slightly. As a result virtual, power supply, V_{DDV}, feeding the bandgap circuit will have a relatively fixed voltage. The voltage is decoupled through the capacitor C_V, while the diode M_{N0} further clamps the voltage. We use this voltage to power the BGR circuit which provides overall high power supply rejection. Figure 12.15 shows the simulation result with the power supply variation of the reference voltage. We see a 1.8 mV variation in the reference voltage for a V_{DD} variation from 0.5 to 0.6 V. This variation is caused by the transistor M_{PC} not completely operating in saturation region. The variation in the V_{REF} from 0.6 to 1.5 V is 1 mV. The DC

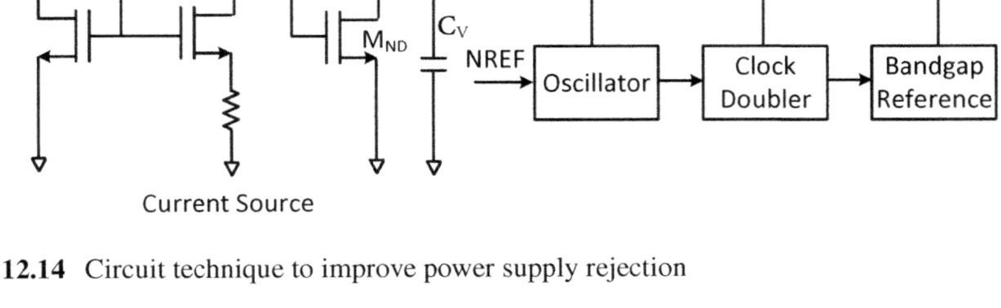

Fig. 12.14 Circuit technique to improve power supply rejection

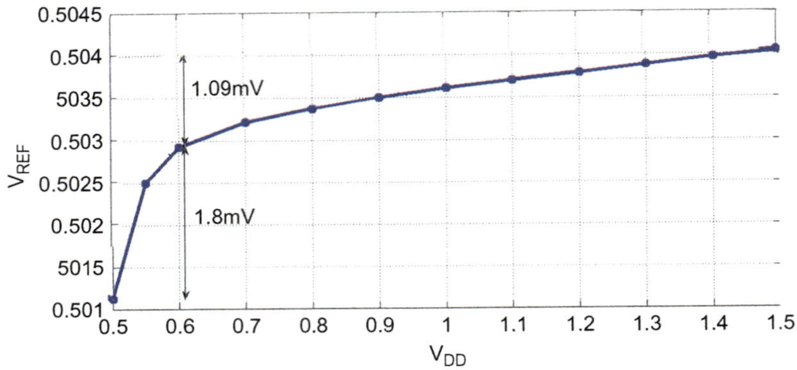

Fig. 12.15 Simulation result of the reference voltage with the variation of V_{DD}

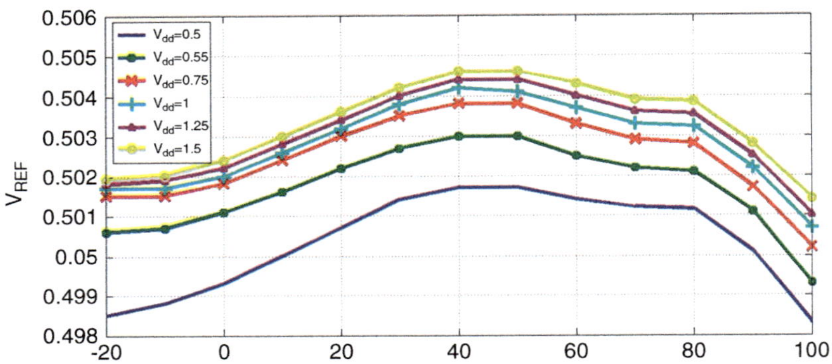

Fig. 12.16 Overall variation of the bandgap reference circuit temperature and power supply

PSRR of this circuit is -60 dB improving significantly over previously reported work [12]. A high-voltage implementation of the circuit was presented in [13]. The reference voltage varies by 1 mV for a V_{DD} variation from 600 mV to 5.5 V achieving a DC PSRR of -70 dB. Note that the PSRR of the bandgap circuit is worst at DC because of the low-pass response of the switched capacitor network.

The overall performance of the charge-pump-based bandgap reference with improved power supply rejection circuit is presented in Fig. 12.16. The circuit also utilizes two minor improvements. First, the V_{BE} scaling circuit uses MIM capacitor instead of nMOS capacitor which improves the temperature stability. Secondly, the scaling capacitor (C_{a1} in Fig. 12.9) in the V_{BE} circuit is not discharged to ground, instead both the top and the bottom plates are connected back to V_{BE} which helps in reducing the power consumption. The PSRR improvement circuit incurs additional power consumption and the circuit now consumes 48 nW at 0.5 V. Table 12.1 presents the comparison of the charge-pump-based reference circuit with other reference voltages in the literature. The charge-pump-based reference

Table 12.1 Comparison of the bandgap circuit with the state of the art

	[14]	[15]	[10]	[11]	[9]	[12]	This work
Power consumption	300 nW	52.5 nW	1.85 μW	20 μ W	170 nW	32 nW	48 nW
Area (mm^2)	0.055	0.0246	0.1	0.4	0.07	0.0264	NA
3σ variation (%)	21	4.8	5.8	1.5	3	2	2
Temp variation (ppm/°C)	7	114	119	11	40	75	45
PSRR(dB)	−45 @ 100Hz	−56 @100Hz	−57 @DC	−86@DC	− 93 @DC	−42@DC	− 60@DC
Min voltage (V)	1.4	0.7	0.84	1.1	0.75	0.5	0.5
Type	CS	CS	CS	CS	CS	CP	CP
Process	350 nm	180 nm	400 nm	500 nm	130 nm	130 nm	130 nm

voltage achieves lower power consumption, lower area, and lower voltage operation compared to the state-of-the-art bandgap references that are current source (CS) based.

12.5 Conclusions

Table 12.1 compares this work with previous reported state-of-the art low-power bandgap circuits. Our work reports operation from minimum input voltage of 0.5 V improving over 1.5 × from previous reported least operating voltage circuit in [9]. Our power consumption is 32 and 48 nW which is one of the lowest power consumptions for bandgap reference without duty cycling. The work in [9] achieves a power of 170 nW by sampling the reference voltage on a capacitor by periodically turning it on and off. Even lower power can be achieved in our work if duty cycling is employed. The power supply variation is higher in the charge-pump circuit of [12] which is improved using a new PSRR improvement circuit presented in this paper. The lower power consumption, lower voltage operation, and lower area of the charge-pump-based bandgap reference makes it ideally suited for IoT devices.

References

1. Roy, A., Klinefelter, A., Yahya, F.B., Chen, X., Gonzalez-Guerrero, L.P., Lukas, C.J., Kamakshi, D.A., Boley, J., Craig, K., Faisal, M., Oh, S., Roberts, N.E., Shakhsheer, Y., Shrivastava, A., Vasudevan, D.P., Wentzloff, D.D., Calhoun, B.H.: A 6.45 μw self-powered soc with integrated energy-harvesting power management and ulp asymmetric radios for portable biomedical systems. IEEE Trans. Biomed. Circuits Syst. **9**(6), 862–874 (2015)

2. Widlar, R.: New developments in ic voltage regulators. In: 1970 IEEE International Solid-State Circuits Conference. Digest of Technical Papers, vol. XIII, pp. 158–159, Feb 1970
3. Kuijk, K.E.: A precision reference voltage source. IEEE J. Solid-State Circuits **8**(3), 222–226 (1973)
4. Brokaw, A.: A simple three-terminal ic bandgap reference. In: 1974 IEEE International Solid-State Circuits Conference. Digest of Technical Papers, vol. XVII, pp. 188–189, Feb 1974
5. Kinget, P., Vezyrtzis, C., Chiang, E., Hung, B., Li, T.L.: Voltage references for ultra-low supply voltages. In: 2008 IEEE Custom Integrated Circuits Conference, pp. 715–720, Sept 2008
7. Leung, K.N., Mok, P.K.T.: A CMOS voltage reference based on weighted delta;VGS for CMOS low-dropout linear regulators. IEEE J. Solid-State Circuits **38**(1), 146–150 (2003)
8. Allen, P.E., Holberg, D.R.: CMOS Analog Circuit Design. Oxford University Press, New York (2002)
9. Ivanov, V., Brederlow, R., Gerber, J.: An ultra low power bandgap operational at supply from 0.75 v. IEEE J. Solid-State Circuits **47**(7), 1515–1523 (2012)
10. Banba, H., Shiga, H., Umezawa, A., Miyaba, T., Tanzawa, T., Atsumi, S., Sakui, K.: A CMOS bandgap reference circuit with sub-1-v operation. IEEE J. Solid-State Circuits **34**(5), 670–674 (1999)
11. Sanborn, K., Ma, D., Ivanov, V.: A sub-1-v low-noise bandgap voltage reference. IEEE J. Solid-State Circuits **42**(11), 2466–2481 (2007)
12. Shrivastava, A., Craig, K., Roberts, N.E., Wentzloff, D.D., Calhoun, B.H.: 5.4 a 32nw bandgap reference voltage operational from 0.5v supply for ultra-low power systems. In: 2015 IEEE International Solid-State Circuits Conference – (ISSCC) Digest of Technical Papers, pp. 1–3, Feb 2015
13. Roberts, N.E., Craig, K., Shrivastava, A., Wooters, S.N., Shakhsheer, Y., Calhoun, B.H., Wentzloff, D.D.: 26.8 a 236 nw–56.5 dbm-sensitivity bluetooth low-energy wakeup receiver with energy harvesting in 65 nm CMOS. In: 2016 IEEE International Solid-State Circuits Conference (ISSCC), pp. 450–451, Jan 2016
14. Ueno, K., Hirose, T., Asai, T., Amemiya, Y.: A 300 nw, 15 ppm/°c, 20 ppm/v CMOS voltage reference circuit consisting of subthreshold mosfets. IEEE J. Solid-State Circuits **44**(7), 2047–2054 (2009)
15. Osaki, Y., Hirose, T., Kuroki, N., Numa, M.: 1.2-v supply, 100-nw, 1.09-v bandgap and 0.7-v supply, 52.5-nw, 0.55-v subbandgap reference circuits for nanowatt CMOS lsis. IEEE J. Solid-State Circuits **48**(6), 1530–1538 (2013)

Part III
Sub-1V and Advanced-Node Analog Circuit Design

The third part of this book is dedicated to recent developments in the field of ultra-low-voltage (1 V) design and/or in advanced nodes. The first three papers deal with technology innovations, the fourth one about future A/D converter developments, while the fifth and the sixth ones report two low-voltage design cases.

The first paper from Andreia Cathelin (STM) presents the planar 28 nm FD-SOI Technology, which appears advantageous for analog/RF/millimeter-wave and high-speed mixed-signal circuits, by taking full advantage of ultra-wide voltage range body biasing tuning. Concrete design examples are given in order to highlight the main FD-SOI design features.

As comparison, the second paper from Alvin L.S. Loke (Qualcomm) deals with analog/mixed-signal design in 16/14 nm FinFET technologies. The compact 3-D FinFET structure offers superior short-channel control that achieves digital power reduction and adequate device performance, while analog/mixed-signal designers must adapt to new design constraints. The challenges and considerations faced when porting analog/mixed-signal designs to FinFET are given.

The third paper from Lukas Dörrer (Intel) presents Intel Tri-Gate transistors (FinFET) that further shrink MOSFET technologies and have been a disruptive semiconductor innovation offering lower area, lower supply voltage, and lower power consumption. This paper presents and compares measurements and designs implemented in the 14 nm FinFET and in a planar 28 nm technology.

The fourth paper from Michael Flynn (University of Michigan) discusses about future developments of SAR ADC which received strong impulse in recent years. A new charge injection-based SAR architecture shrinks the die area and achieves outstanding energy efficiency. The SAR-assisted pipeline structure enables record efficiency for >12bit pipeline ADCs. The ring amplifier replaces the OTA in a pipeline so that pipeline ADCs can achieve outstanding efficiency in advanced nodes.

In the fifth paper, Rachit Mohan (imec) presents a low-voltage design case for biomedical applications, where low-power and low-cost sensor readout devices are demanded. The proposed solution uses time-domain operation to overcome the trade-off between cost, power consumption, and accuracy or the dynamic range capability.

In the sixth paper, Masoud Babaie (Delft University) proposes as second design case a 4.4 mW-TX, 3.6 mW-RX fully integrated Bluetooth Low-Energy Transceiver for IoT Applications optimized for 28 nm CMOS.

Chapter 13
FDSOI Technology, Advantages for Analog/RF and Mixed-Signal Designs

Andreia Cathelin

13.1 General Considerations Regarding FDSOI Technology

The race on the More Moore integration scale has brought several major limitations for efficient process integration starting from the 40 nm technology node, in CMOS planar solutions. It had appeared that the transistor channel was more and more difficult to control in terms of electrostatics, and a lot of process engineering methods (e.g., silicon strain) have been needed in order to provide transistors with good carrier speed and decent electrical characteristics. Starting from the 28 nm node, the obvious solution for transistors with increased electrical performances was the use of fully depleted channel devices. Two integration paths have been chosen in the semiconductor industry for these fully depleted devices: fully depleted silicon on insulator CMOS (FDSOI) and FinFET CMOS devices. The fundamental carrier semiconductor equations are similar; nevertheless, the process integration is very different. This paper will focus on planar FDSOI CMOS technology features as integrated by STMicroelectronics [1, 2] and its specific features for analog, RF, and mmW integrations.

Figure 13.1 gives a generic cross section of a FDSOI CMOS device. We call this technology Ultrathin Body and BOX (UTBB) FDSOI CMOS, as the active device is integrated atop an ultrathin layer of buried oxide (BOX). The transistor active conduction area (also called silicon film) is very thin as well. In the 28 nm UTBB FDSOI technology from ST, the BOX thickness is 25 nm and the film layer is 7 nm. This *planar* topology's direct implications are the following: thanks to the

A. Cathelin (✉)
STMicroelectronics, Crolles, France
e-mail: andreia.cathelin@st.com

P. Harpe et al. (eds.), *Hybrid ADCs, Smart Sensors for the IoT, and Sub-1V & Advanced Node Analog Circuit Design*, DOI 10.1007/978-3-319-61285-0_13

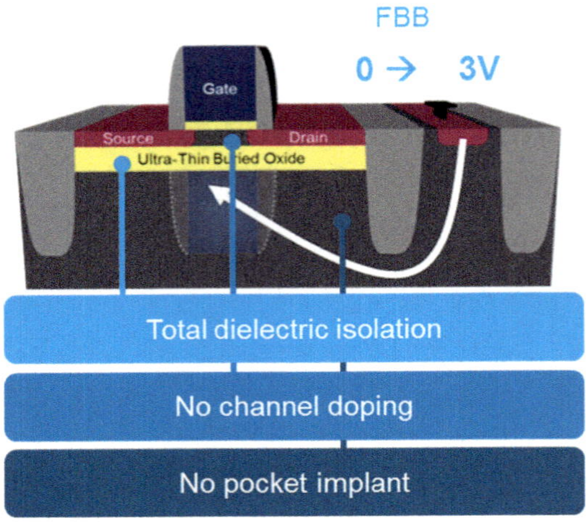

Fig. 13.1 Generic cross section of an UTBB FDSOI transistor

SOI BOX layer, the transistor gets total dielectric isolation. No channel doping is needed as, thanks to the thin silicon film, the channel is fully depleted. Also enabled by this topology, no pocket implants are needed for the source and drain, which enhances naturally the analog/RF transistor's behavior. Another implication of the thin layer consisting the BOX is the fact that the front side transistor's electrostatics can be controlled through the area underneath the BOX, also called transistor body. By applying a voltage on the transistor body (hence underneath the BOX), one can change or modulate the threshold voltage of the main (front side) transistor. We can see this device as well as a planar dual-gate device: the front gate is the regular one (like in bulk technology), and the second one comes from the body tie, with the BOX as the backside gate oxide. As the thickness report of the front and back gate oxides is about 10, we can state the front side transistor's transconductance Gm is ten times bigger than the one of the backside gate.

In terms of manufacturing, the 28 nm FDSOI CMOS process from STMicroelectronics is sharing most of process steps with the equivalent 28 nm *bulk* technology from ST. It is a modified bulk 28 nm high-K metal gate LP process using the same back end of the line and same gate module. Several process steps, specifically channel implants, halo implants, and masking levels are removed compared to the traditional 28 nm bulk technology because of the undoped FDSOI channel. There is less than 20% change with respect to a classical CMOS bulk flow; the two extra specific steps are related to the hybridation and raised source/drain epitaxy. At this node, more than 10% of the process steps and seven masks are saved, resulting in an overall manufacturing process cost saving of 10% [3].

13.2 Electrical Features of Active Devices in 28 nm FDSOI CMOS Technology and Comparison with an Equivalent Bulk Technology

13.2.1 Different Types of Active Devices in FDSOI Technology and Body Biasing Effect on the Threshold Voltage

In order to get a good understanding of the different types of FDSOI transistor's specific features, let's first start with a small reminder from planar *bulk* CMOS transistors.

Figure 13.2 gives a cross section of NMOS and respective PMOS transistors in a regular *bulk* CMOS technology. The potential threshold voltage (V_T) modulation can be obtained by the body bias tuning through the application of a voltage on the respective transistor's body ties. In such technologies, this is very limited by the threshold voltage of the parasitic source/drain diodes toward the respective transistor body. As this voltage is around 0.6 V, the effective safe body biasing voltage range spans only from 0 to 0.3 V (modulus). The body factor in bulk technologies is also quite limited (45 mV/V); hence, the total possible variation of the threshold voltage is limited in bulk technologies to only few tens of mV.

Figure 13.3 presents cross sections of the respective NMOS and PMOS transistors in the 28 nm UTBB FDSOI technology. The darker side of the respective NWells corresponds to the deep-NWell layer.

Regarding the regular VT (RVT) transistors (see the bottom part of the figure), one can observe the specific UTBB FDSOI topology, with the raised sources and drains and the thin BOX isolating the front side transistor for its body. If we get interest on the VT modulation through body bias biasing, one can see that now the only limiting parasitic diodes are those between deep NWell to P-substrate for NMOS and respectively embedded PWell to deep NWell for PMOS devices. These are Zener type of diodes, with an opening voltage around modulus of 3 V. We hence can do a very efficient body biasing on these RVT devices, called here reverse body biasing (RBB), the effective biasing voltage range being from roughly 0 to 3 V (modulus).

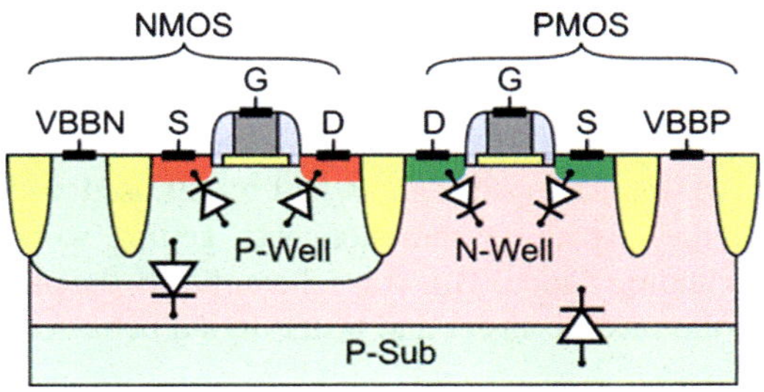

Fig. 13.2 Cross section of CMOS transistors in a regular bulk technology

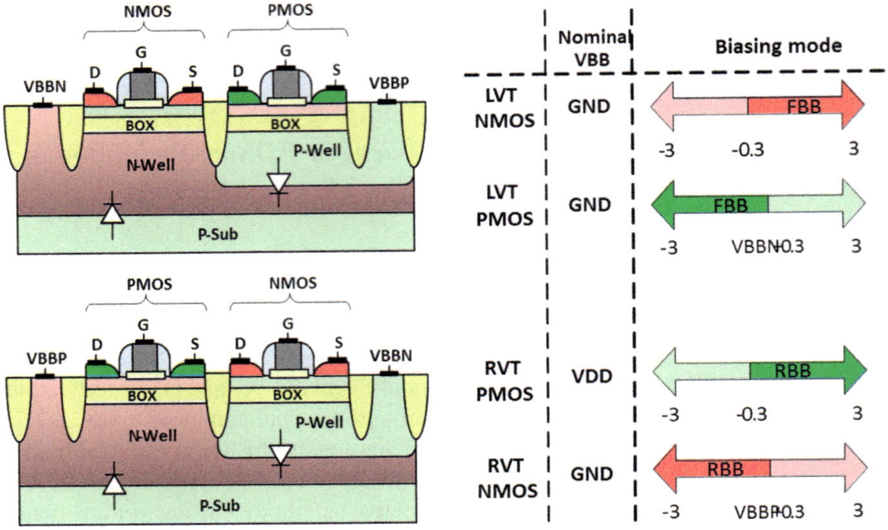

Fig. 13.3 Cross section of 28 nm UTBB FDSOI CMOS transistors: *Top* – Low VT (LVT) transistors; *Bottom* – Regular VT (RVT) transistors

Let's now focus on the low VT (LVT) devices on the upper part of Fig. 13.3. These devices are also called "flipped-well" devices, as in order to obtain the lower VT characteristics, the process engineering has proposed a solution when the NMOS devices lay on an NWell body, and respectively the PMOS on a PWell body. In an equivalent way, we can apply forward body biasing (FBB) on these devices, with also an equivalent body voltage variation from roughly 0 to 3 V (modulus).

The body factor for both families of devices (RVT and LVT) is much larger in FDSOI than in an equivalent body node, here in 28 nm being 85 mV/V. This argument together with the very wide body biasing range result in an unprecedented wide variation of the threshold voltage (VT) of around 250 mV, as depicted in Fig. 13.4.

13.2.2 FDSOI Transistors' Analog Features

When we were following the historical nm downscaling of the CMOS bulk road map, the analog designers had to get used with the fact that the analog behavior of the respective transistors was getting worse and worse with the downscaled technology node. This was inherent from the planar CMOS bulk transistor topology, in the race for faster and faster digital behavior and/or low power. Some foundries,

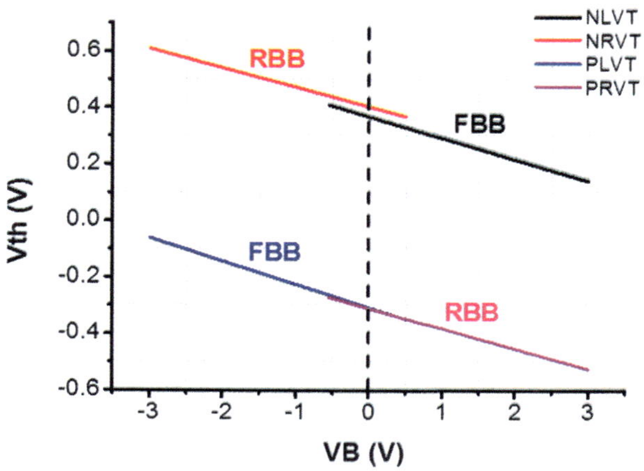

Fig. 13.4 Threshold voltage (VT) variation with respect to body biasing voltage for RVT and LVT devices in 28 nm FDSOI technology

like ST, had solved that in the 65nm node by introducing a specific analog transistor called HPA (high-performance analog) which was overcoming this problem by eliminating the transistor pockets. In the 28 FDSOI planar technology, thanks to the thin film structure, we do not need transistor pockets, and hence we can recover nice native analog behavior.

FDSOI hence brings several advantages to analog designers in terms of efficient short channel devices, improved analog performances, and lower noise variability. For comparison, in the several following figures, we show comparison of the 28 nm FDSOI technology with its equivalent 28 nm LP *bulk* technology, both from STMicroelectronics.

Figure 13.5 shows major improvements of FDSOI vs. bulk solution regarding analog gain and VT matching parameter. For example, in 28 nm FDSOI, an LVT NMOS device of size 1 μm/100 nm can show an excellent analog performance of DC gain of 80 and a sigma (VT) of only 6 mV.

Figure 13.6 shows that FDSOI provides higher Gm for a given current, with respect to the equivalent 28 nm LP bulk node. This, combined with the lower parasitic capacitances coming inherently from the SOI insulation, permits to achieve higher operation bandwidths for a given current consumption or – if working at constant bandwidth – lower power consumption.

The variability in planar FDSOI technologies is improved with respect to an equivalent LP bulk node, thanks to the simpler manufacturing process steps. This helps a lot as well for the noise behavior, as it can be seen in Fig. 13.7. As an example, for an LVT NMOS of size 1 μm/120 nm biased at a 1 μA drain current, we get 1.5 dB lower 1/f noise in FDSOI than in bulk. This is a typical value of the main branch current for a LNA in low GHz frequencies.

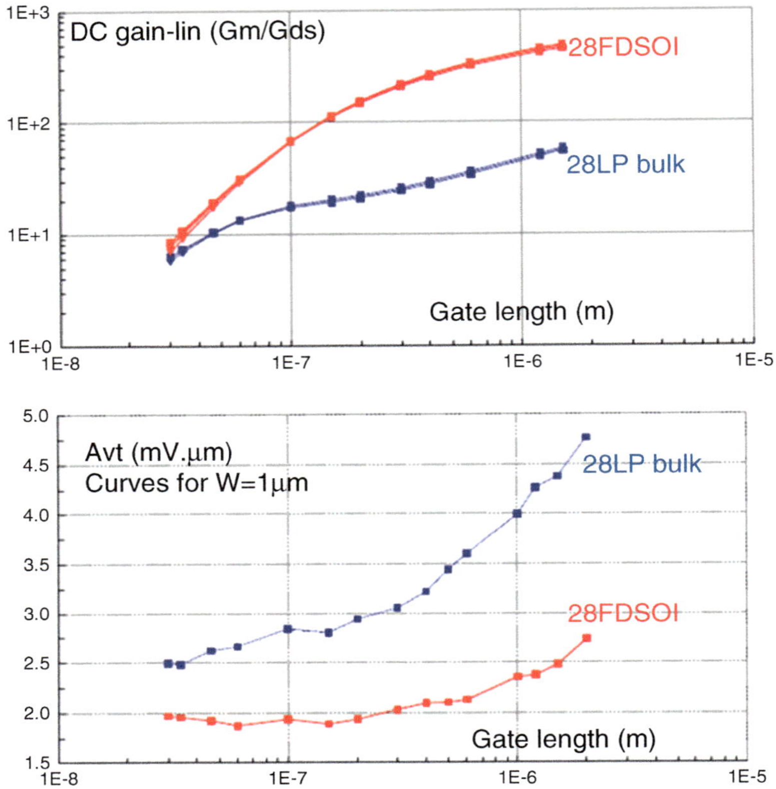

Fig. 13.5 Analog gain (Gm/Gds) and matching (AVT) for NMOS LVT devices in 28 nm FDSOI technology (*red*) and comparison with 28 nm LP bulk (*blue*)

13.2.3 FDSOI Transistors RF and mmW Features

Thanks to the deep submicron lithography, this technology node provides very fast transistors. The intrinsic devices (FEOL plus Metal1 contact) in the LVT flavor, for example, here NMOS, show f_T and f_{max} superior to 300 GHz (Fig. 13.8).

We can hence distinguish two types of dimensioning and biasing strategies, depending on the operation frequency.

If we talk about RF operation frequency below 10 GHz, we can then work with a transistor length of 100 nm. Performances such as a maximum available gain MAG = 12 dB and NFmin 0.5 dB can be obtained for a current density: 125 μA/μm (here $W = 1$ μm).

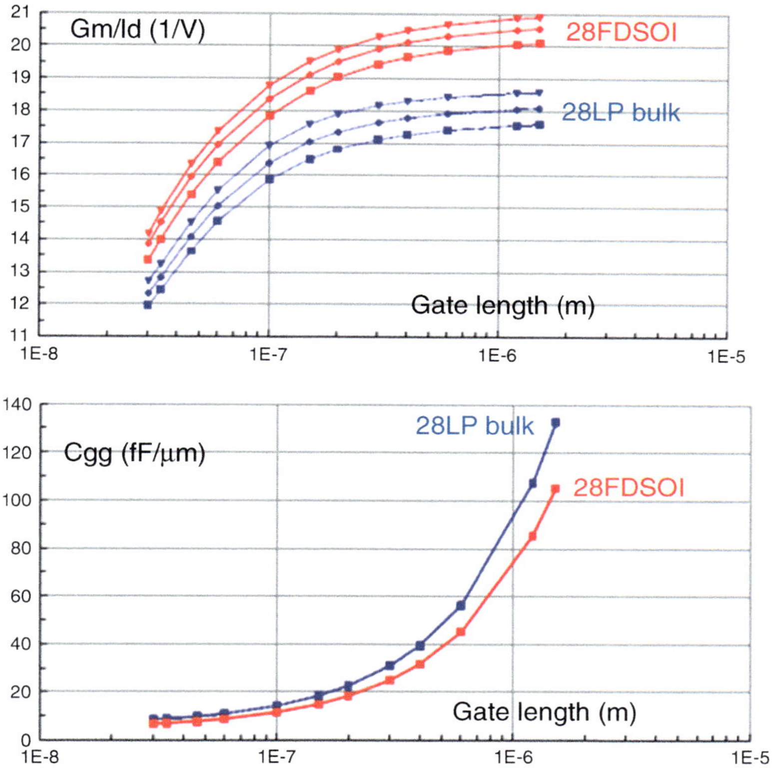

Fig. 13.6 Improved analog performance (Gm/Id and total gate capacitance Cgg) for NMOS LVT devices in 28 nm FDSOI technology (*red*) and comparison with 28 nm LP bulk (*blue*)

Going higher in frequency will then request working with the minimum transistor length of 30 nm. Here, for example, for a 60 GHz operation frequency, a MAG = 12 dB and NFmin 1.3 dB can be obtained, when working at a current density of 200 μA/μm. This value is 33% lower than in 28LP bulk.

Deep submicron CMOS has the counterpart of very low and dense back end of line, with a large number of metal layers. This can be seen as a limiting point for mmW design; nevertheless, the eight metal layers of the discussed technology permit to obtain very decent values for the integrated passive devices. This is enabled by the operation in a low parasitics environment coming with the SOI technologies. Several examples can be cited here: an inductor of $L = 0.5$ nH with a Q factor of 18 at 10 GHz, a varactor of $C = 50$fF with a Q factor of 20 at 20 GHz and a 50 Ohm transmission line of 08 dB/mm losses at 60 GHz.

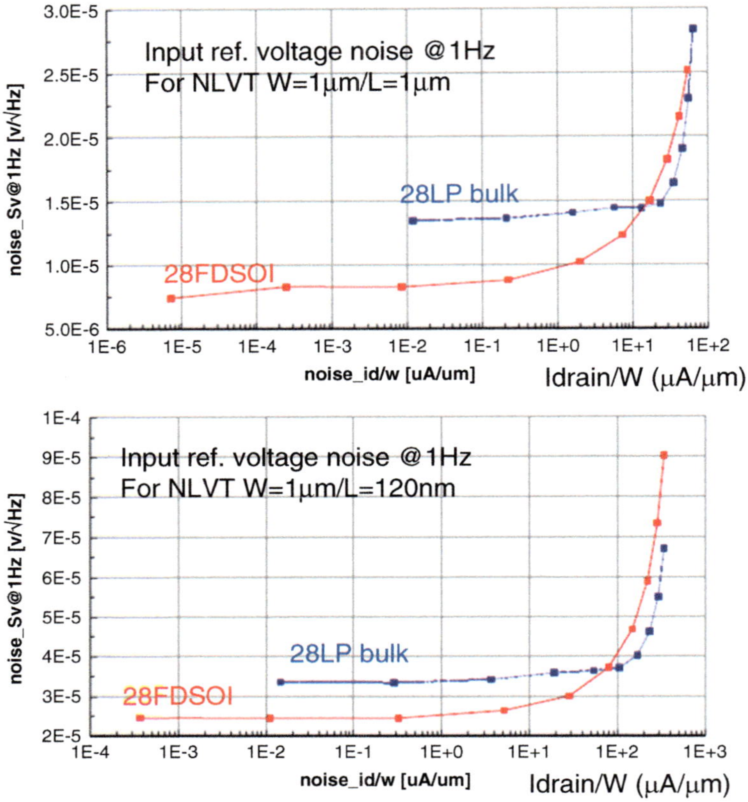

Fig. 13.7 Noise behavior for NMOS LVT devices in 28 nm FDSOI technology (*red*) and comparison with 28 nm LP bulk (*blue*)

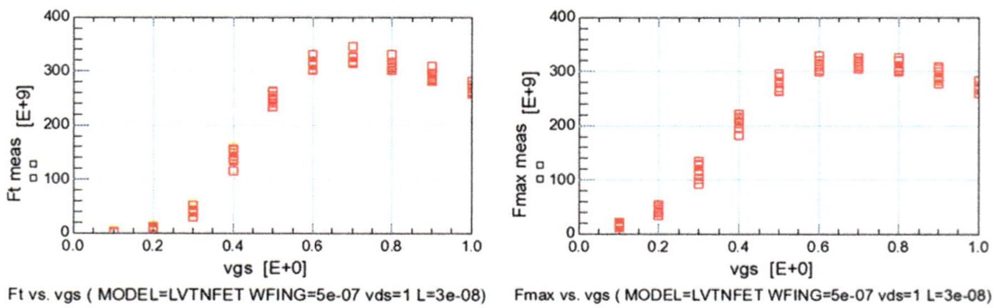

Fig. 13.8 High-frequency behavior (f_T and f_{max}) of LVT NMOS 0.5 μM/30 nm in 28 nm FDSOI CMOS

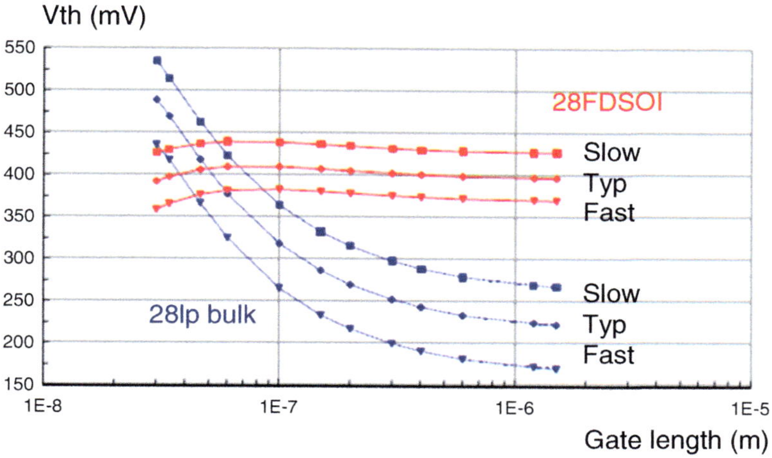

Fig. 13.9 VT process corners for LVT NMOS devices, comparison between 28 nm FDSOI CMOS and 28 nm LP bulk

13.2.4 FDSOI Transistors' Mixed-Signal Features

In terms of process variability, we experience in FDSOI tighter process corners and less random mismatch than competing processes at the same lithography node. The benefits become immediate in terms of simpler design process and shorter design cycle, hence improved yield or improved performance at given yield. Figure 13.9 illustrates this for the VT parameter.

Body biasing allows for VTH reduction by tuning, as depicted in Fig. 13.10 [4]. This result is an unprecedented quality of analog switches. One can observe the exceptional flat behavior of the CMOS switch resistance in the case of the FDSOI integration using body biasing, its absolute value being much lower than the one in bulk. We experience here compounding benefits: a smaller resistance yields a smaller switch with a more compact layout, hence with lower parasitics, which finally gives an even smaller switch. This feature is key for high-performance data converters and other switched-capacitor circuits.

Lower junction capacitances as those experienced in FDSOI make a substantial difference in high-speed circuits. They permit drastic reduction of self-loading in gain stages and a significant reduction of switches self-loading. This yields a twofold benefit: not only incremental improvements, but mostly they allow the designer to use circuit architectures that would be infeasible/inefficient in bulk technologies.

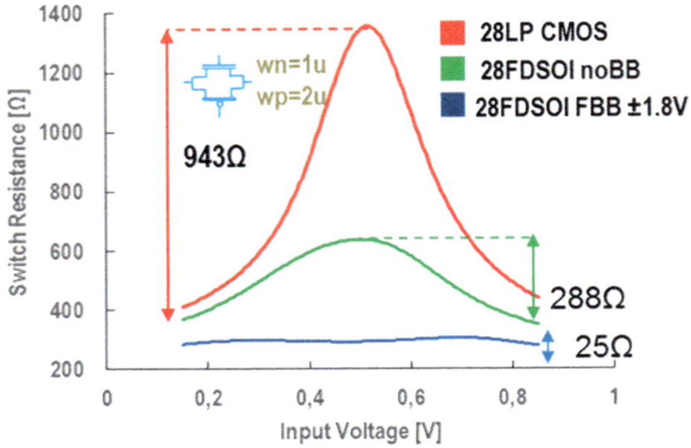

Fig. 13.10 LVT CMOS switch resistance variation wrt input voltage, comparison between 28 nm FDSOI CMOS and 28 nm LP bulk

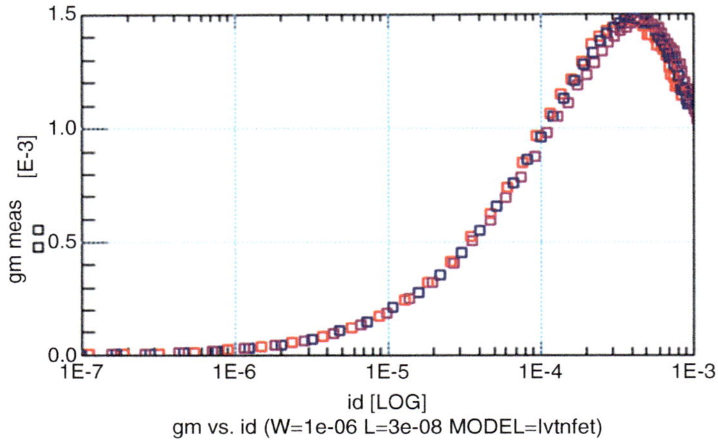

Fig. 13.11 28 nm FDSOI LVT CMOS 1 μm/30 nm transistor measured Gm for different drain currents, Vbody varies from 0 to 2 V, Vds = 1.1 V

13.2.5 FDSOI Transistors' Feature Variation with the Body Biasing Tuning

The semiconductor physics in FDSOI predict that the main design transistor's parameters (such as Gm or f_T) do not depend on the body biasing variation, for an operation at constant current.

The following two figures illustrate this aspect by providing measurement curves for different devices (Figs. 13.11 and 13.12).

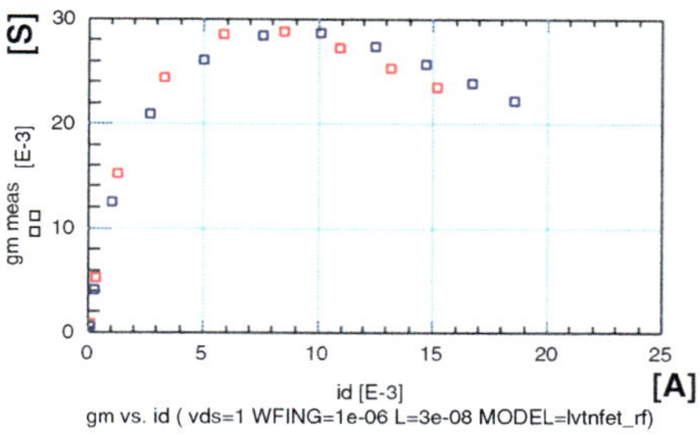

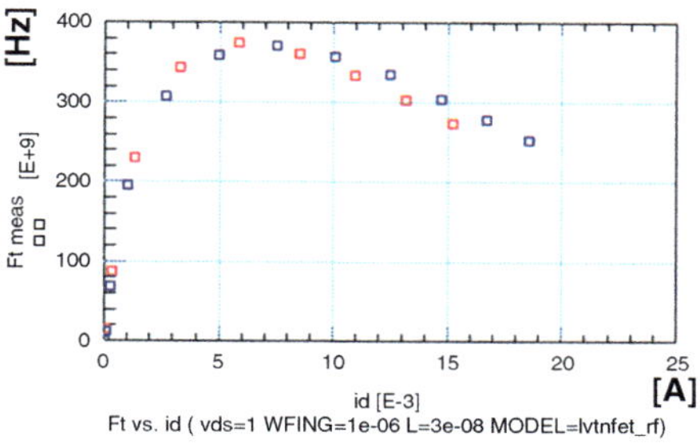

Fig. 13.12 28 nm FDSOI LVT CMOS 1 μm × 20 fingers/30 nm transistor* measured Gm and f_T for different drain currents, Vbody varies from 0 to 2 V, Vds = 1.1 V. ()* intrinsic device (FEOL plus Metal 1)

13.3 Design Example of an Analog/RF Building Bloc with FDSOI Specific Features

This section presents a typical bloc used in most of wireless communication IC's, and it highlights the main benefits that such a design can take from a 28 nm FDSOI integration. We hence discuss in this section the integration of analog filters with several 100's MHz bandwidth. In the real life of a SoC, they suffer from process, voltage, temperature (PVT), and aging variations, which affect system operation and directly influence overall applicative behavior. All such blocs need to be tuned or trimmed inside the SoC, in order to strictly control independently several of its parameters: cutoff frequency, linearity, and noise, all these for an optimal power consumption.

Since more than 10 years now, in deep submicron CMOS processes, it is very practical and straightforward to implement inverter-based analog functions: They yield to simple and compact solutions, which nicely scale with technology nodes. The example here discusses about an analog low-pass Gm-C filter with cutoff frequencies in the range of hundreds of MHz. The typical implementation of such topology is realized with fixed capacitors, then the filter parameters are varied by tuning the different filter Gm's; see Fig. 13.13. Each filter transconductor Gm is composed of several invertors operated in the gain region (analog operation with biasing point on the middle of the out/in transfer function), as, for example, presented in [5].

In traditional bulk CMOS technologies, the only tuning way of such inverter-based topologies is to define as unique tuning knob the local Vdd of the analog operated inverters. A dedicated LDO with controlled output voltage is, for example, implemented between the global system Vdd and the bloc local Vdd, generating this tuning value. This is generating extra power consumption and reduced voltage headroom (hence linearity) for the bloc to be controlled. In general, with only one tuning knob for such blocs and several parameters to tune (cutoff frequency, linearity, etc.),

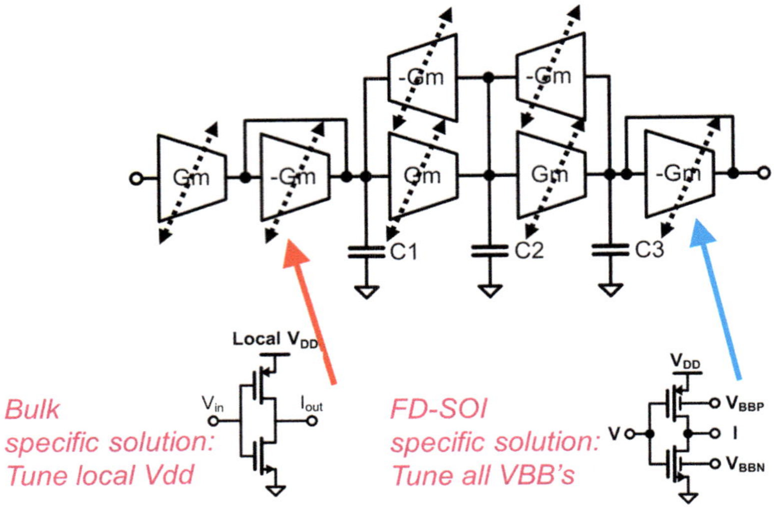

Fig. 13.13 Typical tuning methods in bulk and FDSOI CMOS integration for analog Gm-C filters

the designers have to take a very large power consumption margin in order to be able to satisfy the system level specifications in all operation conditions.

In FDSOI CMOS technologies, as presented in the previous section, there are two available and independent new tuning knobs in the system: the local bodies of the NMOS and respective PMOS transistors. Thanks to the very wide tuning voltage range of these individual body biases knobs, there is an independent variation of the respective transistors' threshold voltages, hence generating at system level a variation of the system level parameters, and this over a very wide range. Moreover, this permits also an *independent* tuning of the different system level parameters (e.g., cutoff frequency and linearity).

Another consideration can be made here regarding the parasitic influence of a tuning control loop on the main signal path operation of a system. In classical bulk CMOS implementation, the loop control signal is somewhere either on the signal path or with direct parasitic influence on the signal path. In these FDSOI tuning systems, the tuning control signal are on the body ties, which are isolated by a buried oxide layer with respect to the signal path main operation, hence all parasitic signals on these signals have much less influence on the main signal path operation.

All these will be illustrated in a simple intuitive way in the following paragraphs.

In a bulk CMOS implementation, the inverter transconductance variation (hence the global filter transconductor variation) is obtained by the local Vdd variation, as depicted in Fig. 13.14. In this example, the capacitor values are fixed, meaning that the global transconductor variation is directly proportional to the one of the filter cutoff frequency. Unfortunately, this transconductance variation will also imply a variation of the linearity; hence, a larger design margin will have to be taken in order to cover these two variations.

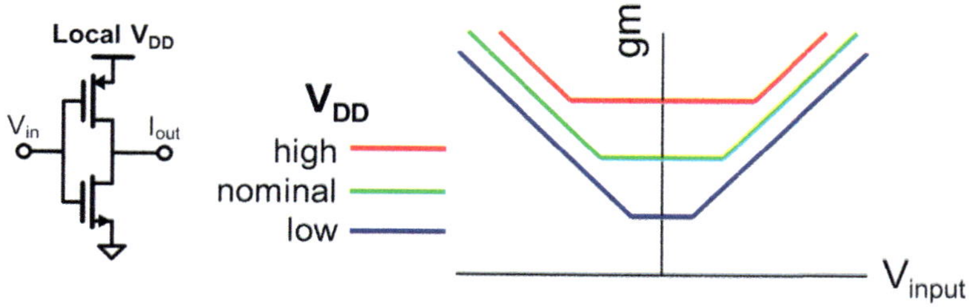

Fig. 13.14 Bulk CMOS operation: inverter-based design tuning through local Vdd variation

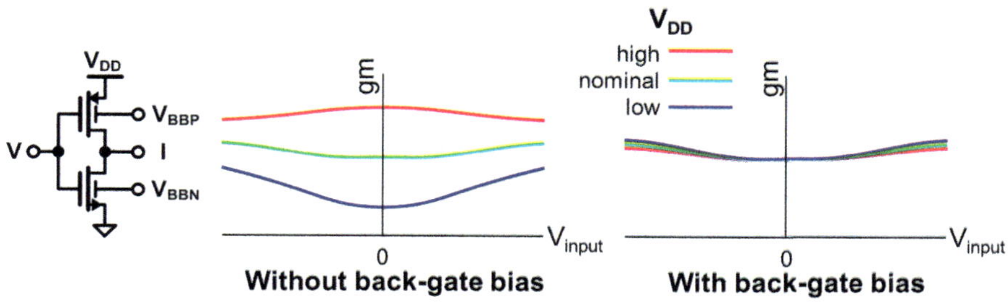

Fig. 13.15 FDSOI CMOS operation: inverter-based design tuning through independent Vbody_N and respective Vbody_P variation

In Fig. 13.15, we show an FDSOI CMOS typical implementation for this filter tuning strategy. The inverters constituting the Gm-C filter have all a global Vdd, which is the global (system) Vdd, and in any ways it is not used for tuning means. Two independent tuning knobs are materialized here by the body ties of the NMOS, respective PMOS transistors. Without body biasing, any variation on the global Vdd will induce a variation in the transconductance and the linearity of the inverters, hence the filter. Thanks to these new independent tuning knobs, a combination of the two tuning voltages can always be found in order to get the desired values for both transconductance and linearity. The very wide tuning voltage range available on these body ties (see Sect. 13.2.1) permits to lower the design margins that are needed on such design, hence the typical power consumption.

This FDSOI tuning concept has been implemented in a third-order Gm-C low-pass filter, as presented in [6]. The measured performances as well as the comparison with the state of the art are given in Fig. 13.16.

This RF low-pass Gm-C filter using CMOS inverters has been successfully tuned by back gate instead of supply, moreover with no signal path interference. This supply regulator-free operation is energy efficient and also able of low voltage operation (down to VDD = 0.7 V).

There are two main categories of analog filters, each with advantages and drawbacks, these considered for an equivalent input referred noise level: the active-RC filters show generally excellent linearity, with the counterpart of large power

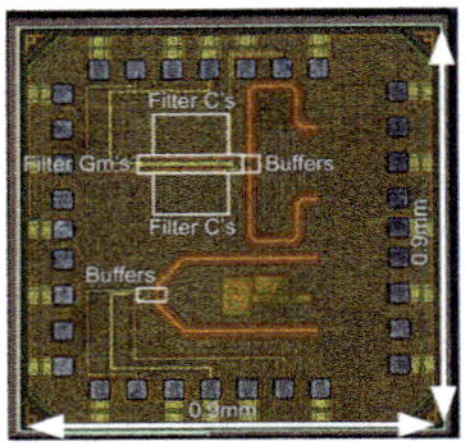

[2] Houfaf, *et al.*, *ISSCC 2012*
[5] Saari, *et al.*, *TCAS-I 2009*
[6] Mobarak, *et al.*, *JSSC 2010*
[7] Kwon, *et al.*, *TMTT 2009*

	This work				[2]	[5]	[6]	[7]
Technology	28nm FD-SOI CMOS				65nm CMOS	65nm CMOS	0.13um CMOS	0.18um CMOS
Order	3				3	5	2	3
Supply voltage [V]	0.7	0.8	0.9	1	1	1.2	1.2	1.8
Cut-off freq. [MHz]	454	454	457	459	4700	275	200	300
Input ref. noise [$nV_{rms}/\sqrt{Hz}$]	5.9	6.1	6.1	5.9	6.6	7.8	35.4	5
in-band IIP3 [dBVp]	1.2	4.0	4.0	2.4	-3	-12.5	4	6.9
Power diss. [mW]	4.0	4.6	5.2	5.6	19	36	21	72
SFDR/BW [dB/Hz]	109	110	110	109	105	98	100	113
NSNR [dB]	**137**	**139**	**138**	**137**	125	111	117	131

Fig. 13.16 28 nm FDSOI Gm-C low-pass filter, chip photomicrograph, measured performances, and comparison with the state of the art

consumption and limited operation in high frequency; the Gm-C filters show excellent frequency behavior and straightforward implementation from passive LC prototypes but have traditionally limited linearity performance. This FDSOI Gm-C implementation shows nevertheless competitive linearity. When compared to similar circuit in 65 nm bulk [7], at the same noise level, we get twice the linearity for a power level divided by four. When compared to best-in-class filters (active-RC) [8], at same noise level and cutoff frequency, we get competitive linearity for a power level divided by 14.

This 28 nm FDSOI CMOS solution implemented in ST's technology exhibits best in class compromise noise linearity power, thanks to the excellent analog/RF process intrinsic features.

13.4 Design Example of a Millimeter-Wave Building Bloc with FDSOI Specific Features

This section presents a millimeter-wave representative design in 28 nm FDSOI CMOS technology. We have chosen to present an integrated power amplifier for 60 GHz WiGig applications, as in such 60 GHz transceivers half of the power is burned by the power amplification section. For CMOS mmW-integrated power amplifiers (PA), there is also a trade-off to be solved between linearity and power-added efficiency at −8 dB back-off (this is the point with the maximum operation probability in such OFDM-based standard). The goal of this design example is to leverage this compromise and propose high linearity PAs with, at the same time, performant efficiency.

In order to satisfy with a same bloc the linearity and efficiency constraints, classical architectures rely on the Doherty PA topology. This structure relies on two different power amplification units that are operated one at a time, in order to privilege one or the other system constraints. The parallelization of these structures

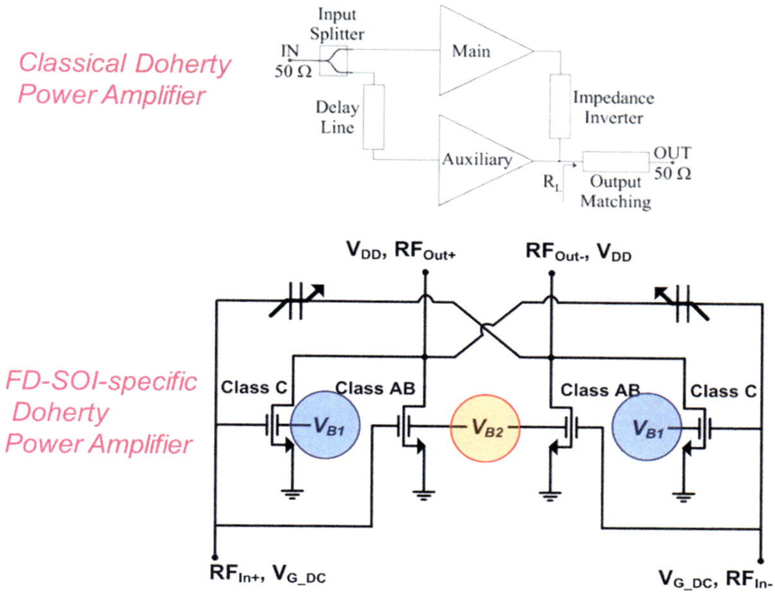

Fig. 13.17 28 nm Classical and FDSOI revisited Doherty power amplification topology

implies also passive power adaptation structures at the input and output, which inherently bring passive losses, hence natural lower efficiency. Such structures are typically implemented in classical CMOS integration.

In a specific FDSOI CMOS integration, we have decided to revisit this classical Doherty PA architecture, as depicted in Fig. 13.17 [9]. Two different class power amplifiers are as well connected in parallel, but here they are materialized by two differential pairs. The parallelization of such structures is straightforward, with no lossy passive elements; hence, the starting point for energy efficiency is already improved. Each PA cell (i.e., differential pair) is biased in a different class (here class AB and C), and each individual biasing point is changed by *gradually* varying the body voltage of each differential pair. We hence get the ability of gradually changing the overall class of the PA (mix of class AB and class C), thanks to the wide range of the forward body biasing voltage. Moreover, we get a new equivalent class of PA, which at any instant is composed of *x%* class AB operation and *y%* class C. At the two-class operation extremities, we get either Class-C at zero body bias or Class-A in maximum forward bias.

The compression of the Class-AB transistors is compensated by the gain expansion of the Class-C transistors, increasing the PA compression point without DC current penalty. The Class-AB devices are sized to carry the RMS power of the modulated signal and determine the average power consumption, while Class-C devices pass the peaks. Static body bias and dynamic modulated signals induce drain-gate nonlinear capacitance variations, tracked by MOS neutralization

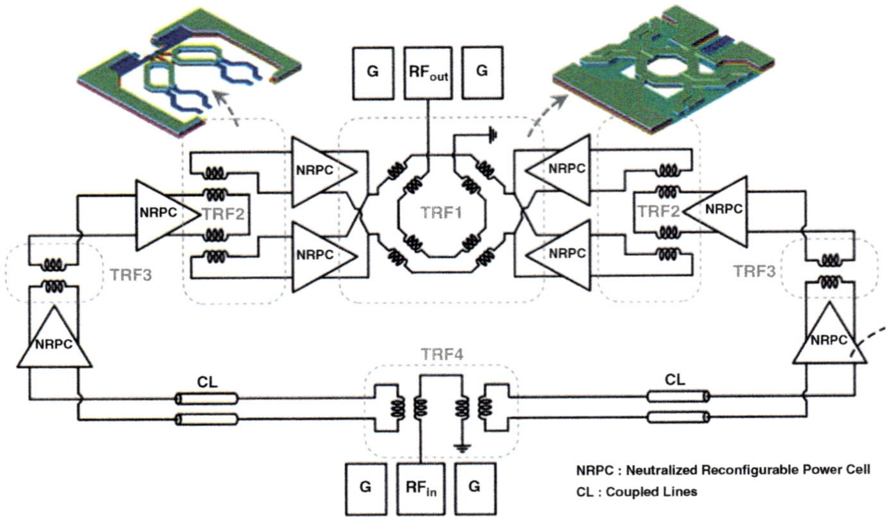

Fig. 13.18 28 nm FDSOI 10ML CMOS implementation of the WiGig 60 GHz Power Amplifier

capacitors across process, temperature, bias, and signal conditions. This robustness allows to safely use the full neutralized power cell flexibility to trade power gain and consumption for linearity, between "high gain mode" (all forward bias, Class-A) and "high linearity mode" (dual body bias, optimized Class-AB Class-C combination).

This new design topology, uniquely enabled by FDSOI wide range body biasing capabilities, permits to optimize in the same time power efficiency and linearity in such power amplification cells.

The total PA consists of three-stage transformer-based power amplification cells, each active cell as depicted before. This PA has been implemented in a 10 ML process option of the ST's 28 nm FDSOI CMOS technology; see Fig. 13.18. The VLSI-like dense and low BEOL still permits to obtain very competitive passives design, thanks to the SOI substrate feature.

At the output, a parallel-series eight-way power combiner TRF1 sums four differential reconfigurable power cells and performs differential to single-ended conversion. While the distributed active transformer (DAT) transforms 50Ω output to four 7Ω input ports with a coupling factor close to 0.87, the access lines are used to create eight ports presenting the optimal large signal load impedance for the power devices. This compact topology achieves an insertion loss of only 1 dB at 60 GHz. The 1:2 differential TRF2 and 1:4 TRF4 power splitters are based on a DAT structure and implement wideband matching networks. The two secondary windings are orthogonally placed to reduce parasitic magnetic coupling. The 1:2 transformer TRF3 performs impedance matching between the two driver stages.

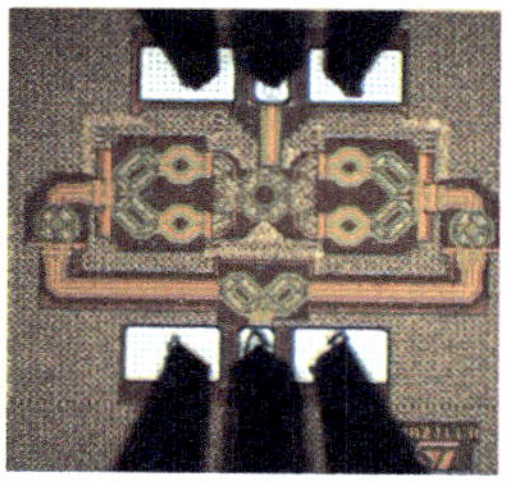

	This work			S. Kulkami *ISSCC 2014*	D. Zhao *JSSC 2013*	D. Zhao *JSSC 2012*	E. Kaymaksut *RFIC 2014*	A. Siligaris *JSSC 2010*
Technology	28nm UTBB FD-SOI			40nm	40nm	40nm	40nm	65nm PD-SOI
Operating mode	High gain	High linearity		NA	Low/High power	NA	NA	NA
Supply voltage [V]	1.0	1.0	0.8	0.9	1.0	1.0	0.9	1.8
Freq. [GHz]	61	60	60	63	61	60	77	60
Gain [dB]	35	15.4	15.1	22.4	16.8 / 17	26	9	16
P_{SAT} [dBm]	18.9	18.8	16.9	16.4	12.1 / 17	15.6	16.2	14.5
P_{1dB} [dBm]	15	18.2	16.2	13.9	9.1 / 13.8	15.6	15.2	12.7
PAE_{max} [%]	17.7	21	21	23	22.2 / 30.3	25	12	25.7
PAE_{1dB} [%]	9	21	21	18.9	14.1 / 21.6	25	11.1	22.6
$PAE_{6dB\ backoff}$ [%]	1.5	8	7.5	3	- / 4.7	5.8	3.5	2.7
P_{DC} [mW]	331	74	58	88	56 / 75[#]	117	126	77.4
$P_{DC\ 6dB\ backoff}$ [mW]	332	124	84	94	56 / 78[#]	120	140	79
$100 \times P_{1dB}/P_{DC}$	9.6	89	72	28	14.5 / 32[#]	31	26	24
Active area [mm²]	0.162			0.081	0.074	0.33	0.1	0.573[*]
ITRS FOM [W.GHz²]	161,671	1,988	1,198	6,925	641 / 2,832	13,009	236	1,038

ITRS FOM = $P_{SAT} \cdot PAE_{max} \cdot Gain \cdot Freq^2$ [*]: with pads [#]: estimated

Fig. 13.19 28 nm FDSOI 10 ML CMOS WiGig PA, chip photomicrograph, measured performance, and comparison with the state of the art

Figure 13.19 gives the measured performance of this mmW PA, for three extreme operation cases, knowing that an infinite number of operation conditions can be met when the two different body bias voltages are varied from 0 to 2 V (FBB conditions).

This power amplifier is fully WiGiG compliant, when taking into consideration linearity and frequency operation range (all four bands of the standard). It has introduced a new PA architecture which permits to continuously reconfigure power cells. This continuous operation class tuning is enabled by the very wide body bias voltage range.

Moreover, in the high gain mode, it exhibits the highest ITRS FOM, improving by an impressive factor of 10 the previous state of the art. In the high linearity mode, it breaks the linearity/consumption trade-off. It permits also a low-voltage high-efficiency operation (Vdd_min = 0.8 V).

13.5 Overview of FDSOI Specific Tuning and Trimming Techniques Using Body Biasing for Analog/RF Designs

Before concluding, this section gives an overview of the potential body biasing enabled tuning and trimming methods, identified as of today.

By taking advantage of the unique very wideband body biasing (BB) voltage range available in FDSOI technologies, the state of the art proposes several unique techniques bringing uncontested chip energy saving and revisiting system performances.

The first method consists in generating and making available on chip a body bias voltage variable over time and process, voltage, and temperature (PVT) variations. This permits to:

- Cancel system level PVT effects by continuously tuning transistors' respective VT. Several design examples can be found in references [6, 10, 11].
- Reconfigure circuit/bloc/system depending on application operation mode. Design examples can be found in [9] (at bloc level) and [11] (at system level).
- Propose new energy-efficient design techniques for tunable blocs via body tie, as in [12].

The second method consists in generating and making available on chip a fixed body bias voltage. In such cases, the design can efficiently:

- Enable operation at ULV (0.5 V) and at the same time increase circuit speed. Some design examples are [13, 14].
- Minimize switches on-resistance value and excursion for energy-efficient and high-speed switched-capacitors circuits (e.g., ADC), as in [4, 15].

For sure, each type of such on-chip body bias generators has to be carefully considered, in terms of power consumption, ripple, and noise on the generated control voltages. They may have a constant over time operation or duty cycled, depending on the system operation. And finally, they can address smaller or larger islands of transistors to be tuned or controlled.

13.6 Conclusion and Perspectives

This paper has presented a short overview of planar UTBB FDSOI technologies and their application for analog, RF, mmW, and mixed-signal designs.

As a summary, here are the major arguments of such technologies to focus on for analog/RF designs:

- Make massive usage of the body biasing techniques which enable transistors' VT as tuning knob and this over an unprecedented very wide tuning range
- Take profit of the very good analog performances. They permit designs with lower power consumption and which can safely operate at $L > L_{min}$ for design margin.

For RF to mmW design, atop the previously mentioned aspects, we should take into consideration the deep submicron technology features for the active devices (excellent f_T, f_{max}). The back end of line in an FDSOI environment permits to obtain performant passive devices, despite the very dense VLSI constraints.

For mixed-signal and high-speed designs, the major key parameters are the improved variability, the remarkable CMOS switches performance, and the reduced parasitic capacitances.

And finally, the UTBB FDSOI technologies open a new era in terms of innovative energy-efficient circuits and systems. They find excellent application for IoT implementations. Ultralow Power SoCs take benefit of efficient ultralow-voltage digital performances [16]; the full mixed-signal integration is then enabled by efficient analog, RF [11], and mmW operation. Finally, the FDSOI implementation exhibit power and performance flexibility, thanks to the new tuning knobs brought in by very wide voltage range body biasing.

References

1. Planes, N., Weber, O., Barral, V., Haendler, S., Noblet, D., Croain, D., Bocat, M., Sassoulas, P., Federspiel, X., Cros, A., Bajolet, A., Richard, E., Dumont, B., Perreau, P., Petit, D., Golanski, D., Fenouillet-Beranger, C., Guillot, N., Rafik, M., Huard, V., Puget, S., Montagner, X., Jaud, M.-A., Rozeau, O., Saxod, O., Wacquant, F., Monsieur, F., Barge, D., Pinzelli, L., Mellier, M., Boeuf, F., Arnaud, F., Haond, M.: 28 nm FD-SOI technology platform for high-speed low-voltage digital applications. In: Proceedings of Symposium VLSI Technology (VLSIT), pp. 133–134 (2012)
2. Arnaud, F., Planes, N., Weber, O., Barral, V., Haendler, S., Flatresse, P., Nyer, F.: Switching energy efficiency optimization for advanced CPU thanks to UTBB technology. In: IEEE International Electron Devices Meeting (IEDM) Dig., pp. 3.2.1–3.2.4 (2012)
3. Jacquet, D., Hasbani, F., Flatresse, P., Wilson, R., Arnaud, F., Cesana, G., Di Gilio, T., Lecocq, C., Roy, T., Chhabra, A., Grover, C., Minez, O., Uginet, J., Durieu, G., Adobati, C., Casalotto, D., Nyer, F., Menut, P., Cathelin, A., Vongsavady, I., Magarshack, P.: A 3 GHz Dual Core Processor ARM Cortex TM -A9 in 28 nm UTBB FD-SOI CMOS With Ultra-Wide Voltage Range and Energy Efficiency Optimization. IEEE J. Solid-State Circuits. **49**(4), (2014)
4. Kumar, A., Debnath, C., Narayan Singh, P., Bhatia, V., Chaudhary, S., Jain, V., Le Tual, S., Malik, R.: A 0.065mm2 19.8mW single channel calibration-free 12b 600MS/s ADC in 28nm UTBB FDSOI using FBB. In: ESSCIRC Conference 2016: 42nd European Solid-State Circuits Conference, pp. 165–168 (2016)
5. Nauta, B.: A CMOS transconductance-C filter technique for very high frequencies. IEEE J. Solid-State Circuits. **27**(2), 142–153 (1992)
6. Lechevallier, J., Struiksma, R., Sherry, H., Cathelin, A., Klumperink, E., Nauta, B.: A forward-body-bias tuned 450 MHz Gm-C 3rd-order low-pass filter in 28nm UTBB FD-SOI with >1dBVp IIP3 over a 0.7-to-1V supply. In: 2015 IEEE International Solid-State Circuits Conference – (ISSCC) Digest of Technical Papers, pp. 1–3 (2015)
7. Houfaf, F. et al.: A 65nm CMOS 1-to-10GHz Tunable Continuous-Time Lowpass Filter for High-Data-Rate Communications. In: IEEE ISSCC Digest of Techical Papers, pp. 362–364 (2012)
8. Kwon, K., et al.: A 50–300-MHz highly linear and low-noise CMOS gm-C filter adopting multiple gated transistors for digital TV tuner ICs. IEEE Trans. Microwave Theory Techn. **57**(2), 306–313 (2009)
9. Larie, A., Kerhervé, E., Martineau, B., Vogt, L., Belot, L.: A 60GHz 28nm UTBB FD-SOI CMOS reconfigurable power amplifier with 21% PAE, 18.2dBm P1dB and 74mW PDC. In: 2015 IEEE International Solid-State Circuits Conference – (ISSCC) Digest of Technical Papers, pp. 1–3 (2015)
10. Danilovic, D., Milovanovic, V., Cathelin, A., Vladimirescu, A., Nikolic, B.: Low-power inductorless RF receiver front-end with IIP2 calibration through body bias control in 28 nm UTBB FDSOI. In: 2016 IEEE Radio Frequency Integrated Circuits Symposium (RFIC), pp. 87–90 (2016)
11. de Streel, G., Stas, F., Gurné, T., Durant, F., Frenkel, C., Cathelin, A., Bol, D.: Sleep Talker: A ULV 802.15.4a IR-UWB Transmitter SoC in 28-nm FDSOI Achieving 14 pJ/b at 27 Mb/s With Channel Selection Based on Adaptive FBB and Digitally Programmable Pulse Shaping. IEEE J. Solid-State Circuits. **PP**(99), 1–15 (2017)

12. Sourikopoulos, I., Frappé, A., Cathelin, A., Clavier, L., Kaiser, A.: A digital delay line with coarse/fine tuning through gate/body biasing in 28nm FDSOI. In: ESSCIRC Conference 2016: 42nd European Solid-State Circuits Conference, pp. 145–148 (2016)
13. Fanori, L., Mahmoud, A., Mattsson, T., Caputa, P., Rämö, S., Andreani, P.: A 2.8-to-5.8 GHz harmonic VCO in a 28 nm UTBB FD-SOI CMOS process. In: 2015 IEEE Radio Frequency Integrated Circuits Symposium (RFIC), pp. 195–198 (2015)
14. Lahiri, A., Gupta, N.: A 0.0175 mm2 600 μW 32kHz input 307 MHz output PLL with 190 psrms jitter in 28 nm FD-SOI. In: ESSCIRC Conference 2016: 42nd European Solid-State Circuits Conference, pp. 339–342 (2016)
15. Le Tual, S., Narayan Singh, P., Curis, C., Dautriche, P.: A 20 GHz-BW 6b 10GS/s 32 mW time-interleaved SAR ADC with Master T&H in 28 nm UTBB FDSOI technology. In: 2014 IEEE International Solid-State Circuits Conference Digest of Technical Papers (ISSCC), pp. 382–383 (2014)
16. Zimmer, B., Lee, Y., Puggelli, A., Kwak, J., Jevtić, R., Keller, B., Bailey, S., Blagojević, M., Chiu, P.-F., Le, H.-P., Chen, P.-H., Sutardja, N., Avizienis, R., Waterman, A., Richards, B., Flatresse, P., Alon, E., Asanović, K., Nikolić, B.: A RISC-V vector processor with simultaneous-switching switched-capacitor DC–DC converters in 28 nm FDSOI. IEEE J. Solid-State Circuits. **51**(4), 930–942 (2016)

Chapter 14
Analog/Mixed-Signal Design in FinFET Technologies

Alvin L. S. Loke, Esin Terzioglu, Albert A. Kumar, Tin Tin Wee, Kern Rim, Da Yang, Bo Yu, Lixin Ge, Li Sun, Jonathan L. Holland, Chulkyu Lee, Deqiang Song, Sam Yang, John Zhu, Jihong Choi, Hasnain Lakdawala, Zhiqin Chen, Wilson J. Chen, Sreeker Dundigal, Stephen R. Knol, Chiew-Guan Tan, Stanley S. C. Song, Hai Dang, Patrick G. Drennan, Jun Yuan, P. R. Chidambaram, Reza Jalilizeinali, Steven J. Dillen, Xiaohua Kong, and Burton M. Leary

14.1 Introduction

The mobile system-on-chip (SoC) has emerged as the principal economic driver to extend CMOS scaling. Since the finFET debuted in manufacturing at the 22-nm node [1], most advanced fabless designs have bypassed planar 20 nm in favor of foundry finFET offerings introduced at 16 and 14 nm [2]. With its superior short-channel control, the fully depleted tri-gate finFET enables a smaller device footprint with dynamic and leakage power savings and competitive performance. Qualcomm's first finFET product (Snapdragon™ 820) was built in 14-nm CMOS and has been in high-volume production since 2015. Its next-generation mobile processor (Snapdragon™ 835) recently became the world's first product built in 10-nm CMOS [3, 4]. Continuing at this aggressive pace, 7-nm finFET products are expected to be commercialized by late 2018.

Analog/mixed-signal (AMS) subsystems are essential in SoCs. Functions such as clocking, I/O connectivity, and core voltage/frequency scaling require a smorgasbord of AMS components including phase-locked loops (PLLs), wireline transceivers, data converters, regulators, thermal sensors, and bandgap references. We address the general considerations faced as we port AMS designs to a finFET node. We also summarize the key scaling innovations preceding the finFET to demonstrate how 16-/14-nm design is complicated by far more than just the new device structure. These are process-induced mechanical strain, high-permittivity

A.L.S. Loke (✉) • E. Terzioglu • A.A. Kumar • T.T. Wee • K. Rim • D. Yang • B. Yu • L. Ge
L. Sun • J.L. Holland • C. Lee • D. Song • S. Yang • J. Zhu • J. Choi • H. Lakdawala • Z. Chen
W.J. Chen • S. Dundigal • S.R. Knol • C.-G. Tan • S.S.C. Song • H. Dang • P.G. Drennan • J. Yuan
P.R. Chidambaram • R. Jalilizeinali • S.J. Dillen • X. Kong • B.M. Leary
Qualcomm Technologies, Inc., 5775 Morehouse Drive, San Diego, CA 92121-1714, USA
e-mail: alvin.loke@ieee.org

P. Harpe et al. (eds.), *Hybrid ADCs, Smart Sensors for the IoT, and Sub-1V & Advanced Node Analog Circuit Design*, DOI 10.1007/978-3-319-61285-0_14

gate dielectric and metal gate (HKMG), multiple and spacer-based patterning, and a substantially more complex middle-end-of-line (MEOL). As with previous SoC nodes, AMS designers continue to adapt to process technology prioritized to the scaling needs of logic and SRAM which dominate the die area and thus dictate the wafer cost.

14.2 Technology Considerations

14.2.1 Short-Channel Limitations in Planar MOSFET

With channel lengths approaching 20 nm, short-channel effects in the planar bulk silicon MOSFET have become ever so severe. At a given supply voltage (V_{DD}), building a smaller device with threshold voltage (V_T) low enough for good on-state drive current while preserving a low off-state leakage remains the perennial challenge, constrained by subthreshold operation. Here, the MOSFET behaves as a BJT whose base voltage, analogously the source-side surface potential (ϕ_S), is induced from coupling to the gate (C_{ox}), body (C_B), and drain (C_D) as depicted in Fig. 14.1 [5]. Leakage is governed by the subthreshold swing *SS* (ΔV_{GS} required for a decade change in subthreshold current) and degraded by *DIBL* (V_T reduction per ΔV_{DS} of 1 V). Good gate control of the silicon surface below a channel length of 1 μm has largely been maintained through gate dielectric thinning for higher C_{ox} but also through channel engineering. Here, higher well doping levels and retrograded well doping profiles, localized halo implants under the gate edge, and shallow source/drain extensions helped to preserve gate control of the surface depletion by suppressing the impact of the source/drain junction depletion regions. ϕ_s can then closely track the gate voltage to modulate the source-to-channel potential barrier and device subthreshold diffusion current. Unfortunately, ϕ_s coupling to the body and drain has been increasing with channel optimization. The resulting tug-of-war

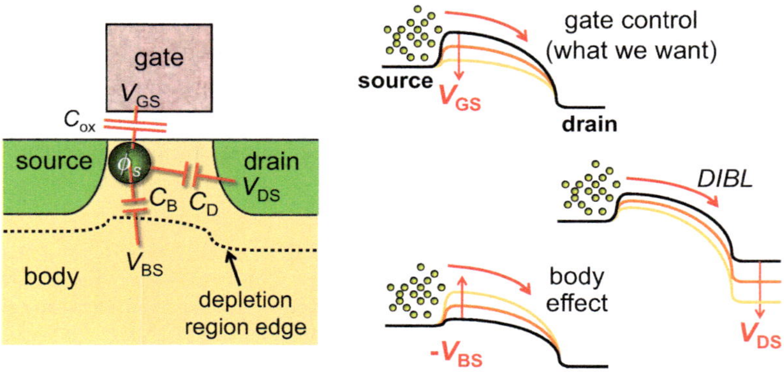

Fig. 14.1 Subthreshold tug-of-war to control ϕ_s

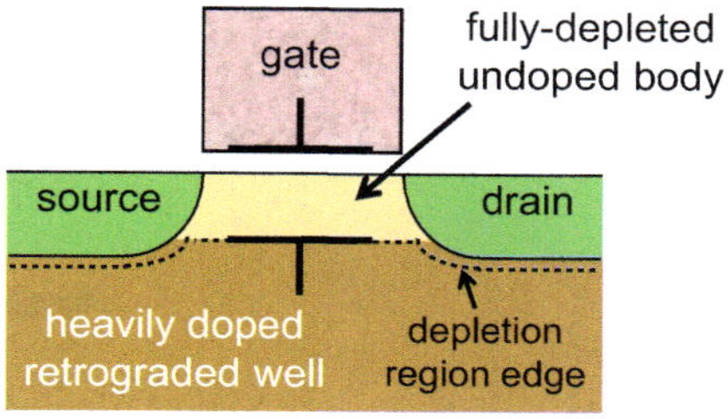

Fig. 14.2 Concept of fully depleted MOSFET

between gate, body, and drain weakens gate modulation of ϕ_s which leads to an increase in both *SS* and *DIBL*. Even in a competitive 32-nm technology, *SS* and *DIBL* as high as 100 mV/decade at 25 °C (vs. the ideal 60-mV/decade in a BJT) and 200 mV/V respectively have already been reported [6]. With the existing device architecture, *SS* and *DIBL* have limited further scaling of V_{DD} and V_T for a given performance and leakage.

14.2.2 Fully Depleted MOSFET Basics

A fully depleted (FD) MOSFET enables V_{DD} and V_T reduction for uncompromised lower-power operation. Conceptually, when the well doping profile of a bulk planar MOSFET is extremely retrograded [7], the undoped body surface becomes fully depleted as it is devoid of ionized dopant fixed charge. This makes the device resemble a parallel plate capacitor with source and drain attached to the plate dielectric as shown in Fig. 14.2. In subthreshold, gate charge is balanced by opposite polarity charge in the heavily doped well region to establish a vertical electric field through the undoped body. This field subsequently induces energy band bending beneath the gate dielectric. At higher V_{GS}, more band bending will eventually create a conductive surface channel to support the on-state drift current. In fact, the channel would be formed in a similar fashion should the body be uniformly doped, a simplifying condition typically assumed in introductory device textbooks [8]. So, fundamentally, a distribution of body dopants at or near the silicon surface is not essential for field-effect action in a MOS system.

Because electric fields from the gate and drain cannot terminate in the charge-free fully depleted region but must instead terminate in the heavily doped region, ϕ_s is weakly coupled to the body and drain, i.e., C_B and C_D are substantially smaller than C_{ox}. The resulting stronger gate control reduces both *SS* and *DIBL* [9, 10]. V_T and V_{DD} can therefore be lowered for a given performance (I_{Dsat} or I_{eff} [11]) and leakage (I_{off}) target to reduce the V_{DD}^2-dependent dynamic switching power. See Fig. 14.3. As an example, the 22-nm tri-gate finFET in [1] demonstrates a *SS* of 70 mV/decade and *DIBL* of 50 mV/V.

The fully depleted structure of Fig. 14.2 can also be realized in a silicon-on-insulator (SOI) substrate [12, 13]. Here, the undoped body is simply replaced by

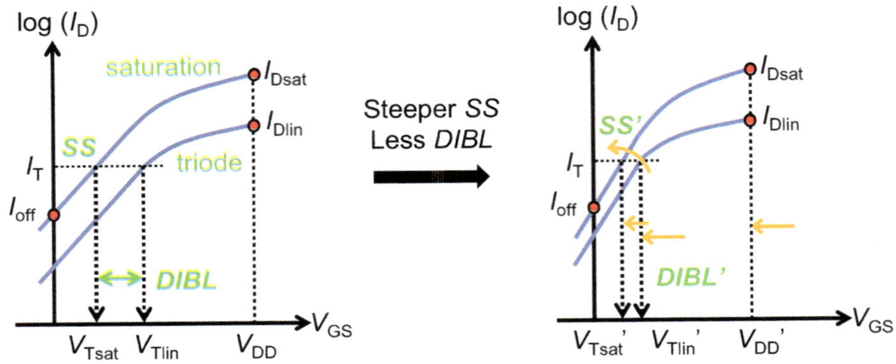

Fig. 14.3 Enabling lower V_{DD} with reduced SS and DIBL

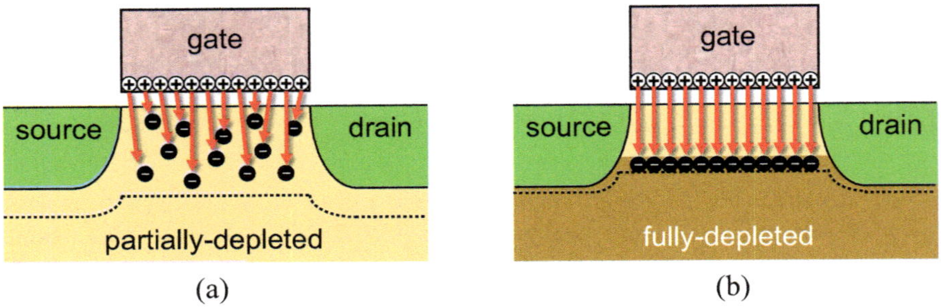

Fig. 14.4 Electric fields in (**a**) partially depleted and (**b**) fully depleted MOSFET

thin undoped SOI and buried oxide layers. Gate charge is balanced by mirror charge in the substrate beneath the buried oxide. Unlike partially-depleted SOI, FD-SOI is not prone to V_T hysteresis because there is no quasi-neutral floating body in the SOI layer. At the 22-nm node, FD-SOI can achieve *SS* of 80–85 mV/decade and *DIBL* of 85–90 mV/V as demonstrated in [13]. FD-SOI does require a more expensive wafer substrate but circumvents the bulk substrate challenge of growing an undoped surface epitaxially at high temperature while suppressing dopant outdiffusion from the heavily doped well region [14].

With the surface ideally undoped, the fully depleted structure enjoys better channel mobility given no ionized impurity scattering. In addition, V_T variation due to random dopant fluctuation (RDF) [15] is eliminated [16]. In a conventional "partially depleted" MOSFET (Fig. 14.4a), dopants in the body vary not only in quantity but also in location due to the stochastic nature of ion implantation. Consequently, there will be variation in the lengths of the electric field lines from the gate to the ionized dopants which integrates to variation in surface band bending and

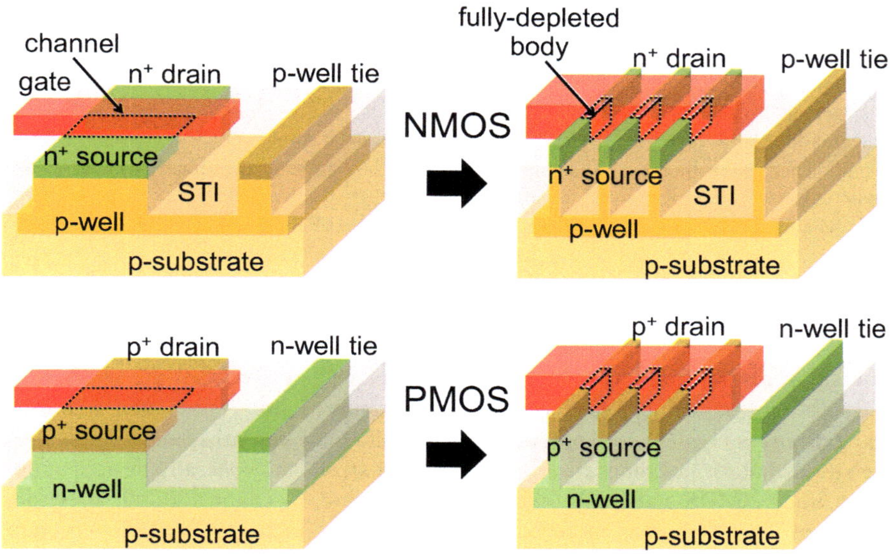

Fig. 14.5 Planar MOSFET transition to fully depleted finFET

V_T. However, in a fully depleted MOSFET (Fig. 14.4b), the lengths of the electric field lines exhibit less variation to realize less V_T variation.

The sensitivity to electric field profile does, however, make fully depleted structures especially sensitive to variations in geometry. For instance, V_T depends on variation in the undoped body thickness in FD bulk, the SOI and buried oxide thicknesses in FD-SOI, and the width, height, and shape of a bulk finFET.

A fully depleted device can be realized as a two- or three-dimensional structure built in bulk or SOI [1, 13, 14, 17]. The bulk tri-gate finFET [1] of Fig. 14.5 has by far become the dominant flavor in high-volume production.

14.2.3 Innovations Leading to FinFET

The 16/14-nm node incorporates several earlier process advances, summarized in Table 14.1, to improve device performance and extend scaling.

14.2.3.1 Mechanical Stressors

Process-induced mechanical strain is employed to boost electron and hole mobilities [18]. Since silicon is piezoresistive, as little as 1% lattice strain can reduce carrier

Table 14.1 Summary of technology innovations leading to finFET

Technology innovation	Foundry debut	Reason required
Mechanical stressors	40 nm	Mobility boost for more FET current
HKMG replacement metal gate integration	28 nm (HK-first) 20 nm (HK-last)	Higher C_{ox} for more FET current and channel control
Multiple and spacer-based patterning	20 nm	Sub-80-nm pitch lithography without EUV
Middle-end-of-line	20 nm	Contact FET diffusion and gate with tighter contacted poly pitch (CPP)

effective mass to amplify mobility several times especially in short channels. The strain results from surrounding the channel with regions that are stressed to 1–3 GPa using techniques such as epitaxial SiGe source/drain, stress memorization, and gate stress [18–20]. To appreciate the magnitude of this stress, device channels are stressed at several times the yield strength of steel! The generally desired strain is tensile in NMOS and compressive in PMOS along the channel length direction. However, the situation is more complex in a three-dimensional finFET structure where the top and sides of the fin exhibit (100) and (110) normals, respectively, and mobility response to stress is highly anisotropic.

Channel stress in each transistor is influenced by its source/drain volume and CPP, and is also perturbed by the stress of other devices in the same active area, surrounding dielectric isolation, and adjacent devices as shown in Fig. 14.6 [4, 21–24]. For instance, the device current per fin depends on the number of contiguous fins and fingers. The three-dimensional finFET structure has also given rise to new complex stress interactions such as the gate-cut stress effect [4]. Here, cutting the gate between adjacent groups of fins disrupts the mechanical support offered by a continuous gate and consequently modulates the stress of fins near the cut. These layout-dependent effects can be responsible for nontrivial post- vs. pre-layout simulation discrepancies that make design closure especially time-consuming and must also be carefully considered for proper device matching. Stress proximity effects are by no means new, but their impact continues to grow with each new node as stressors become stronger. As early as for 130-nm CMOS, the BSIM4.3.0 compact model began capturing the effect of mechanical compression of the active area due to the surrounding shallow trench isolation (STI) [25].

Channel stress can also be impacted by chip-package interactions caused by die attachment to the package [26]. Shown in Fig. 14.7, carrier mobility can vary by as much as 10% especially near the die edge and corners where mostly AMS content typically reside. These long-range gradients can compromise the matching of distributed bias references, data converters, and I/O impedance termination.

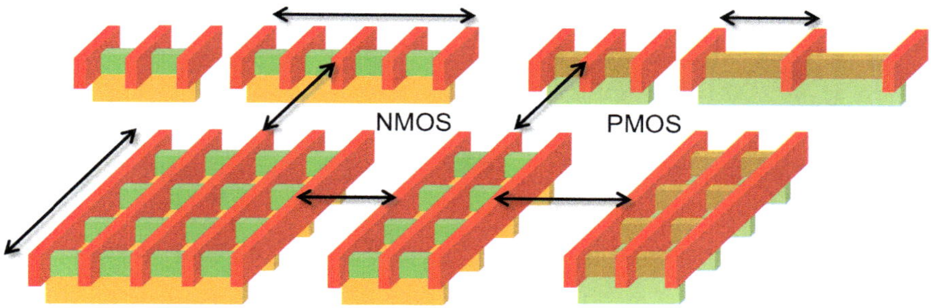

Fig. 14.6 Layout-dependent effects due to stress of surrounding isolation and devices

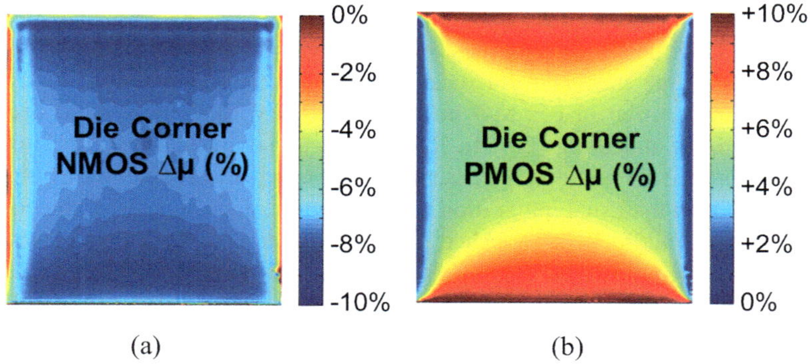

Fig. 14.7 Mobility variation in (**a**) NMOS and (**b**) PMOS resulting from die attachment to package [26]

14.2.3.2 High-Permittivity Gate Dielectric and Metal Gate (HKMG)

Starting at 45 nm [20], HKMG replaced the polysilicon/nitrided oxide gate stack for better short-channel gate control in order to overcome gate tunneling leakage [27] and polysilicon depletion [28]. The higher C_{ox} comes from using a higher permittivity (K) hafnium oxide dielectric (HK) capped by a thin but more conductive metal gate (MG) layer. The remaining gate volume is filled with a lower resistivity metal to reduce gate resistance (R_G). Both HK and MG layers are deposited using atomic layer deposition (ALD) for precise control of thickness and material stoichiometry. The higher K facilitates a thicker gate dielectric to suppress tunneling leakage while maintaining or increasing C_{ox}. The device V_T is controlled by choosing the appropriate work function (ϕ_M) of the MG layer which in turn sets the flatband voltage (V_{FB}). A minimum of two ϕ_M materials is required to replace n^+ and p^+ polysilicon whose work functions are near the conduction and valence band edges, respectively. Instead of using well implants subject to RDF, some HKMG integration schemes employ additional MG materials and HK dipoles to offer multiple V_T flavors with reduced mismatch [3, 17, 29].

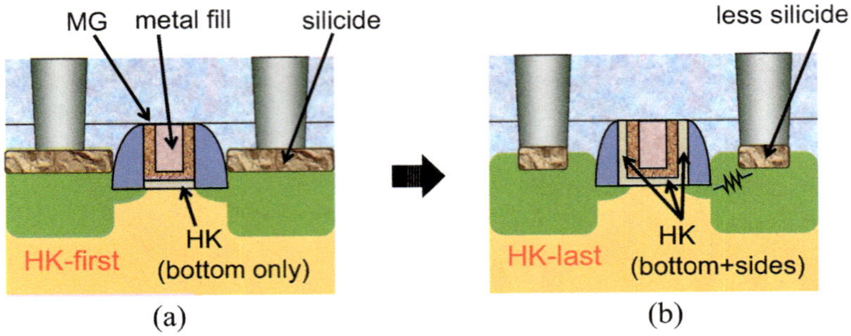

Fig. 14.8 HKMG integration variants: (**a**) HK-first and (**b**) HK-last

As the MG/HK interface is extremely delicate and prone to thermally aggravated ϕ_M instabilities, the gate is formed using a damascene replacement metal gate (RMG) integration that places MG in direct contact with the HK dielectric only after source/drain annealing [20]. In RMG, the polysilicon gate is deposited and patterned after the STI and well-implant modules, and the source/drain junctions are annealed as in the usual integration flow. However, once deposited, the contact-level dielectric is additionally polished to expose the top surface of the polysilicon for complete gate removal. The resulting gate trench is subsequently filled with MG and metal fill, and excess metal outside the trench is removed with another chemo-mechanical polish (CMP).

HKMG was introduced in two iterations of integration: HK-first [20] and HK-last [6]. In HK-first integration, the HK dielectric is deposited early in the process flow just prior to deposition of the sacrificial polysilicon gate (Fig. 14.8a). During RMG, only the polysilicon is removed, leaving behind the underlying HK. This approach is unfortunately prone to front-end process exposure that thickens the HK dielectric at the gate edge, resulting in less gate control. Consequently, one node later, the industry migrated to the HK-last integration where the HK deposition is postponed until just prior to MG deposition (Fig. 14.8b). The late incorporation of the HK dielectric improves gate-to-channel coupling.

RMG introduces several new process constraints. Because CMP is integral to the RMG process, it necessitates reducing the maximum allowed gate length (L_{max}) to contain the extent of metal dishing. Furthermore, the migration from HK-first to HK-last integration imposes several key changes. The silicide module must be postponed until after the contact etch step as silicide cannot tolerate temperatures associated with HK deposition and anneal [30]. This resequencing limits silicide formation to only the bottom of the contact opening in contrast to pre-HKMG and HK-first integration where silicide straps the entire diffusion. To mitigate the resulting device performance degradation from series resistance, trench contacts lining the entire device width have become standard. Lastly, the unsilicided polysilicon precision resistor (with silicided ends) became unavailable and has been replaced by a thin-film MEOL resistor.

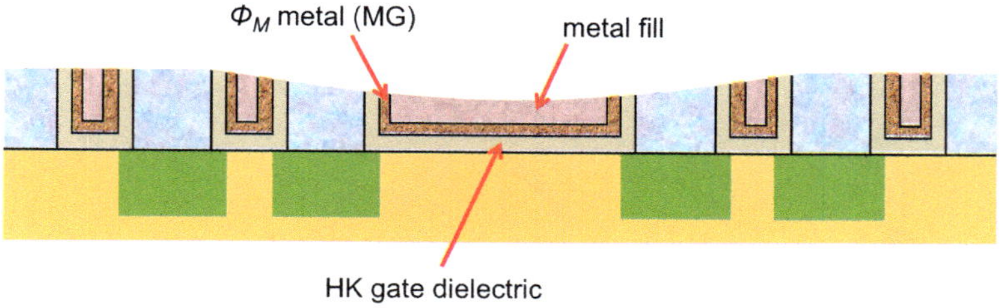

Fig. 14.9 Gate density-induced mismatch

HKMG integration introduces several new sources of V_T mismatch. Shown in Fig. 14.9, one source is gate height variation caused by MG CMP dishing [31]. Because gate charge cannot be completely contained inside the thin MG layer due to the limited MG conductivity but spills into the metal fill, the effective ϕ_M is also influenced by the metal fill work function. Another source of variation is MG grain orientation. ϕ_M depends on the material texture of the deposited polycrystalline MG film which exhibits a distribution of different grain orientations [32].

A new layout-dependent effect known as the metal boundary effect (MBE) has been observed in HKMG [33]. When a single gate is composed of two dissimilar ϕ_M materials for more compact layout (e.g., inverter NMOS and PMOS), interdiffusion of the ϕ_M metals will occur. Therefore, devices close to a boundary between these ϕ_M metals may experience some V_T shift. MBE can be mitigated by increasing the device separation and even eliminated by unsharing the gate, both at the expense of layout bloat.

14.2.3.3 Multiple and Spacer-Based Patterning

Several novel techniques using conventional 193-nm ArF immersion lithography have been introduced to enable even more aggressive layout rules. The 13.5-nm extreme ultraviolet (EUV) infrastructure will only be available starting at the 7-nm node. See Fig. 14.10.

Introduced at 45 nm for SRAM bit cell size reduction [20], orthogonal cut masks help to reduce line end-to-end spacing (Fig. 14.10a). To print below the single-exposure pitch limit of 80 nm, pitch splitting was introduced at 20 nm for contacts and the lowest metal layers [34]. It involves printing alternate lines with a first exposure (Mask A) and the remaining unprinted lines with a second exposure (Mask B) (Fig. 14.10b). Here, multiple patterning adds the complexity of layout metal coloring as well as new rules to cope with misalignment between the two masks. Pitch splitting can be extended by incorporating additional exposures to achieve even finer pitch. Another sub-resolution technique is spacer-based patterning, also known as self-aligned double-patterning or sidewall image transfer. This technique is used not only for patterning fins [1] but also increasingly for

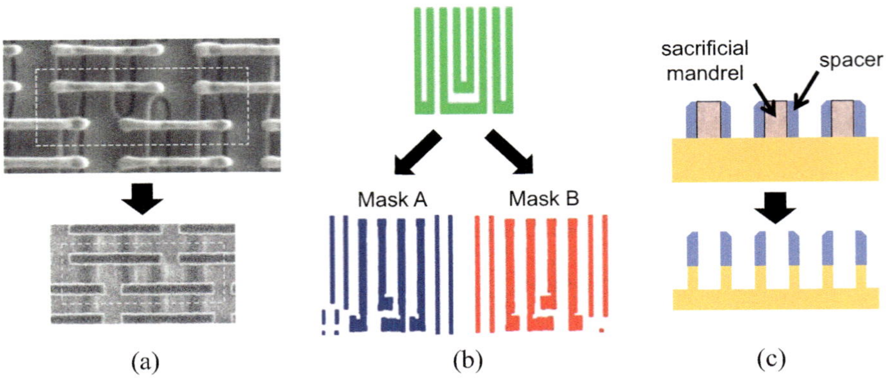

Fig. 14.10 Lithography innovations: (**a**) cut mask [20], (**b**) pitch splitting [34], and (**c**) spacer-based patterning [1]

patterning short-channel gates with CPP below 80 nm. Spacer-based patterning is also used in conjunction with conventional lithography in creative schemes to print fine interconnect features [35]. Spacers are formed on the sidewalls of a sacrificial mandrel grating that is subsequently removed, leaving behind a sea of spacers at half the mandrel pitch to pattern the underlying material (Fig. 14.10c). Spacer-based patterning is limited to a pitch of 40 nm but can be applied recursively. In self-aligned quadrature patterning, the spacers produced from the original mandrel (Spacer 1) serve as the mandrel for forming a second set of spacers (Spacer 2) at one quarter the pitch of the original mandrel.

14.2.3.4 Middle-End-Of-Line (MEOL)

Tighter CPP has resulted in a complex and far costlier MEOL to contact the devices in 16/14 nm [26, 36]. See Fig. 14.11a. This contrasts with the 28-nm MEOL which consists of only a single mask module to form both diffusion and gate contacts. The MEOL is typically the most problematic module to yield. As such, despite a higher R_G penalty especially as the gate traverses over the fins (Fig. 14.11b) [37], starting at the 22 nm node, self-aligned source/drain contacts (SACs) with full dielectric encapsulation of the gate are built to eliminate overlay-related contact-to-gate shorts [1]. See Fig. 14.12. Furthermore, diffusion contacts and gate contacts are formed independently to overcome process difficulties associated with a single module for both sets of contacts. Finally, an additional via level is inserted to bridge these contacts to Metal-1. The combination of finer geometries and additional contact interfaces (whose quality always dominates the overall contact resistance) have added substantially more resistance in the MEOL [26].

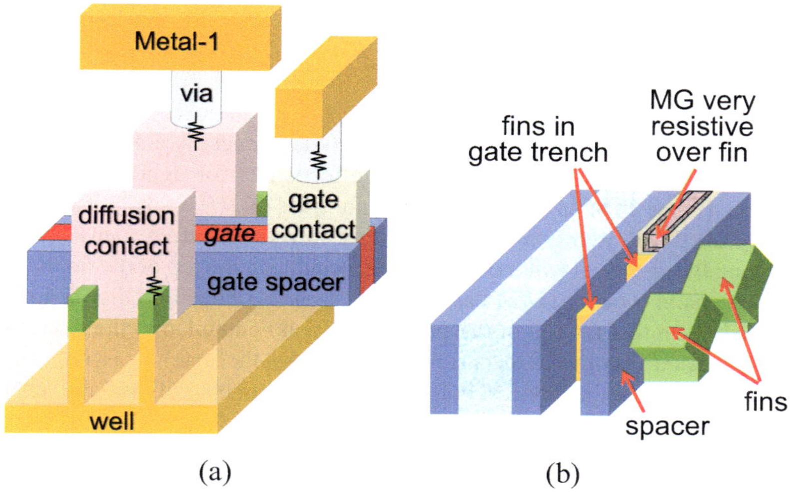

Fig. 14.11 (**a**) Complex MEOL to contact finFET [26], (**b**) corresponding high R_G

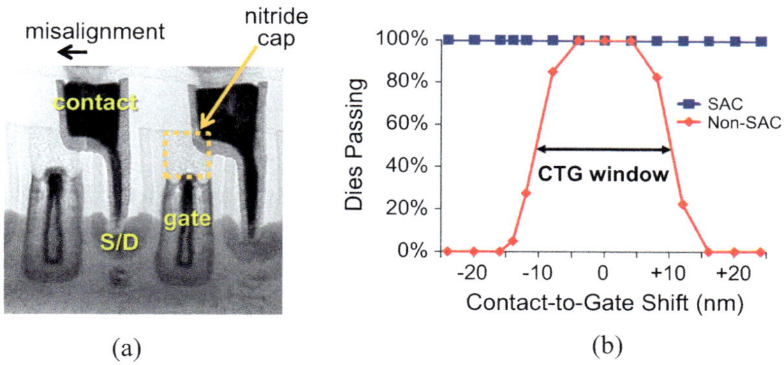

Fig. 14.12 (**a**) Self-aligned contact and (**b**) its resilience to misalignment [1]

14.3 Design Considerations

Porting a design to finFET requires some design re-optimization to address the new device structure and aforementioned technology constraints. To little surprise, these process complexities have spawned many more restrictive layout design rules which have significantly increased the design closure effort.

14.3.1 General

14.3.1.1 Channel Width Quantization

Fin dimensions (height H_{fin}, width W_{fin}, and pitch) are generally uniform throughout the die to reduce lithography and etch load effects, i.e., for tight process control, forcing the tri-gate channel width W_{eff} to be multiples of $2H_{\text{fin}} + W_{\text{fin}}$. This quantity is an estimate for design convenience because in reality, current density actually varies along the electrical width of the device, peaking at the top of the fin where the onset of inversion occurs and decreasing along the sides of the fin. This phenomenon is expected given that the top of the fin is most weakly coupled to the device body node (undepleted portion of the fin not wrapped by the gate). It is important to realize that H_{fin} is not the physical height of the entire fin but only the amount of fin protrusion above the STI oxide.

W_{eff} quantization is a challenge for logic and SRAM design and has, for instance, led to the growing use of SRAM assist techniques [38] where variable gate and bitline voltages are used to modulate the strength of bit cell devices with finer resolution. However, its impact on analog design is minimal as the gate transconductance, g_m, under typical biasing conditions is granular enough (10–100 μA/V per fin) for sufficient design flexibility.

14.3.1.2 FinFET Advantages

Compared to planar CMOS, device density is the clear winner in a finFET node with W_{eff} being 1.5–3 times the width projected onto the wafer plane. Given the superior gate control in a fully depleted structure, the body effect in a finFET is practically nonexistent ($\Delta V_T < 10$ mV for $|\Delta V_{BS}| = V_{DD}$) [39]. Hence, there is no stack penalty such as in the NMOS pulldown network of a NAND gate. Furthermore, reduced *DIBL* means more effective drive current (I_{eff}) for better digital performance [9] and more ideal analog behavior with 3× better intrinsic gain [40]. Defined as the average of drain current (1) at $V_{GS} = V_{DD}$ and $V_{DS} = V_{DD}/2$ and (2) at $V_{GS} = V_{DD}/2$ and $V_{DS} = V_{DD}$, I_{eff} is an increasingly preferred metric for logic performance because it is a more accurate average than I_{Dsat} of the switching current in an inverter. Without RDF, V_T variation is also reduced by about 30% [40], much to the benefit of the SRAM minimum supply voltage $V_{\min}$. These benefits may reopen opportunities for precision analog approaches that became unfeasible in recent planar CMOS nodes.

14.3.1.3 FinFET Disadvantages

The compact finFET structure comes at the cost of more parasitics. Source/drain resistance is dramatically higher as currents must funnel from wide trench contacts into the narrow fins. The extrinsic source/drain resistance is particularly high given

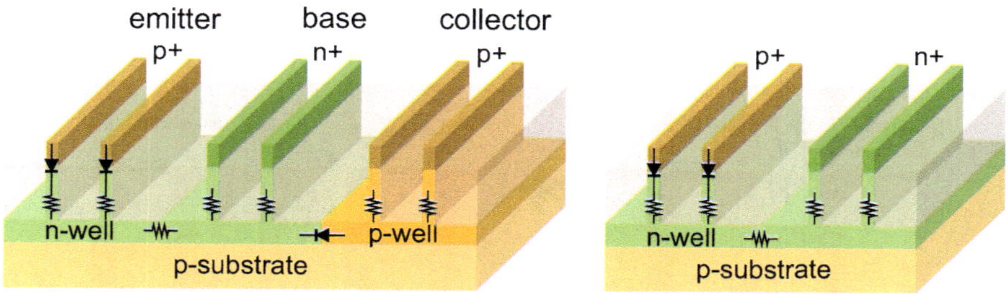

(a) (b)

Fig. 14.13 (**a**) PNP BJT and (**b**) STI ESD diode in finFET technology

limited silicidation. C_{GS} and C_{GD} are also higher due to gate coupling to the trench diffusion contacts and to the epitaxial source/drain fill between fins. As W_{fin} is a small fraction of the fin pitch, junction capacitance to the wells is reduced, but the vertical well resistance is much higher. This enforces stricter well-tie and latch-up layout rules and increases series resistance in analog and ESD diodes (see Fig. 14.13). Higher R_G especially in short-channel gates also exacerbates non-quasistatic behavior, making the design of circuits such as RF low-noise amplifiers especially challenging. Finally, self-heating is worse given the higher area density of device currents, elevating both device and metallization reliability concerns.

14.3.2 Analog/Mixed-Signal

14.3.2.1 Transistor Parasitics

AMS designs, for the most part, can be ported to 16/14 nm with expected node-to-node adjustments. However, some designs are increasingly difficult due to growing parasitics and accumulating layout constraints from continued scaling.

Techniques to reduce resistance are becoming vitally important even at the expense of increased capacitance. For example, the double-source layout of Fig. 14.14 is now a common technique to combat high contact resistance and resulting supply droop especially in circuits such as I/O transmitters and clock drivers that need to drive large loads.

Worse parasitic capacitance is also troublesome. For example, in Fig. 14.15a, higher C_{GD} (Miller) coupling in a low-dropout (LDO) regulator with a PMOS pass element translates to worse high-frequency supply-noise rejection. In another example (Fig. 14.15b), higher C_{GS} can inject substantial V_{ref} kickback noise in say an LPDDR receiver commonly implemented as a PMOS differential amplifier with one input tied to the V_{ref} threshold. Such non-idealities can be mitigated by incorporating more capacitive clamping which obviously costs more area as well as longer circuit start-up time. Slower circuit wake-up is increasingly undesirable,

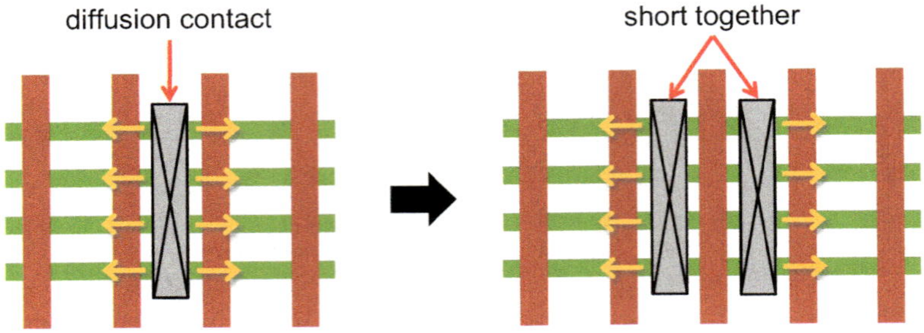

Fig. 14.14 Double-source layout to halve contact resistance

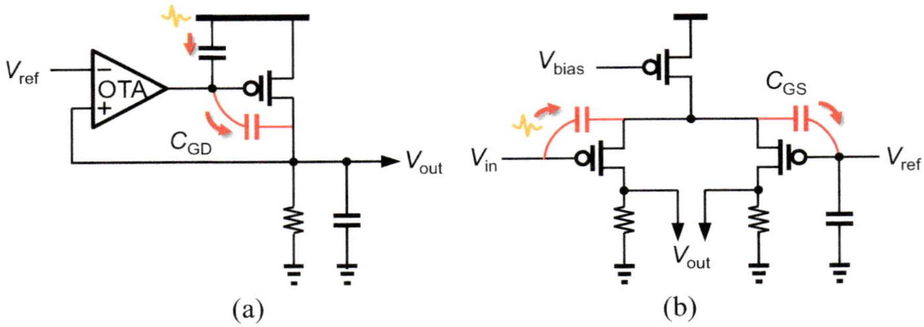

Fig. 14.15 (**a**) LDO regulator with PMOS pass element and (**b**) LPDDR receiver

especially in mobile ICs, as many AMS subsystems support on demand burst modes for power saving. As designers become more familiar with finFETs though, clever solutions such as anti-kickback circuits [41] are emerging.

14.3.2.2 Stacked FET

The stacked FET of Fig. 14.16 has become ubiquitous for building current sources as L_{max} continues to shrink. The desired higher r_{out} is realized through resistive source degeneration of the top device in the stack operating in saturation. Area bloat is incurred by the intermediate diffusions, but r_{out} into the gigahertz range is also worse as these diffusions electrically short to ground, degrading analog metrics such as intrinsic gain and common-mode noise rejection.

14.3.2.3 PNP (Analog Diode)

Conventional bandgap references and temperature sensors use diode-connected PNP BJTs and face similar non-idealities in a finFET technology.

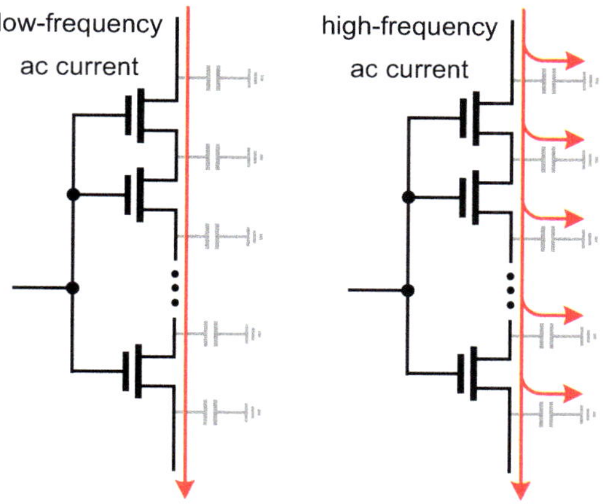

Fig. 14.16 Stacked MOSFET r_{out} at low and high frequencies

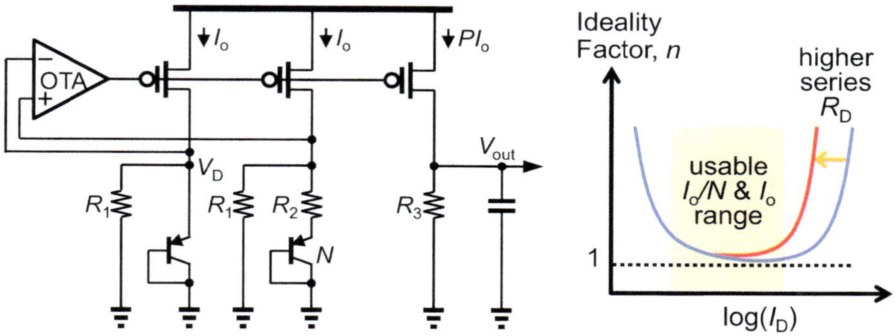

Fig. 14.17 Low-voltage bandgap reference

Figure 14.17 illustrates the typically used low-supply variant of a bandgap reference [42]. Higher base-to-emitter diode series resistance (R_D) narrows the allowable bias current window with relatively constant ideality factor (n) where ΔV_{BE} can be expressed as the logarithm of the PNP current ratio (N). Smaller values of N must therefore be used to generate the proportional-to-absolute-temperature (PTAT) ΔV_{BE} more accurately. Even then, some ideality degradation is expected in a finFET vs. pre-HKMG planar PNP because base-emitter current can flow vertically along the STI edge where it is subject to surface recombination. In pre-HKMG nodes, the ideality factor was improved by using the silicide block oxide (no longer available in a finFET node) to provide an enclosure ring around the emitter that kept current away from the active area edge.

The diode forward voltage (V_D) has also increased in 16/14 nm from heavier fin doping beneath the STI level to suppress short-channel subsurface punchthrough leakage. Given that the core logic V_{DD} is already reduced in a finFET node, this bandgap reference topology may require a separate higher supply to address the limited voltage headroom imposed by V_D.

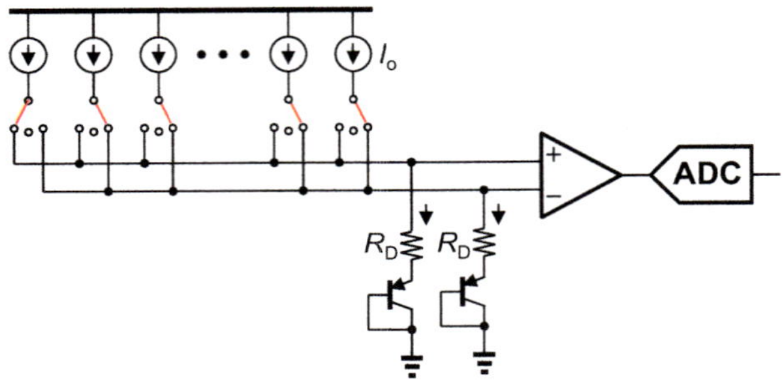

Fig. 14.18 Conventional diode-based temperature sensor

The conventional temperature sensor with $N + 1$ identical current sources is shown in Fig. 14.18. Here, we generate two ratioed currents, NI_o and I_o, that flow into a pair of identical PNPs and then sense and digitize the resulting ΔV_{BE}. Resolution is improved by employing dynamic element matching to eliminate random mismatch effects, i.e., repeat measurements by interchanging the current sources as well as the PNPs [43].

For better temperature accuracy, the effect of R_D must be eliminated. To do this, we measure ΔV_{BE} twice: first with a current ratio of N:1 to obtain $\Delta V_{BE,N}$ and then with a different current ratio of M:1 (where $M < N$) to get $\Delta V_{BE,M}$ [44].

$$\Delta V_{BE,N} = \frac{nkT}{q} \ln N + (N - 1)\, I_0 R_D \tag{14.1}$$

$$\Delta V_{BE,M} = \frac{nkT}{q} \ln M + (M - 1)\, I_0 R_D \tag{14.2}$$

We then solve (14.1) and (14.2) to extract the absolute temperature:

$$\frac{nkT}{q} = \frac{(N - 1)\, \Delta V_{BE,M} - (M - 1)\, \Delta V_{BE,N}}{(N - 1) \ln M - (M - 1) \ln N} \tag{14.3}$$

14.3.2.4 Varactor

The accumulation-mode varactor of Fig. 14.19 remains available in a finFET node primarily for supply noise decoupling. It is commonly leveraged for fine-tuning the frequency of an LC-VCO. The thick-oxide flavor is frequented for lower gate leakage. Several factors must be considered. Because the finFET is a fully depleted device, less ΔV_G is required to transition from accumulation to inversion in the small-signal capacitance–voltage (C_G vs. V_G) characteristic. This steeper transition translates to the benefit of higher PLL VCO gain (dC_G/dV_G) but, depending on circuit topology, may require more careful biasing to bias the varactor into its

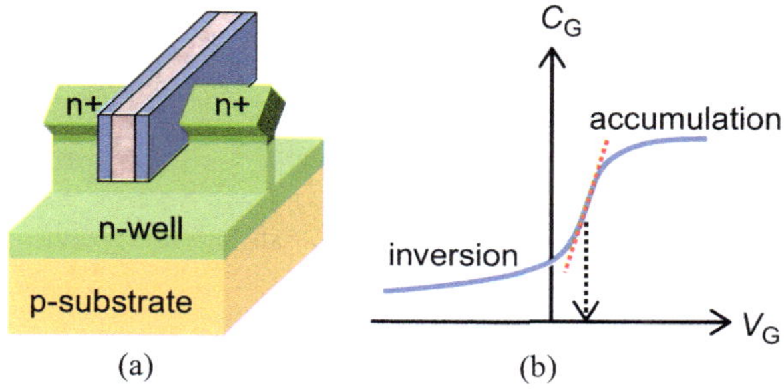

Fig. 14.19 (**a**) Varactor in finFET technology and (**b**) C_G–V_G characteristic

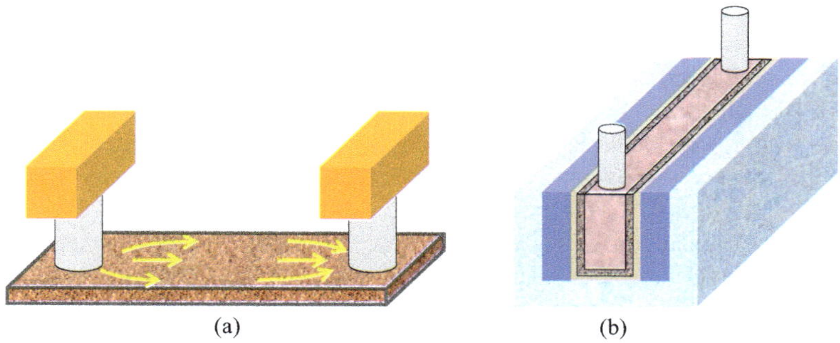

Fig. 14.20 Resistor options: (**a**) MEOL resistor and (**b**) gate resistor

narrower voltage window of useful tuning. In fact, for this very reason, quarter gap as opposed to band-edge gate work function materials are required in fully depleted devices to properly target V_T from being too low [45]. Lastly, as C_{GS} and C_{GD} are higher in a finFET-based varactor, tuning range will inevitably be compromised.

14.3.2.5 Passive Elements (Resistors, Capacitors, and Inductors)

In a finFET node, the precision resistor of choice for AMS applications is the thin-film MEOL resistor shown in Fig. 14.20a. It replaced the polysilicon resistor in the transition to HK-last integration. The MEOL resistor is composed of a thin refractory metal compound that is deposited and subtractively etched prior to the metallization module. It is specifically built for analog/mixed-signal usage, so its integration is decoupled from the finFET. The MEOL resistor is necessarily thin to obtain a small material grain structure that is electrically dominated by surface and grain boundary (as opposed to bulk) scattering in order to realize a lower temperature coefficient of resistance. Unlike the polysilicon resistor which has well-defined, low-resistance silicided resistor ends, the MEOL resistor is more prone to current spreading near its contacts. The finFET gate of Fig. 14.20b can also be leveraged as a resistor. Unfortunately, it is not well controlled and is plagued

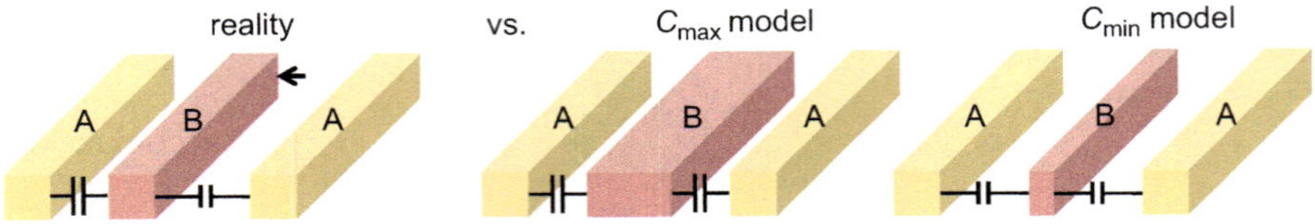

Fig. 14.21 Nonphysical modeling of misalignment in double-patterned metal

by many sources of variation including RMG CMP and the SAC gate recess etch. Moreover, its width is limited by the transistor L_{max}.

Linear capacitors are still implemented using metal-oxide-metal (MOM) fingers in the interconnect stack. The area density of MOM capacitors is one of only a handful of scaling consequences that benefits analog design. Designers have to be cautious though of technology-imposed modeling subtleties. For example, corner models of the lowest metal layers may account for double-patterning misalignment in a nonphysical way shown in Fig. 14.21 where Mask B has been misaligned to the left relative to Mask A.

Planar metal-insulator-metal (MIM) capacitors built with a high-permittivity dielectric like hafnium oxide are sometimes available. As they require additional processing, MIM capacitors are usually only justified in more expensive ICs such as high-performance servers [1] that can command a premium profit margin.

The migration to finFET impacts inductor design minimally as the highest thick layers of interconnect metal used for forming planar spirals generally do not scale so as to preserve low enough resistance to mitigate voltage droop in supply distribution. Subtle degradation of inductor Q will occur as interconnect scaling at lower metal layers requires more dummy fill surrounding the inductor for tighter CMP pattern density control to minimize accumulation of topography.

14.3.2.6 Thick-Oxide Devices and I/O Supply Scaling

Although V_{DD} scaling has helped to reduce core power tremendously in finFET nodes, the supply voltage for sub-gigahertz general-purpose I/Os (GPIOs) remains mostly at 1.8 V which imposes several technology and design challenges. FinFETs with thicker gate dielectrics are increasingly difficult to build because a tighter fin pitch requires more aggressive ALD MG fill capability for the thicker oxide [46]. Also, the larger separation between core and I/O voltages leads to more complex voltage level shifter designs. Historically, the GPIO voltage has scaled from 5.0 V to 3.3 V and eventually 1.8 V. System ecosystem consensus is needed to lower the GPIO supply voltage yet again, but little cost motivation exists for designs staying in cheaper legacy nodes. However, high-performance memory interfaces like LPDDR4X, which require a higher DRAM die supply, are migrating to lower and more SoC-friendly signaling voltages to enable the SoC I/O supply to scale [41].

14.3.2.7 Limitation of V_T-Based Design Principles

Supply voltages have scaled to the point where traditional V_T-based analog design principles can at best be loosely applied, finFET nodes being no exception. Given limited voltage headroom, transistors operating in saturation have been biased with gate overdrive ($V_{GS} - V_T$) and saturation margin ($V_{DS} - V_{DSAT}$) as low as 50 mV going as far back as the 45-nm node. These voltage levels are basically drowned in the ΔV_{GS} required to transition from subthreshold to weak inversion, making impossible a clear demarcation between "off" and "on" regions of operation. This ambiguity has spawned a host of practical though less convenient design metrics such as current efficiency (g_m/I_D) [47], inversion coefficient [48], and r_{out}-based saturation margin [49] for optimum analog biasing.

V_T is a cumbersome quantity to define. Based on the inversion condition $\phi_s = -\phi_b$ where ϕ_b is the bulk potential, the traditional V_T definition is not only electrically immeasurable but more fundamentally inapplicable to a fully depleted device where ϕ_b vanishes if the body is undoped. The BSIM-CMG ϕ_s-based model defines V_T as V_{GS} at which the superthreshold drift current matches the subthreshold diffusion current as the channel forms [50]. Although this definition is theoretically unique, this condition also cannot be measured. As a result, foundries typically measure and report the constant-current V_T [51], defined as V_{GS} corresponding to the threshold current.

$$I_T = I_0 \frac{W_{eff}}{L_{eff}} \tag{14.4}$$

I_0 is a foundry-specific parameter, and with typical values in the 10–100 nA range, the device is typically still operating in subthreshold at the constant-current V_T condition. Because I_0 is somewhat arbitrary, the constant-current V_T should be used as a reference point along the I_D–V_{GS} curve for direct model-to-silicon comparison and not for computing gate overdrive and saturation margin reliably.

14.4 Conclusion

With SoC finFET technologies already in production for several years, AMS designs have clearly migrated to finFET nodes without showstoppers. AMS designers are pressed to understand process technology even more than ever in order to anticipate its impact on design. HKMG, MEOL, and finFET parasitics as well as their layout-related effects and constraints have already increased AMS design effort significantly. As we march toward the remaining few CMOS nodes, this design landscape will stay on course as logic and SRAM needs continue to dictate technology priorities.

Acknowledgments We wish to extend sincere thanks to Yanxiang Liu, Michael Brunolli, Ray Stephany, Andy Wei, Bich-Yen Nguyen, Jung-Suk Goo, Shawn Searles, Dennis Fischette, Larry Bair, Jia Feng, Simon Wong, and Marcel Pelgrom for valuable discussions. The first author is especially indebted to Tin Tin, Theo, and Josephene for their love, support, and encouragement.

References

1. Auth, C., et al.: A 22nm high performance and low-power CMOS technology featuring fully-depleted tri-gate transistors, self-aligned contacts and high density MIM capacitors. In: IEEE Symposium on VLSI Technology Technical Digest, Honolulu, HI, June 2012, pp. 131–132
2. Wu, S.-Y., et al.: A 16nm finFET CMOS technology for mobile SoC and computing applications. In: IEEE International Electron Devices Meeting Technical Digest, Washington, DC, Dec. 2013, pp. 224–227
3. Cho, H.-J., et al.: Si finFET based 10nm technology with multi Vt gate stack for low power and high performance applications. In: IEEE Symposium on VLSI Technology Technical Digest, Honolulu, HI, June 2016, pp. 12–13
4. Yang, S., et al.: 10 nm high performance mobile SoC design and technology co-developed for performance, power, and area scaling. In: IEEE Symposium on VLSI Technology Technical Digest, Kyoto, Japan, June 2017, pp. 70–71
5. King Liu, T.-J.: Bulk CMOS scaling to the end of the roadmap. In: IEEE Symposium on VLSI Circuits, Short Course, Honolulu, HI, June 2012
6. Packan, P., et al.: High performance 32nm logic technology featuring 2nd generation high-k + metal gate transistors, In: IEEE International Electron Devices Meeting Technical Digest, Baltimore, MD, Dec. 2009, pp. 659–662
7. Yan, R.-H., Ourmazd, A., Lee, K.F.: Scaling the Si MOSFET: From bulk to SOI to bulk. IEEE Transactions on Electron Devices. **39**(7), 1704–1710 (July 1992)
8. Muller, R.S., Kamins, T.I.: Device electronics for integrated circuits, 2nd edn. Wiley, New York (1996)
9. Wei, L., Boeuf, F., Antoniadis, D., Skotnicki, T., Wong, H.-S.P.: Exploration of device design space to meet circuit speed targeting 22nm and beyond. In: Proceedings of the International Conference on Solid State Devices and Materials, Miyagi, Japan, Sep. 2009, pp. 808–809
10. Skotnicki, T.: CMOS technologies – trends, scaling and issues. In: IEEE International Electron Devices Meeting, Short Course, San Francisco, CA, Dec. 2010
11. Na, M.H., Nowak, E. J., Haensch, W., Cai, J.: The effective drive current in CMOS inverters. In: IEEE International Electron Devices Meeting Technical Digest, San Francisco, CA, Dec. 2002, pp. 121–124
12. Colinge, J.-P.: Silicon-On-Insulator Technology: Materials to VLSI. Kluwer Academic Publishers, Norwell (1991)
13. Cheng, K., et al.: Fully depleted extremely thin SOI technology fabricate by a novel integration scheme featuring implant-free, zero-silicon-loss, and faceted raised source/drain. In: IEEE Symposium on VLSI Technology Technical Digest, Kyoto, Japan, June 2009, pp. 212–213
14. Fujita, K., et al.: Advanced channel engineering achieving aggressive reduction of V_T variation for ultra-low power applications. In: IEEE International Electron Devices Meeting Technical Digest, Dec. 2011, pp. 749–752
15. Stolk, P.A., Widdershoven, F.P., Klaassen, D.B.M.: Modeling statistical dopant fluctuations in MOS transistors. IEEE Transactions on Electron Devices. **45**(9), 1960–1971 (Sep. 1998)
16. Asenov, A.: Suppression of random dopant-induced threshold voltage fluctuations in sub-0.1-μm MOSFET's with epitaxial and δ–doped channels. IEEE Transactions on Electron Devices. **46**(8), 1718–1724 (Aug. 1999)

17. Seo, K.-I., et al.: A 10nm platform technology for low power and high performance application featuring FINFET devices with multi work function gate stack on bulk and SOI. In: IEEE Symposium on VLSI Technology Technical Digest, Honolulu, HI, June 2014
18. Chan, V., Rim, K., Ieong, M., Yang, S., Malik, R., Teh, Y.W., Yang, M., Ouyang, Q.: Strain for CMOS performance improvement. In: Proceedings of the IEEE Custom Integrated Circuits Conference, San Jose, CA, Sep. 2005, pp. 667–674
19. Liu, Y., et al.: NFET effective work function improvement via stress memorization technique in replacement metal gate technology. In: IEEE Symposium on VLSI Technology Technical Digest, Kyoto, Japan, June 2013, pp. 198–199
20. Auth, C., et al.: 45nm high-k + metal-gate strain-enhanced transistors. In: IEEE Symposium on VLSI Technology Technical Digest, Honolulu, HI, June 2008, pp. 128–129
21. Faricelli, J.: Layout-dependent proximity effects in deep nanoscale CMOS. In: Proceedings of the IEEE Custom Integrated Circuits Conference, San Jose, CA, Sep. 2010
22. Garcia Bardon, M. et al.: Layout-induced stress effects in 14nm & 10nm finFETs and their impact on performance. In: IEEE Symposium on VLSI Technology Technical Digest, Kyoto, Japan, June 2013, pp. 114–115
23. Lee, C., et al.: Layout-induced stress effects on the performance and variation of finFETs. In: IEEE International Conference on Simulation of Semiconductor Processes and Devices, Washington, DC, Sep. 2015, pp. 369–372
24. Sato, F., et al.: Process and local layout effect interaction on a high performance planar 20nm CMOS. In: IEEE Symposium on VLSI Technology Technical Digest, Kyoto, Japan, June 2013, pp. 116–117
25. Xi, X., et al.: BSIM4.3.0 MOSFET Model – User's Manual. Regents University California, Berkeley (2003)
26. Terzioglu, E.: Design and technology co-optimization for mobile SoCs. In: International Conferance on IC Design & Technology, Leuven, Belgium, June 2015
27. McPherson, J.: Reliability trends with advanced CMOS scaling and the implications on design. In: Proceedings of the IEEE Custom Integrated Circuits Conference, Sep. 2007, pp. 405–412
28. Wong, P.: Beyond the conventional transistor. IBM Journal of Research and Development. **2–3**(46), 133–168 (Mar. 2002)
29. Kang, C.Y., et al.: The impact of La-doping on the reliability of low Vth high-k/metal gate nMOSFETs under various gate stress conditions. In: IEEE International Electron Devices Meeting Technical Digest, San Francisco, CA, Dec. 2008
30. Hou, C.: A smart design paradigm for smart chips. In: IEEE International Solid-State Circuits Conference Technical Digest, San Francisco, CA, Feb. 2017, pp. 8–13
31. Yang, S., et al.: High-performance mobile SoC design and technology co-optimization to mitigate high-K metal gate process variations. In: IEEE Symposium on VLSI Technology Technical Digest, Honolulu, HI, June 2017
32. Dadgour, H.F, Endo, K., De, V.K., Banerjee, K.: Modeling and analysis of grain-orientation effects in emerging metal-gate devices and implications for SRAM reliability. In: IEEE International Electron Devices Meeting Technical Digest, San Francisco, CA, Dec. 2008
33. Hamaguchi, M., et al.: New layout dependency in high-K/metal gate MOSFETs. In: IEEE International Electron Devices Meeting Technical Digest, Washington, DC, Dec. 2011, pp. 579–582
34. Dorsch, J.: Changes and challenges abound in multi-patterning lithography. Semiconductor Manufacturing & Design Community, www.semi.org/en/node/54491, Feb. 2015
35. Woo, Y., Ichihashi, M., Parihar, S., Yuan, L., Banna, S, Kye, J.: Design and process technology co-optimization with SADP BEOL in sub-10nm SRAM bitcell. In: IEEE International Electron Devices Meeting Technical Digest, Washington, DC, Dec. 2015, pp. 273–276
36. Rashed, M., et al.: Innovations in special constructs for standard cell libraries in sub 28nm technologies. In: IEEE International Electron Devices Meeting Technical Digest, Washington, DC, Dec. 2013, pp. 248–251
37. Wu, W., Chan, M.: Gate resistance modeling of multifin MOS devices. IEEE Electron Device Letters. **27**(1), 68–70 (Jan. 2006)

38. Wang, Y.: Embedded memory design in CMOS finFET technology. In: IEEE Symposium on VLSI Circuits, Short Course, Honolulu, HI, June 2016
39. Sheu, B.: Circuit design using finFETs. In: IEEE International Solid-State Circuits Conference, Tutorial T4, San Francisco, CA, Feb. 2013
40. Hsueh, F.-L., et al.: Analog/RF wonderland: circuit and technology co-optimization in advanced finFET technology. In: IEEE Symposium on VLSI Technology Technical Digest, Honolulu, HI, June 2016, pp. 114–115
41. Lee, C.-K., et al.: A 5Gb/s/pin 8Gb LPDDR4X SDRAM with power-isolated LVSTL and split-die architecture with 2-die ZQ calibration scheme. In: IEEE International Solid-State Circuits Conference Technical Digest, San Francisco, CA, Feb. 2017, pp. 390–391
42. Banba, H., Shiga, H., Miyaba, T., Tanzawa, T., Atsumi, S., Sakui, K.: A CMOS bandgap reference circuit with sub-1-V operation. IEEE Journal of Solid-State Circuits. **34**(5), 670–673 (May 1999)
43. Pertijs, M.A.P., Huilsing, J.H.: Precision temperature sensors in CMOS technology. Springer, Dordrecht (2006)
44. ADT7461 $\pm 1°$C temperature monitor with series resistance cancellation. ON Semiconductor Publication No. ADT7461/D, Mar. 2014
45. Chang, L., Tang, S., King, T.-J., Bokor, J., Hu, C.: Gate length scaling and threshold voltage control of double-gate MOSFETs. In: IEEE International Electron Devices Meeting Technical Digest, San Francisco, CA, Dec. 2000, pp. 719–722
46. Wei, A., et al.: Challenges of analog and I/O scaling in 10nm SoC technology and beyond. In: IEEE International Electron Devices Meeting Technical Digest, San Francisco, CA, Dec. 2014, pp. 462–465
47. Silveira, F., Flandre, D., Jespers, P.G.A.: A g_m/I_D based methodology for the design of CMOS analog circuits and its application to the synthesis of a silicon-on-insulator OTA. IEEE Journal of Solid State Circuits. **31**(9), 1314–1319 (Sep. 1996)
48. Sansen, W.: Analog CMOS from 5 micrometer to 5 nanometer. In: IEEE International Solid-State Circuits Conference Technical Digest, San Francisco, CA, Feb. 2015, pp. 22–27
49. Feng, J., et al.: Bridging design and manufacture of analog/mixed-signal circuits in advanced CMOS. In: IEEE Symposium on VLSI Technology Technical Digest, Kyoto, Japan, June 2011, pp. 226–227
50. Khandelwal, S., et al.: BSIM-CMG 110.0.0 Multi-Gate MOSFET Compact Model Technical Manual. Regents University California, Berkeley (2015)
51. Loke, A.L.S., Wu, Z.-Y., Moallemi, R., Cabler, C.D., Lackey, C.O., Wee, T.T., Doyle, B.A: Constant-current threshold voltage extraction in HSPICE for nanoscale CMOS analog design. In: Synopsys Users Group (SNUG) Conference, San Jose, CA, Mar. 2010

Chapter 15
Analog Circuits in 28 nm and 14 nm FinFET

Lukas Dörrer, Franz Kuttner, Francesco Conzatti, and Patrick Torta

15.1 Introduction

The introduction of a new technology node is always followed by a discussion between experts about its usability for analog and RF design. By changing to FinFET design, this discussion relates to the fact that the transistor world suddenly is quantized. The designer has the choice of using one, two, or any discrete number of transistors in parallel or stacked. No analog width and length scaling are possible anymore. Nevertheless, it is possible to migrate analog and RF designs to FinFET technologies. We proved this in our testchips and can furthermore claim that the difficulties for analog and RF design in FinFET are not located in this quantization of the transistors but rather in the limited wavelength of the lithography, which leads to double or in future multiple patterning in the metal layers. The drawbacks of the increased metal density are not negligible in terms of parasitic capacitors, resistance, and effort in physical design. The FinFET transistors are also known as Tri-Gate transistors.

15.2 The Intel Tri-Gate FinFET Technology

The FinFETs are processed with a gate which surrounds the channel at three sides as shown in Fig. 15.1. The improved lateral electrostatic control of the Tri-Gate structure leads to the outstanding performance of FinFETs in terms of saturation

L. Dörrer (✉) • F. Kuttner • F. Conzatti • P. Torta
Intel Austria GmbH, Villach, Austria
e-mail: lukas.doerrer@intel.com

P. Harpe et al. (eds.), *Hybrid ADCs, Smart Sensors for the IoT, and Sub-1V & Advanced Node Analog Circuit Design*, DOI 10.1007/978-3-319-61285-0_15

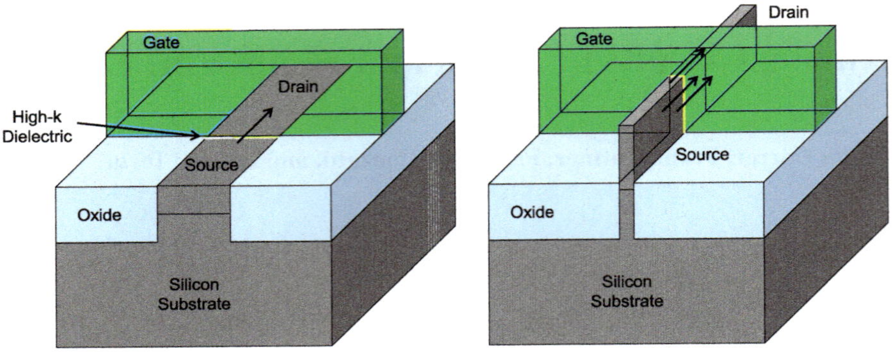

Fig. 15.1 A planar transistor (*left*) and a FinFET transistor (*right*) [1]

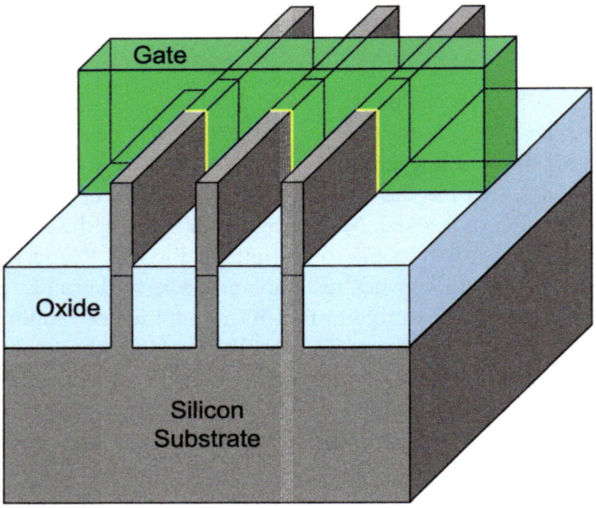

Fig. 15.2 Three FinFETs sharing the same gate [1]

current, leakage current, and other analog parameters, as well as improved short channel effects [1] (Fig. 15.2).

Designing circuits for a FinFET technology is relatively similar to the circuit design for a planar technology. The difference is the parallel connection of fins to get a wanted transistor width and the stacking of devices to increase the effective length. A microscope picture of some transistors implemented in the planar and in the FinFET technologies is shown in Fig. 15.3. Although up to a hundred transistors have to be stacked for a current source, the parasitic capacitance is small enough to not degrade the performance in comparison to planar CMOS (Fig. 15.4).

The main issue for these arrays of fins is the layout generation, because all the fins must to be connected.

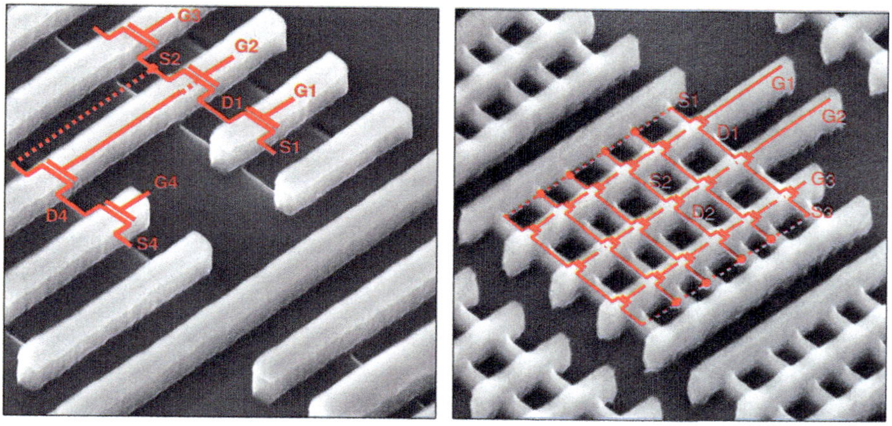

Fig. 15.3 Microscope picture of some transistors in planar (*left*) and in FinFET (*right*) technology [1]

Fig. 15.4 A current mirror implementation with four stacked FinFET transistors

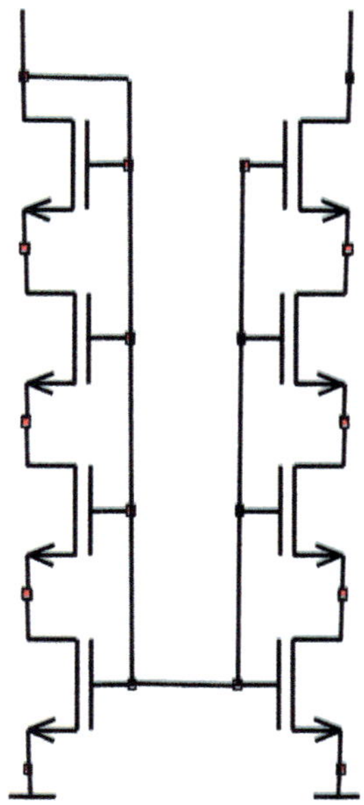

Special care has to be taken at high current or for high frequency circuits. FinFET self-heating (FISH) can be a serious issue because the thermal resistance from the channel of the FinFET to the substrate is much higher than in planar CMOS resulting in increased maximal temperature within the fins [2]. Moreover, the resistances as well as the parasitic capacitances of the lower metal layers are increased because of the small metal pitch.

15.2.1 Photolithography and Double Patterning

Double patterning is a class of photolithography technologies for manufacturing integrated circuits, developed to enhance the feature density. It is necessary for the lower metal layers of 14 nm technology.

As shown in Fig. 15.5, more than one lithographic exposure is required due to the need to first print a series of lines and then cut them. As a consequence of the double patterning, the layout complexity increases for FinFET technologies. A layout is decomposed into two masks (two colors), and therefore the effective pitch is doubled. There are design restrictions caused by double patterning. Double patterning will work best on designs whose critical layers can be split into two separated, very regular patterns which are aligned in a predictable way. This means that producing a layout that looks like a diffraction grating is good, while a design littered with diagonal lines, jogs, and vias between layers may be not effectively split. Double patterning brings alignment issues on to critical layers, rather than between layers as before, with a potential impact on design performance and production yield. Double patterning is expensive because it uses two masks to define a layer that was previously defined with a single one. This means buying more steppers to maintain the manufacturing throughput. For an analog designer, the drawbacks of double patterning are visible in the physical design. Higher metal density and lower metal pitch cause higher parasitic capacitance and resistance. The main drawback is the time required in the layout development and optimization. (For example), in order to lower the parasitic capacitance in 28 nm, a simple shift of metal lines to increase the distance can be done easily, but due process discretization this is harder to achieve in smaller technologies like 14 nm due to the discretization of the process.

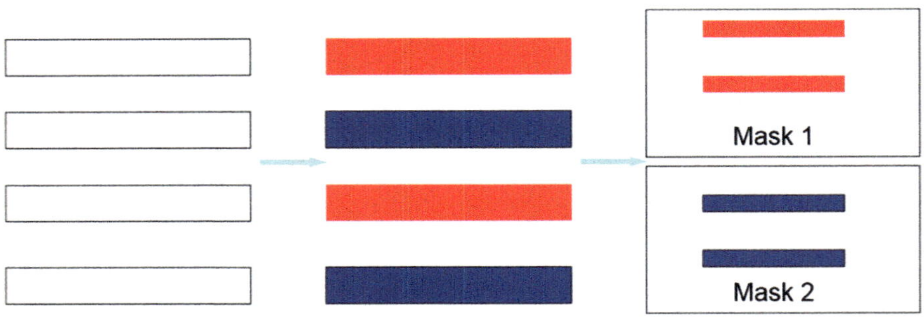

Fig. 15.5 Double patterning requires two different masks for the same layer

15.3 Comparison of Designs

In this section we compare the planar and FinFET designs of a CDAC, a two-step flash ADC and a Sigma Delta ADC. We analyze designs targeting similar constraints and describe how the implemented circuit solutions can be enhanced or limited by the technology.

15.3.1 Digital IQ CDAC

A new architecture for an RFDAC was developed in 28 nm and 14 nm in parallel. In both technologies the RFDAC was implemented and measured on testchips [3]. The RFDAC is a polar/digital IQ combo. It can be operated in polar mode for high efficiency or in IQ mode for high modulation bandwidth, when the DPLL is not able to be modulated at such high bandwidth. The DAC shown in Fig. 15.6 consists of a capacitor array of 10-bit thermometer-coded and 5-bit binary-coded capacitor cells and can directly generate the signal for an RF transmitter.

Figure 15.7 shows the digital nature of the capacitive RFDAC. The main effort in the design is to switch the capacitors with timing errors smaller than picoseconds. The capacitive RFDAC, which is mainly composed by capacitors and logic cells, makes it perfectly suitable for migration to a scaled digital technology.

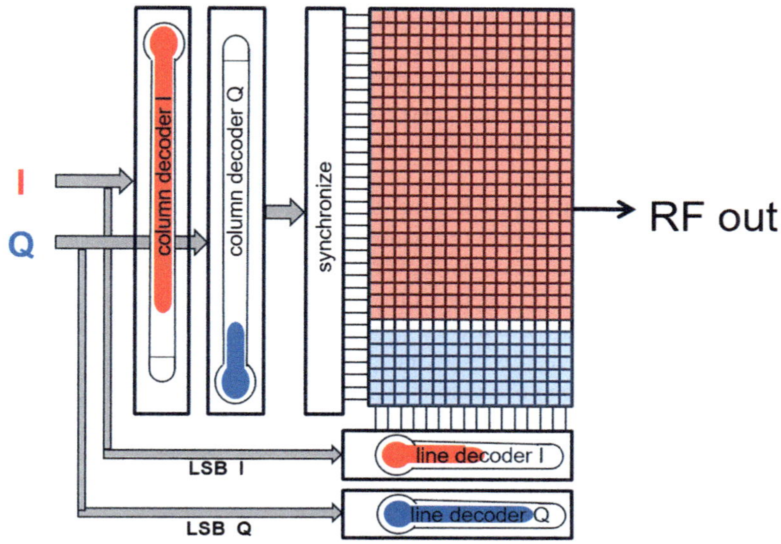

Fig. 15.6 Digital IQ CDAC block diagram

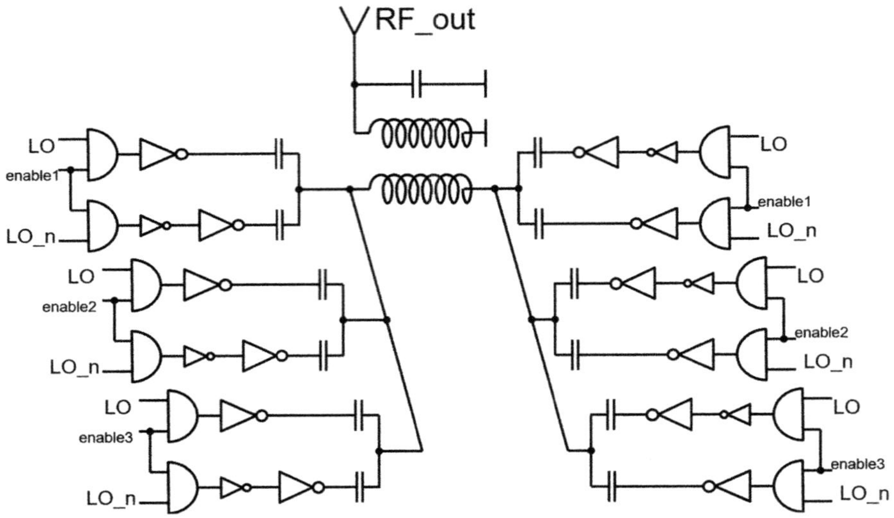

Fig. 15.7 Cells of digital IQ CDAC

A total sine wave output power of 14.2 dBm and 13.5 dBm were measured at 1850 MHz and 850 MHz, respectively. The measurement results in 14 nm and 28 nm are similar. In 14 nm 25% power reduction was possible, which is caused by the digital architecture of the DAC. Although the power delivered to the output has to be the same, losses of switched digital signals are drastically reduced. The area for the DAC is reduced by more than 50%, but the overall area of the transmitter is determined by the matching network at the output. The transformer in the matching network is the main area contributor, and its size is given by the amount of turns for a chosen frequency. Therefore, the total area for the whole transmitter shrinks by approximately 25% (Fig. 15.8).

In IQ mode, an excellent noise floor of -153dBc/Hz has been measured in LTE5 mode (Fig. 15.9). Low error vector magnitudes (EVM) of 3% and 1.5% have been further measured in LTE5 mode at 1850 MHz and 850 MHz, respectively. With digital predistortion, the EVM can be reduced to 0.8% in LTE20 mode. The periodic spurs in Fig. 15.9 are an artifact caused by the testchip RAM in the digital path and thus can be easily avoided. It is expected that the noise floor can be improved by better layout, because power connections were not fully optimized.

15.3.2 Two-Step Flash ADC 8-Bit 4Gs/s

For 5G application fast AD converters with bandwidths in the GHz range are necessary. The basic architecture is described in the literature [4, 5]. The resolution of the AD converter is mainly limited by the power dissipation of the system. For

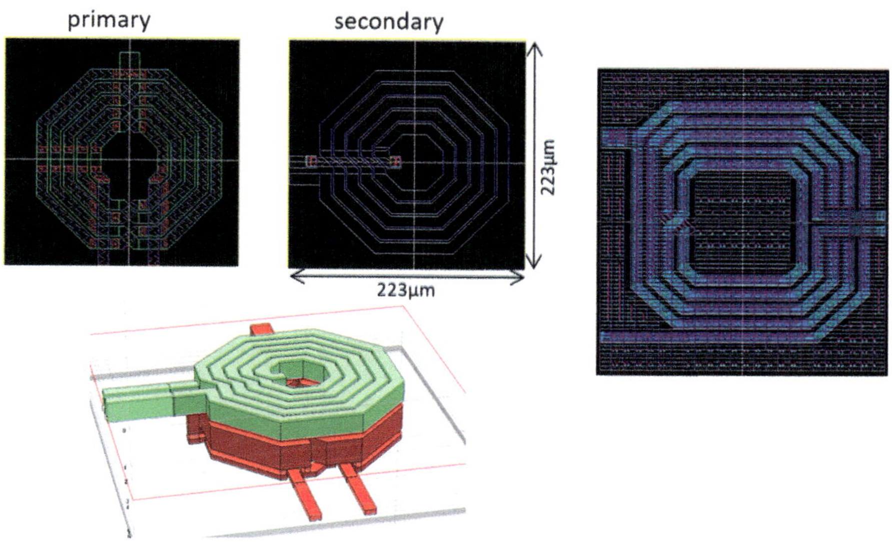

Fig. 15.8 The transformer depends on the matching network and in 28 nm and 14 nm does not shrink substantially

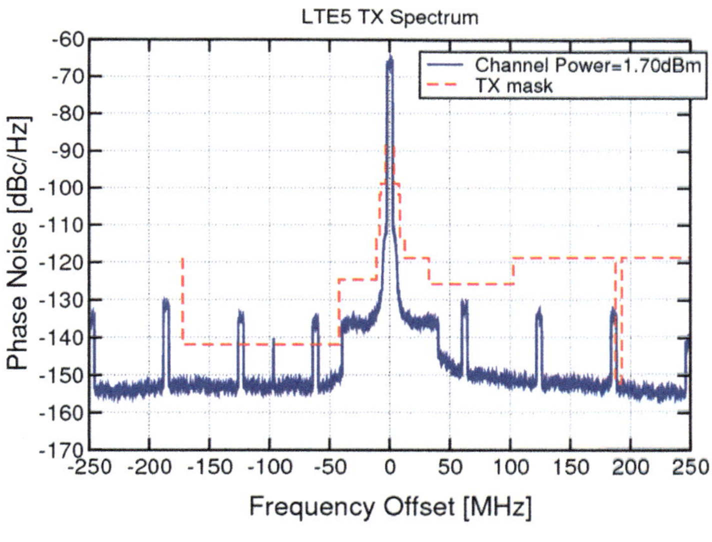

Fig. 15.9 Measurement of the RFDAC in 14 nm LTE5 1950 MHz (band1)

the application 8- to 9-bit resolution is sufficient. A subranging AD converter was developed in 28 nm and 14 nm in parallel; the concept is presented in Fig. 15.10. Exploiting the small technologies made it possible to digitally calibrate the offset of each comparator, and also a new background calibration was developed.

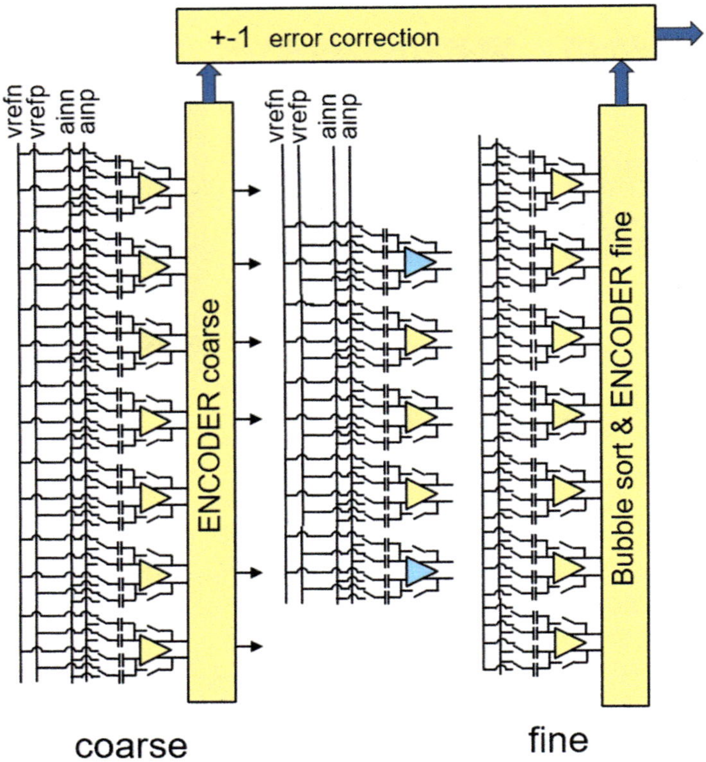

Fig. 15.10 Concept of the subranging converter, coarse and fine stage

The comparators are very simple differential stages followed by a latch which is tuned for low kickback. Figure 15.11 shows the comparator of the coarse stage.

The numbers express the number of used fins or capacitors. For the fine stage, a slice-based design to simplify the layout was used. Instead of sizing every comparator itself, a common size was used and switched in parallel to generate the capacitance and current necessary by noise requirements (Fig. 15.12).

In the preamplifier of 14 nm, it is possible to stack more devices due to better ratio of VDD to threshold voltages. Exploiting the small technologies made it possible to digitally calibrate the offset of each comparator (Figs. 15.13 and 15.14).

The comparators of the coarse stage are first offset calibrated [4]. The remaining errors are corrected by monitoring the difference of the coarse and the fine result. It is obvious that an error occurred whenever an underrange or overrange signal is generated. The implementation in 14 nm is more compact and faster. Simulations showed that for the same power dissipation, the sampling rate can be doubled in 14 nm. So for 15 mW, the ADC can be operated up to 2Gs/s in 28 nm, whereas 4Gs/s is possible in 14 nm. The area shrinks by 50% in 14 nm. Measurements are available in 28 nm, showing 7.2 ENOBs up to Nyquist frequency.

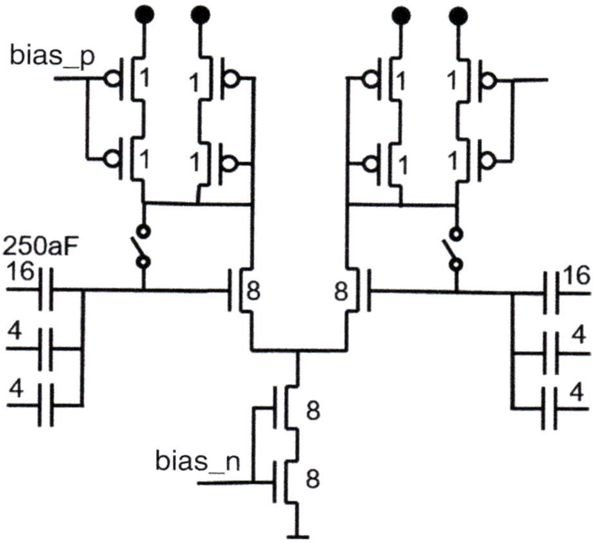

Fig. 15.11 Comparator of the coarse stage

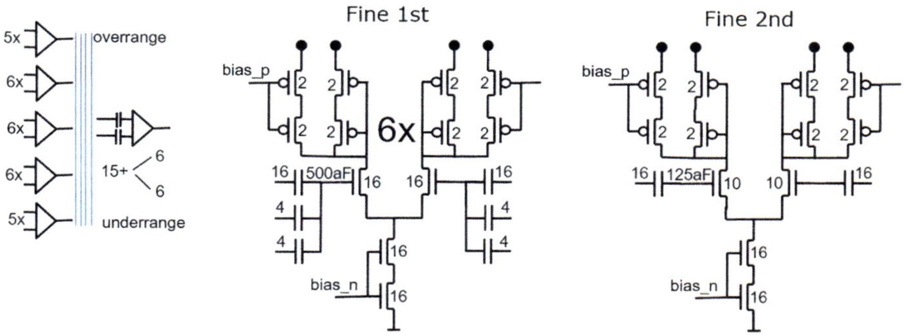

Fig. 15.12 Comparator of the fine stage in 14 nm

15.3.3 Third-Order Continuous Time Sigma Delta ADC

As a last comparison example, a continuous time Sigma Delta ADC has also been developed and measured in the two technologies. This ADC is a good test vehicle to scout and compare the improved performance or the limitations of a technology because its building blocks relate to pure analog, mixed mode, and pure digital circuits. Some literature about similar ADC design can be found in [6, 7].

In this work we compare two similar ADC designs in feedback configuration. The ADC architecture is shown in Fig. 15.15. The ADC can be reconfigured by changing the filter coefficients to cover two bandwidth modes 9 MHz and 47 MHz, respectively. A low-resolution 5-bit flash quantizer and three current steering feedback DACs are used to close the loop and digitize the analog input

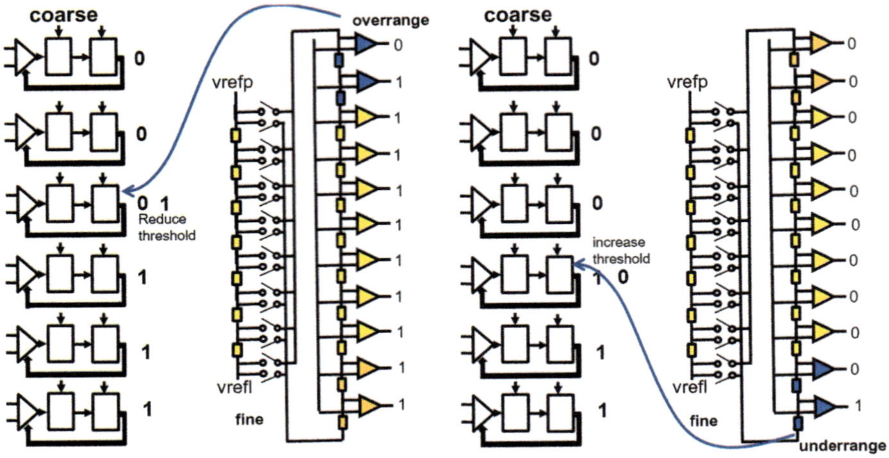

Fig. 15.13 Calibration of coarse comparators

Fig. 15.14 Offset calibration of each comparator

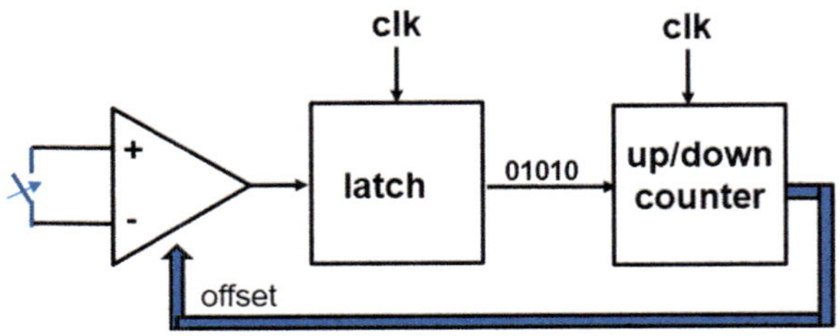

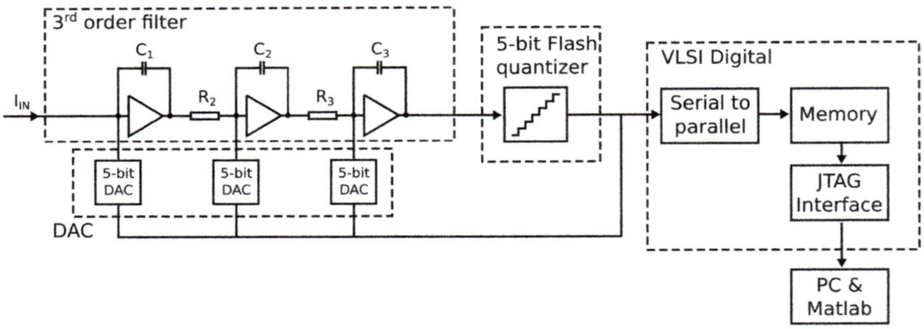

Fig. 15.15 The third-order Sigma Delta ADC architecture used for the technology comparison

signal. In both designs a high-speed digital block is used to collect, parallelize by 4, and store the output data over a 32-bit bus in the 96Kb memory implemented on the chip. A JTAG interface is used to read and evaluate the memory content in a PC.

At block level the filter is composed by three cascaded integrators implementing active amplifiers and one resonator between the third and the second stage.

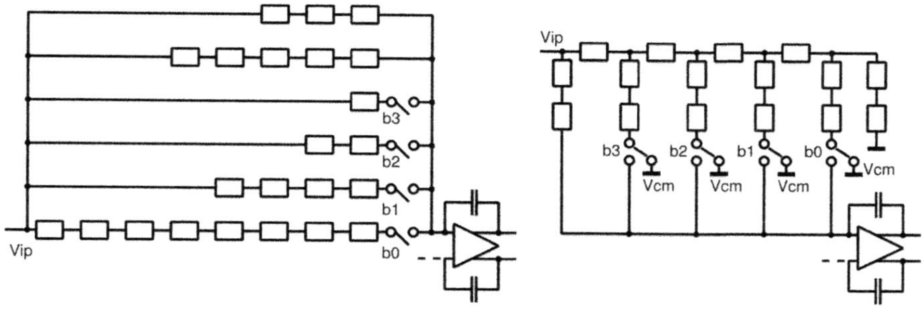

Fig. 15.16 The tuned resistor implementation. The parallel design (*left*) and the more compact R2R implementation (*right*). The latter reduces the area and the parasitic capacitance to ground

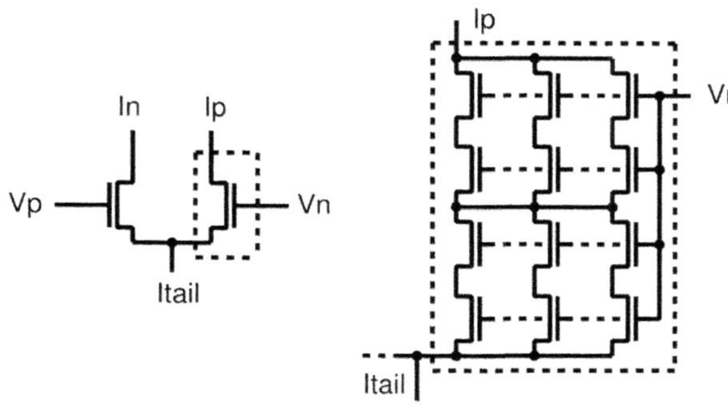

Fig. 15.17 A simple differential pair in planar technology (*left*) becomes a cluster of 20,000 devices in his FinFET implementation (*right*)

The bandwidth mode selection is achieved by capacitor reprogramming. In order to tune the RC product caused by manufacturing mismatch, each resistor in the filter is digitally trimmed. In the planar 28 nm variant, the poly layer has been used to build the unit resistor. In the FinFET 14 nm variant, a dedicated resistor layer introducing substantial parasitics to substrate and ground is used. The increased parasitic capacitance limits the maximum bandwidth of the device. To overcome this limitation, the design was changed from pure parallel to a R2R implementation as shown in Fig. 15.16. Due to its size and layout implications, the resistor usage in the design was minimized.

In the filter the first amplifier has very stringent constraints in terms of noise and linearity. To achieve this the input differential pair is very large. Usually in planar technologies, the two differential pair transistors are split in several smaller transistors placed in parallel and sharing the same length. The actual implementation in 14 nm strives to reduce the amount of devices in the layout and in the design by using characterized subcircuits consisting of several stacked transistors. This introduces new nodes as shown in Fig. 15.17.

The resulting cluster of transistors can be seen as an equivalent device. The saturation condition of the cluster is ensured when the upper transistors are in saturation. Although this condition ensures saturated behavior of the cluster, it can lead to overdesign.

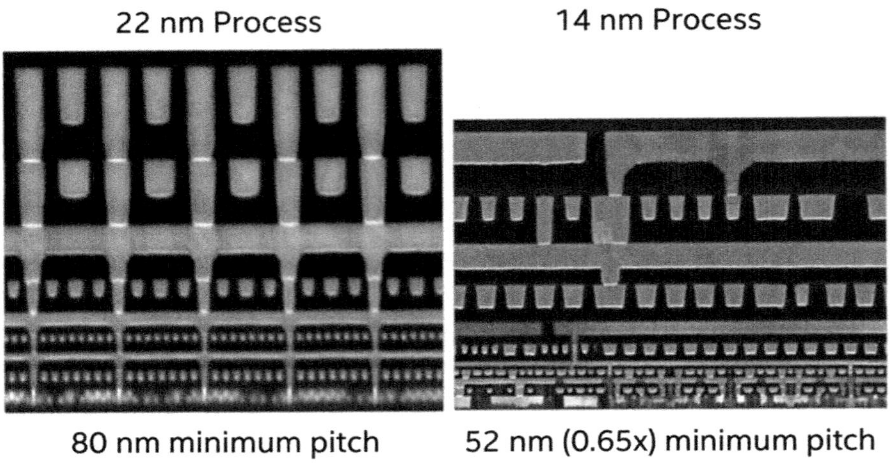

Fig. 15.18 Metal stack section comparison between two Intel FinFET technologies 22 nm and 14 nm [8]

An advantage of having several devices connected in a grid network ensures that in comparison to planar technologies, the current is not flowing into a single wire. This automatically relaxes the reliability requirements due to power dissipation and FinFET self-heating (FISH). A disadvantage of the cluster device is that the amount of elements and the amount of nodes to be simulated increases drastically.

The filter capacitors have been implemented using lateral capacitance. As the pitch between the metal lines reduces with the technology as shown in Fig. 15.18, the area needed to achieve the targeted value greatly reduces. As the filter caps sized for the required lowest bandwidth cover about 40% of the ADC area, the technology contributes substantially to the total ADC area reduction.

In 14 nm the metal interconnects show increased resistance due to the lower width and height needed to contact the smaller transistors. To ensure proper performance of the ADC, it is crucial to route the lines between quantizer, DAC, and filter using low ohmic metals. This means that the higher metals are needed in 14 nm. The current steering DAC cell in 14 nm profits from the lower threshold voltage for the switches compared to the supply as previously mentioned. These are implemented using only NMOS switches and a simplified driving logic (Fig. 15.19). In both technologies the current source is long and made by several transistor in series. This results in more effort in the layout and in the simulation time of the 14 nm design.

The design of the quantizer was also easier in 14 nm due to a better threshold to supply ratio. The implementation in 14 nm is achieving a much higher gain of the cascaded preamplifier plus latch compared to 28 nm because a CMOS input can be used in the preamplifier as shown in Fig. 15.20.

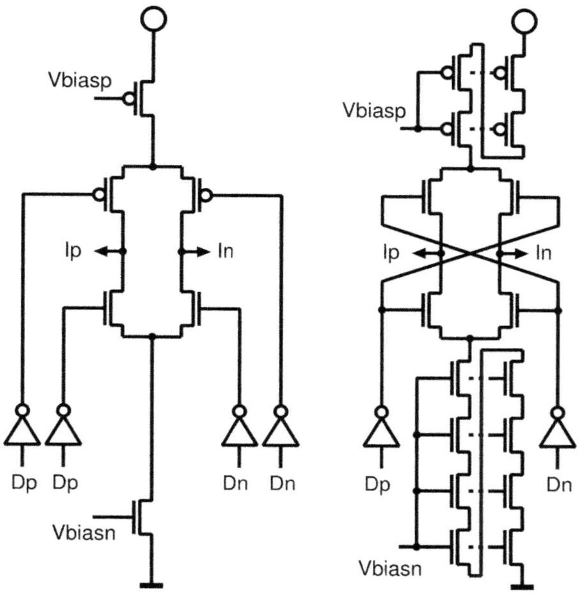

Fig. 15.19 A current steering DAC cell, its driving logic, and the switches implemented in planar technology (*left*) and in FinFET (*right*)

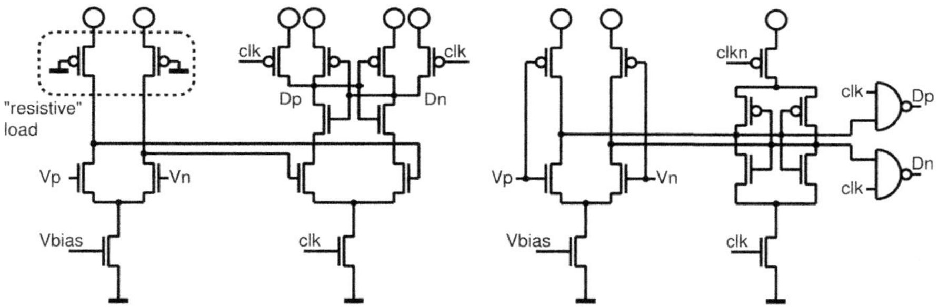

Fig. 15.20 The comparator achieves a higher performance in 14 nm (*right*) compared to 28 nm (*left*) because the preamplifier can be CMOS

The digitized output of the quantizer is collected into a memory in the digital VLSI domain. As expected the digital VLSI block shown in Fig. 15.15 in FinFET 14 nm greatly outperforms in all aspects the planar design.

A comparison of the measured performance is shown in Figs. 15.21 and 15.22 and in Table 15.1.

Although the two ADCs are not identical, they show similar performance and can be thus compared. An even more important fact is that in both designs the simulation results fit well with the measurements. The accurate simulation of the extracted top level view in 14 nm is much more time consuming due to the larger amount of netlist elements that cannot be neglected or simplified/reduced.

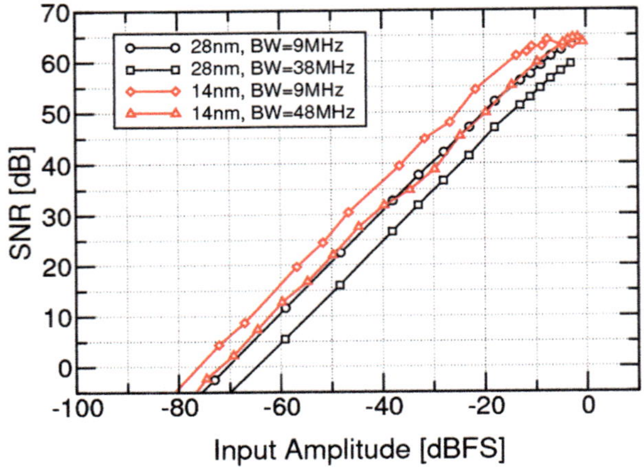

Fig. 15.21 Comparison of the SNR integrated over the specified bandwidth

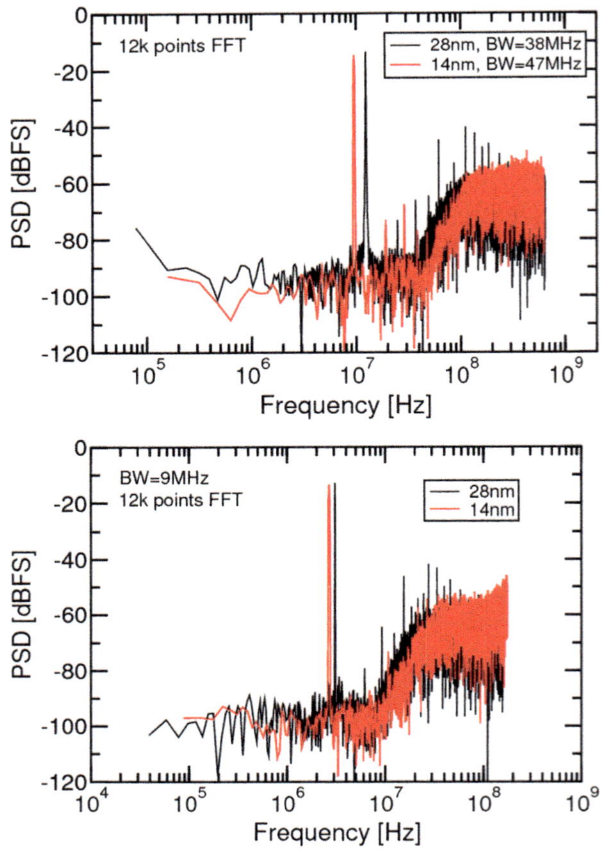

Fig. 15.22 PSD plot of the output data of the two designs

Table 15.1 Performance comparison of the SDADCs

	V_{DD} [V]	BW [MHz]	Current [mA]	DR [dB]
28 nm	1.1	9	6.75	70.7
		38	11.4	65
14 nm	1.0	9	8	73
		47	12	71

15.4 Conclusions

The critical high performance A/D and D/A building blocks for an RF transceiver have been designed in Intel 14 nm Tri-Gate FinFET and 28 nm planar technologies. The competitive advantage in performance has been determined, and hurdles for a commercialization with a competitive execution schedule have been identified.

References

1. Bohr, M., Mistry, K.: Intel's revolutionary 22nm transistor technology, Intel Corporation.www.intel.com/content/dam/www/public/us/en/documents/presentation/revolutionary-22nm-transistor-technology-presentation.pdf, pp. 4–9, May 2011
2. Prasad, C., et al.: Transistor reliability characterization and comparisons for a 14 nm tri-gate technology optimized for system-on-chip and foundry platforms. 2016 IEEE International Reliability Physics Symposium (IRPS), Pasadena, CA, 2016, pp. 4B-5-1-4B-5-8
3. Fulde, M., Kuttner, F., et al.: Digital multimode polar transmitter supporting 40MHz LTE carrier aggregation in 28nm CMOS. ISSCC2017 13.2 D
4. Sandner, C., Clara, M., Santner, A., Hartig, T., Kuttner, F.: A 6-bit 1.2-GS/s low-power flash-ADC in 0.13-μm digital CMOS. IEEE J. Solid-State Circuits. **40**(7), 1499–1505 (2005)
5. Clara, M., Wiesbauer, A., Kuttner, F.: A 1.8 V fully embedded 10 b 160 MS/s two-step ADC in 0.18 μm CMOS. Proceedings of the IEEE 2002 Custom Integrated Circuits Conference (Cat. No.02CH37285), 2002, pp. 437–440. D
6. Dorrer, L., Kuttner, F., Greco, P., Torta, P., Hartig, T.: A 3-mW 74-dB SNR 2-MHz continuous-time delta-sigma ADC with a tracking ADC quantizer in 0.13-μm CMOS. IEEE J. Solid State Circuits. **40**(12), 2416–2427 (2005)
7. Dorrer, L., et al.: A 2.2mW, Continuous-Time Sigma-Delta ADC for Voice Coding with 95dB Dynamic Range in a 65nm CMOS Process. 2006 Proceedings of the 32nd European Solid-State Circuits Conference, Montreux, 2006, pp. 195–198

Additional Reference

8. Borkar, R., Bohr, M., Jourdan, S.: Advancing Moore's Law – The Road to 14 nm. Intel Corporation,www.intel.com/content/www/us/en/silicon-innovations/advancing-moores-law-in-2014-presentation.html, pp. 14–39, August 2014

Chapter 16
Pipeline and SAR ADCs for Advanced Nodes

Michael P. Flynn, Kyojin Choo, and Yong Lim

16.1 Introduction

The energy efficiency of ADCs has improved by orders of magnitude over the past two decades. Even though process scaling degrades the analog characteristics of transistors, by exploiting, scaling the energy efficiency of recently reported ADCs is approaching fundamental limits [1]. These improvements have been achieved through innovative circuit ideas and through the evolution of ADC architectures. In particular, the SAR ADC architecture has benefitted tremendously from scaling. SAR ADCs are among the most efficient stand-alone converters and form excellent building blocks for more complex architectures including pipeline and highly interleaved ADCs. This chapter builds on material that first appears in [2–4].

Section 2 presents a stand-alone SAR ADC that achieves outstanding efficiency and also shrinks the area needed for a SAR ADC. This architecture uses a charge-injection cell-based DAC to avoid problems associated with residual settling in conventional SAR ADCs. Further, reuse of the charge-injection cells allows a very small die area. The small size and efficiency make the charge-injection cell SAR ADC an ideal building block for hybrid and interleaved ADCs.

We concentrate on pipeline ADCs for the remainder of the chapter. These combine sub-ADCs of moderate resolution with high-performance amplifiers to construct a high-resolution pipeline. In Sect. 3, we present a simple argument that in a two-stage pipeline the first stage should have higher resolution. However, this high-resolution first stage is difficult to achieve with flash-based sub-ADCs. The SAR-assisted pipeline ADC allows a high-resolution first stage that enables a very efficient two-stage pipeline.

M.P. Flynn (✉) • K. Choo • Y. Lim
University of Michigan, 1301 Beal Ave, Ann Arbor, MI 48109, USA
e-mail: mpflynn@umich.edu

P. Harpe et al. (eds.), *Hybrid ADCs, Smart Sensors for the IoT, and Sub-1V & Advanced Node Analog Circuit Design*, DOI 10.1007/978-3-319-61285-0_16

We focus on the amplifier part of the pipeline for the remainder of the chapter. In Sect. 4, we argue that a ring amplifier can supersede the workhorse-cascoded telescopic OTA in the switched-capacitor residue amplifier of a SAR-assisted pipeline ADC. In Sect. 5, we present a SAR-assisted pipeline ADC that uses a ring amplifier to achieve outstanding energy efficiency.

16.2 Charge-Injection Cell-Based DAC SAR ADC

SAR ADCs are not only highly effective by themselves but also form critical building blocks of pipeline ADCs and interleaved SAR ADC arrays. Interleaving of SAR ADCs delivers very high sampling speeds and good energy efficiency. However, interleaving of multiple SAR ADCs poses significant challenges due to the large area needed. A particular problem is that interleaving artifacts are exacerbated by die size [5]. Compact and efficient SAR ADCs facilitate highly interleaved ADCs and also serve as efficient building blocks for pipeline and $\Sigma\Delta$ ADCs. One approach to improving SAR ADC performance is the multiple-bit-per-cycle SAR ADC, but this has the disadvantage of significant extra complexity since extra quantizers and capacitor DACs are needed [6, 7]. Furthermore, multiple-bit-per-cycle SAR ADCs need increased die area. The charge-injection cell-based DAC SAR ADC (ciSAR ADC) [3] is a very compact SAR ADC architecture and achieves excellent energy efficiency.

Interrupted settling makes the ciSAR ADC faster, simpler, and more linear for high-speed operation. This is because the ciSAR architecture avoids the distortion suffered by conventional fast SAR ADCs due to insufficient DAC settling. Figure 16.1 illustrates how residual settling compromises the linearity of a conventional SAR ADC. Settling from a previous step continues into the present SAR step, leading to distortion in the conversion. As shown in the example, the residual background settling skews trip points downward. Redundancy in the SAR algorithm can alleviate this problem, but redundancy requires extra SAR steps and more complicated SAR logic. On the other hand, in ciSAR, thanks to interrupted settling, settling for any given SAR step stops *completely* at the end of that step. Returning to the example in Fig. 16.1, we see that with interrupted settling, there is no longer distortion of the trip points.

Modular *charge-injection cells* (CICs), as shown in Fig. 16.2, are the key to interrupted settling. Initially, the input signal is sampled onto two differential integration caps, C_{int+} and C_{int-}. During the SAR operation, the DAC+ and DAC- nodes of these capacitors are connected to the CIC cells and to a comparator. During the binary search, the CICs subtract fixed quanta of charge from the C_{int+} and C_{int-} capacitors. The binary search of the SAR ADC is based on the set-and-down method [8], since the CIC cells can only subtract charge. A unique advantage is that CIC cell can be reused for different SAR steps. Thanks to this reuse, the number of CIC cells can be far fewer than the number of levels in the SAR ADC. For example, in [3] only eight identical CICs are needed for a prototype 6-bit ADC.

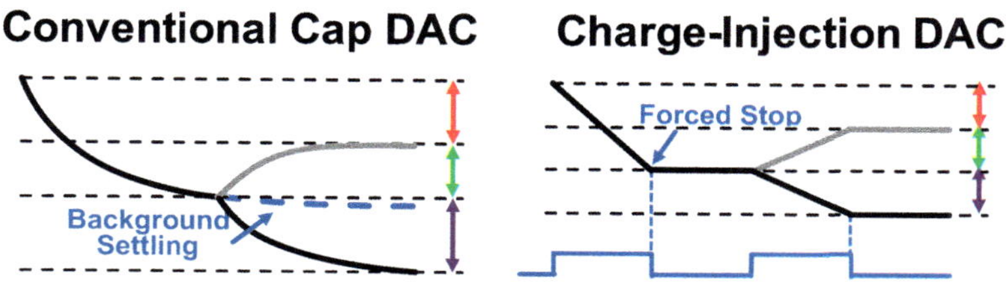

Fig. 16.1 Interrupted settling (*right*) avoids distortion due to limited bandwidth [3]

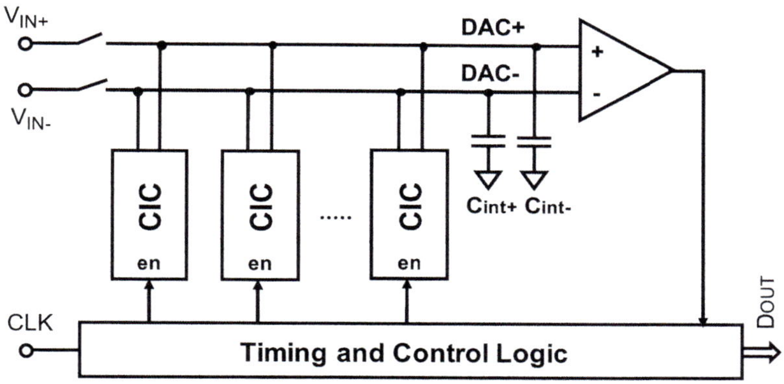

Fig. 16.2 ciSAR architecture with eight CIC cells for a 6-bit ADC [3]

The charge-injection DAC is fundamentally different to a conventional capacitor DAC because it is based on unidirectional transfer of charge instead of bi-directional charge sharing. When enabled, a CIC cell injects a fixed amount of charge onto the integration caps. Unidirectional switches isolate the charge source in the CIC cells from the DAC outputs with high output impedance for unidirectional charge transfer. A CIC only transfers charge from its charge reservoir to the C_{int+} and C_{int-} when it is enabled. Furthermore, the high output impedance makes the transferred charge independent of the voltage on the C_{int+} and C_{int-} capacitors. When a CIC cell is disabled, all charge transfer is halted and settling is interrupted. This interrupted settling eliminates any residual settling in subsequent cycles enabling better linearity, especially in high-speed operation, as shown in Fig. 16.1.

Figure 16.3 shows an implementation of the CIC cell. The CIC cell consists of a charge reservoir, switches, and control logic. The capacitance of M3, operating in triode region, along with parasitic capacitance at its drain node forms the charge reservoir. Initially, M4 resets the voltage in the charge reservoir to ground. During operation, switches M1 and M2 connect the reservoir to either DAC+ or DAC-. These switches operate in saturation for unidirectional transfer. Three logic gates (G1~3) control these transistors based on signals from the SAR control logic and the comparator. Figure 16.3 also shows a timing diagram.

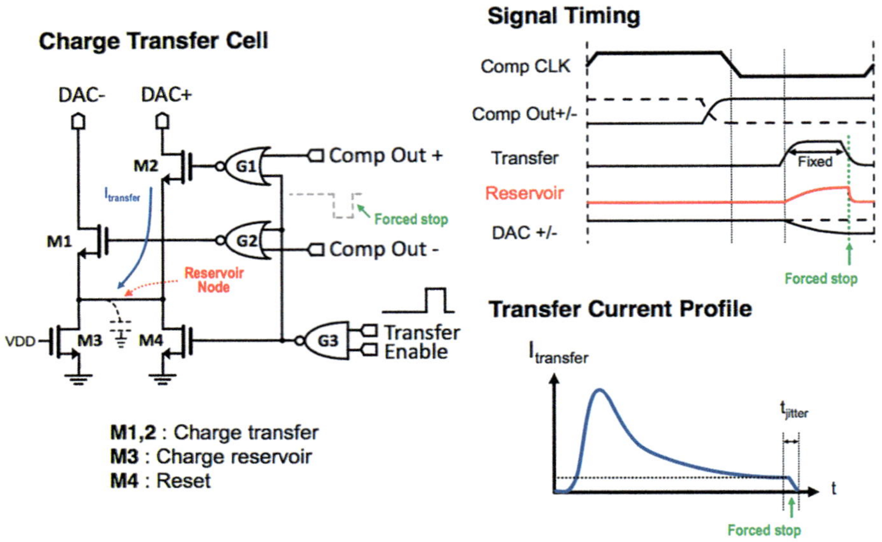

Fig. 16.3 Charge transfer cell, timing diagram, and transfer current profile [3]

The profile of the charge transfer (Fig. 16.3) also facilitates interrupted settling. At the beginning of the charge transfer cycle, one of the charge transfer switches (i.e., either M1 or M2) is strongly on. However, during the transfer, the voltage on the reservoir node rises reducing the gate-source voltage of the conducting NMOS switch (M1 or M2) and causing the current to fall. The current continues to fall until it drops to the level of the small bias current supplied by M3. This falling current profile greatly reduces the sensitivity to jitter in the timing control signal, since the current flow is always small when charge transfer is halted. As CIC cells are only active for a short duration, we use the remaining time to prepare for the next charge transfer.

A prototype ciSAR needs only eight CICs [3]. CIC cells in a ciSAR ADC are reused multiple times in a SAR conversion both to save area and improve linearity. During the MSB cycle, these eight CICs are used twice in two successive phases to deliver 16 units of charge. CIC cell reuse not only halves the DAC area but also halves the driver power for the control signals. CIC cell reuse also improves ADC linearity because the same cells are reused. CIC cell reuse in the prototype [3] only slows down the overall ADC sampling rate by only 15%.

16.3 Combining SAR and Pipeline

16.3.1 The Advantages of the Two-Stage Pipeline

A large first-stage resolution is very beneficial to the performance of a pipeline ADC [2, 9, 10]. A large first-stage resolution reduces power consumption and also attenuates noise and nonlinearity contributions of the subsequent stages to the overall ADC performance [9, 10]. The front-end stages dominate the ADC power consumption because these must be the most accurate and must dissipate more opamp power to achieve sufficiently accurate settling. The accuracy required from subsequent stages decreases exponentially, and so the power dissipated by them is relatively low.

On the other hand, a large first-stage resolution is difficult to implement with a flash-based sub-ADC because of the power and area needed for a large number of accurate comparators. A further challenge is that a high first-stage resolution demands an active front end to reduce aperture and sampling errors related to unavoidable differences in the sampling instants of the MDAC and sub-ADC (i.e., *clock skew* in Fig. 16.4). The SAR-assisted pipeline technique facilitates a large-stage sub-ADC resolution and removes the sampling mismatch between the MDAC and the sub-ADC.

We now consider a simple 3-bit MDAC stage (Fig. 16.4) to examine some of the benefits of a large first-stage resolution [2]. We model the opamp with a

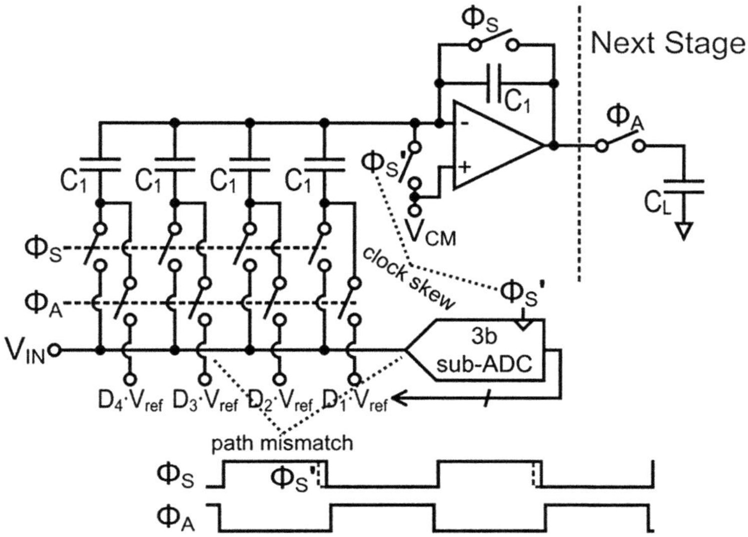

Fig. 16.4 Example of 3-bit SC MDAC [2]

transconductance G_m, and $C_{L,tot}$ is the total output load of the opamp. If we assume a first-order step response, then the output of the first-stage MDAC at the end of hold phase is

$$V_{\text{res}} = V_{\text{ideal}} + V_{\text{err}} \text{ and } V_{\text{err}} = (V_{\text{ideal}} - V_{\text{initial}})\, e^{-\frac{T\beta G_m}{C_{L,\text{tot}}}} \quad (1)$$

where T is the available time for settling and β is the feedback factor.

A simple argument shows the power advantages of an increased first-stage gain. The feedback factor β ($\approx 2^{1\text{-M}}$) is approximately halved with every 1-bit increase in resolution, M, of the first-stage MDAC. A 1-bit increase in the resolution of the first stage also indicates a 1-bit decrease in the required resolution of the subsequent stages. Therefore, the worsened feedback factor, β, is approximately offset by the increased tolerance for settling error, V_{err}. On the other hand, a 1-bit decrease in the required resolution of the subsequent stages also approximately halves the output load capacitance, $C_{L,tot}$. This reduction in $C_{L,tot}$ decreases the required opamp transconductance, G_m, which in turn directly translates into a reduction in the opamp power consumption. However, this power improvement with increasing first-stage MDAC resolution ceases when output self-parasitics of the opamp dominate $C_{L,tot}$.

The linearity of a pipeline ADC improves as the first-stage resolution increases [9]. This is because an increased first-stage resolution lowers nonlinearity due to capacitor mismatch. Furthermore, the large gain of a high-resolution stage decreases the nonlinearity and noise contributions of the subsequent stages.

16.3.2 The SAR-Assisted Pipeline

The SAR-assisted pipeline ADC architecture [2] is an energy-efficient hybrid architecture for moderately high-resolution analog-to-digital conversion. The SAR-assisted pipeline ADC couples two SAR ADCs with a residue amplifier (i.e., Gain), as shown in Fig. 16.5. The two SAR ADCs work as high-resolution sub-ADCs in each of the two pipeline stages. The SAR-assisted pipeline ADC has several advantages compared to both conventional pipeline ADCs with flash sub-ADCs and conventional SAR ADCs [2]. As discussed, we improve the linearity of the overall ADC and reduce opamp power consumption as we increase the resolution of the first-stage sub-ADC [9]. A SAR sub-ADC is very attractive because it uses less power than a flash sub-ADC. Another important benefit of a SAR-based first stage is that SAR sub-ADC and MDAC share the same sampling mechanism, thereby removing the need for a separate front-end sample and hold circuit.

For the same overall ADC resolution, the SAR-assisted pipeline architecture also has advantages compared to the conventional SAR architecture. In particular, for moderately high resolution (e.g., 12 bits), comparator noise performance is challenging in a conventional SAR ADC. On the other hand, the comparator

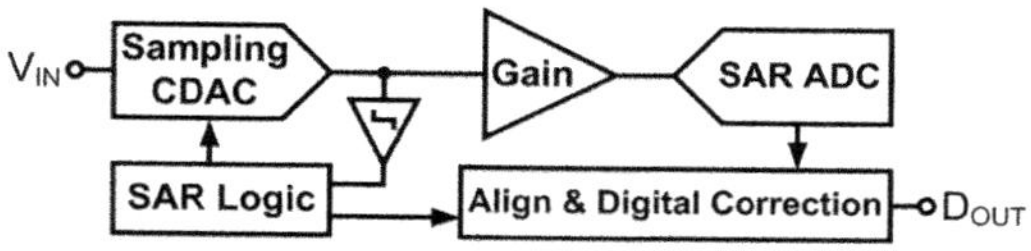

Fig. 16.5 Two-stage SAR-assisted pipeline [2, 4]

noise requirement is greatly relaxed in the SAR sub-ADCs of a comparable-resolution SAR-assisted pipeline ADC. A further advantage is that pipelining removes the speed bottleneck of the conventional SAR architecture. Finally, thanks to redundancy and digital correction, the SAR-assisted pipeline ADC tolerates settling errors of the first-stage SAR CDAC, provided these remain within the error correction range of the stage redundancy.

16.4 The Ring Amplifier

The cascoded telescopic OTA-based SC residue amplifier has been the workhorse of conventional pipeline and SAR-assisted pipeline ADCs [2, 11]. However, the conventional OTA structure consumes a lot of power and suffers from a limited output swing. This restricted output swing forces a stage gain smaller than suggested by the first-stage resolution and the redundancy of the pipeline. Because of this, SAR-assisted pipelines often need to reduce the reference voltages to the second stage [2], but this consumes extra power. Another alternative is to use an R-2R DAC in the second-stage SAR ADC [11]. Dynamic amplifiers are a lower-power alternative to OTAs in a pipeline ADC [12, 13, 15, 16]. Through time-domain integration, a dynamic amplifier offers low power amplification of the residue. An advantage is that this integration filters noise [14]; however, the inaccurate open-loop gain with dynamic amplification requires gain calibration in the pipeline. Not only does calibration increase both the complexity of design and test cost, it also reduces robustness to changes in process, supply voltage, and temperature (PVT) [13].

The ring amplifier [17, 18] is an energy-efficient alternative to an OTA that intrinsically has a high output swing. The high gain of the ring amplifier allows closed-loop operation without the need for calibration of gain. [4] introduced a fully differential ring amplifier enabling a fully differential switched-capacitor stage. Slew-based charging makes ring amplifiers energy efficient. Recent ring amplifiers are robust to PVT variation [4, 18] because they do not need external biasing.

The original ring amplifier [17], Fig. 16.6a, is a three-stage inverter-based amplifier with an offset-canceled first stage. The ring amplifier is stabilized by last stage moving to the subthreshold region as the ring amplifier virtual ground (i.e., V_{IN} in Fig. 16.6) approaches the desired common-mode voltage. This is done with the help of split second-stage inverter amplifiers with separate floating input offset voltages. During auto-zero, the floating input offsets of the second-stage inverters are applied to capacitors C_2 and C_3 via an external bias voltage, V_{OS}.

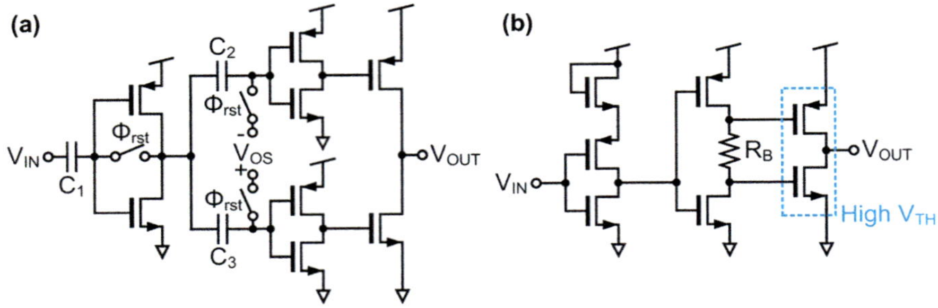

Fig. 16.6 (**a**) Original ring amplifier [17] and (**b**) the self-biased ring amplifier [18]

Operating the third stage in subthreshold results in a high output resistance, thereby forming a dominant pole at the output and stabilizing the amplifier. Ring amplifiers [4, 17, 18] have several intrinsic advantages compared to OTAs. First, even with a low supply voltage, a ring amplifier easily produces high gain from its three cascaded gain stages. Second, as mentioned earlier, slew-based charging is very energy efficient. Third, because the last stage is a simple inverter, ultimately operating in subthreshold, ring amplifiers can handle a near rail-to-rail output signal swing.

The self-biased single-ended ring amplifier introduced in [18] and shown in Fig. 16.6b is more robust to PVT changes and uses less power than the original ring amplifier circuit. The improved robustness and the removal of external biases make this ring amplifier more practical. One of the innovations in [18] is the use of high threshold voltage NMOS and PMOS transistors in the last stage, which extends the stable range since high threshold voltage FETs have an order-of-magnitude higher output resistance for a given gate-source voltage. Another technique that helps stabilize the design is the addition of resistor, R_B, between the gates of third-stage NMOS and PMOS transistors, as shown in Fig. 16.6b. The voltage drop caused by the second-stage inverter current flowing through R_B dynamically applies different voltages to the gates of the last inverter stage, as V_{IN} approaches the desired common-mode voltage. On the other hand, the gates of the NMOS and PMOS transistors of the last stage are still driven rail to rail when V_{IN} is away from the common-mode voltage, ensuring a high slew rate. An advantage compared to the ring amplifier in [17] is that the combined three stages are auto-zeroed for improved PVT tolerance.

The ring amplifiers in [17, 18] are single-ended circuits and therefore inherit the drawbacks of single-ended structures. The well-known disadvantages of single-ended circuits include limited common-mode and supply rejection. Furthermore, single-ended circuits do not reject even order harmonics as differential circuits do. As shown in Fig. 16.7, a pseudo-differential structure along with a common-mode feedback (CMFB) circuit [17, 18] to some extent alleviates these problems. The switched-capacitor CMFB in Fig. 16.7 consists of the common-mode sensing

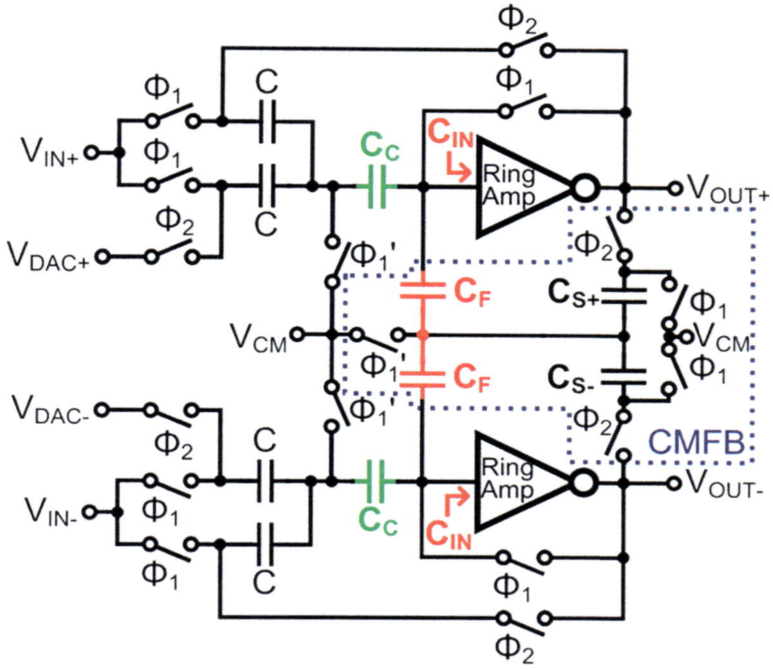

Fig. 16.7 Pseudo-differential MDAC gain stage with two ring amplifiers in [18]

capacitors C_{S+} and C_{S-} and feedback capacitors, C_F. V_{CM} is the common-mode voltage reference. A limitation is that this pseudo-differential CMFB reduces the effective gain of the ring amplifier because C_F forms a capacitive divider at the input of the ring amplifier. The effective gain is reduced from the nominal ring amplifier gain, A_V to $A_V \bullet C_C/(C_C + C_F + C_{IN})$, where C_C is the auto-zero offset storage capacitor and C_{IN} is the input parasitic capacitance of the ring amplifier.

As shown in Fig. 16.8, [4] introduces a fully differential ring amplifier that avoids the problems of the single-ended ring amplifier structure. In the fully differential ring amplifier, a single differential pair replaces the first stages of a pair of single-ended ring amplifiers [18]. Reuse of current by the NMOS and PMOS differential pairs increases the transconductance, thereby reducing the dominant thermal noise of the ring amplifier. To further save power, when not needed, the first stage is powered down via an enable signal Φ_{EN}.

Effective biasing and CMFB are important for reliable operation of the ring amplifier. Biasing and CMFB are shown in Fig. 16.8. The auto-zero forces the ring amplifier input and output voltages to be close to values that lead to the highest amplifier gain. There are separate CMFB loops to set the common mode of the first stage and the common mode of the overall ring amplifier. During the auto-zero phase, a CMFB loop, consisting of PMOS devices M4, M5, and M6

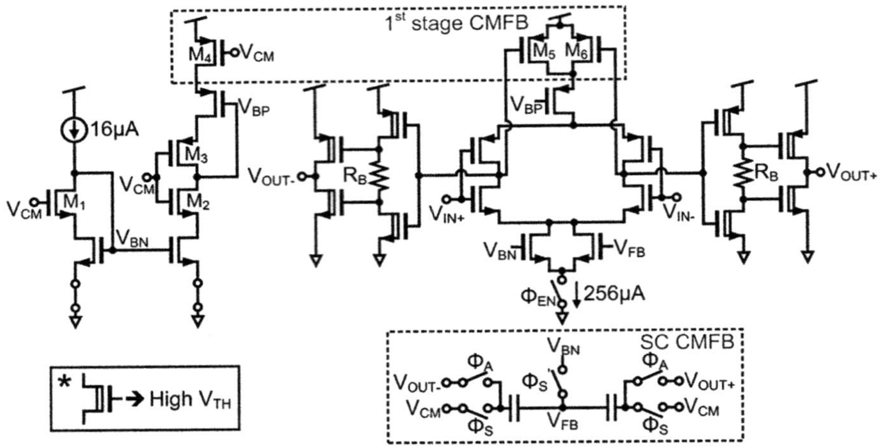

Fig. 16.8 Fully differential ring amplifier, along with bias and CMFB [4]

operating in triode [19], coarsely regulates the output common mode of the first stage. A separate switched-capacitor CMFB circuit forces the output common mode of the entire ring amplifier to V_{CM} during the amplification phase.

The second and third stages of the ring amplifier are based on inverters. Similar to the single-ended self-biased ring amplifier [18], resistors, R_B, apply (Fig. 16.8) dynamically offset voltages to the PMOS and NMOS gates of the third stage. Furthermore, high-threshold voltage devices in the second stage increase gain. This is needed because dynamic biasing can cause the second-stage transistors to operate in triode region. Triode operation greatly reduces both the second-stage gain and also the gain of the entire ring amplifier. As with the third stage, the use of high threshold voltage transistors extends the output voltage range for which the second-stage transistors are operating in saturation. The simulated small-signal gain for a 65 nm CMOS prototype ring amplifier is greater than 80 dB for an output swing range from 0.1 V to 1.1 V and a 1.2 V supply.

16.5 SAR-Assisted Pipeline ADC with Ring Amplifier

A prototype 50MS/s 13-bit ring amplifier-based SAR-assisted pipeline ADC [4], shown in Fig. 16.9, employs a 6-bit first-stage SAR ADC and an 8-bit second-stage SAR sub-ADC. Different to conventional SAR-assisted pipeline ADCs [2, 11–16], an advantage is that the wide output range of the differential ring amplifier permits a full 32× gain residue stage. The wide output swing relaxes the noise constraints on the second-stage sub-ADC and therefore saves power. The overall ADC accepts a 2.4 $V_{\text{pk-pk diff}}$ (i.e., rail to rail) input. One bit of stage redundancy allows the

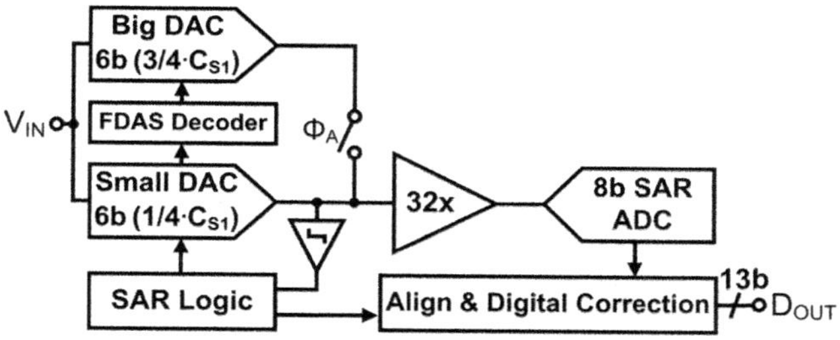

Fig. 16.9 50MS/s 13-bit SAR-assisted pipeline ADC with ring amplifier [4]

pipeline to tolerate first-stage sub-ADC errors. The output range of the residue is 0.3–0.9 V for ideal first-stage CDAC and comparator. The additional output range of the amplifier facilitates the 1-bit redundancy.

To reduce the switching energy of the first stage, SAR CDAC is split into two separate capacitor DAC arrays, *Big DAC* and *Small DAC*, as shown in Fig. 16.9. Splitting the CDAC into two separate capacitor arrays also reduces the INL and DNL errors due to the CDAC capacitor mismatch. The total differential sampling capacitance of the first-stage CDAC is 4 pF to satisfy the 13-bit kT/C noise requirement. Taking advantage of the fact that the 6-bit first-stage SAR sub-ADC needs only to meet 6-bit kT/C noise performance, *Small DAC*, which is part of the first-stage SAR ADC, uses only a quarter of the sampling capacitance, to reduce the SAR DAC power consumption. Merged capacitor switching (MCS) [20] further reduces the energy consumption of the SAR DAC. Asynchronous SAR operation [21] eliminates the need for a high-frequency ADC clock and reduces errors due to comparator metastability.

Both *Big DAC* and *Small DAC* sample the same input signal. *Big DAC* contains the remaining three quarters of the sampling capacitance and is only needed during residue generation. Based on the decision of the SAR, energy-efficient switching of *Big DAC* is achieved with the floated detect-and-skip (FDAS) CDAC switching technique, derived from [22]. Once the first-stage SAR conversion is complete, the residues of the Big and Small DACs are merged together and passed to the 32× residue amplifier.

Figure 16.10 shows a simplified single-ended depiction of the residue gain structure – the actual implementation is fully differential. Φ_A controls the amplification phase, and Φ_S and $\Phi_{S'}$ are sampling/auto-zero phase control signals. Auto-zeroing ensures that the output swing of the ring amplifier is fully utilized. A relatively large (4 pF) offset storage capacitor, C_{AZ}, minimizes folding of the auto-zero noise [23]. However, the fact that the sampled voltage on C_{AZ} stays constant means that the large C_{AZ} capacitance does not have a detrimental effect on power consumption. Furthermore, this large C_{AZ} capacitance has the advantage of stabilizing the ring amplifier during the auto-zero. This is because C_{AZ} presents a large load to the ring amplifier during auto-zero, thereby reducing both the dominant pole frequency and the slew rate.

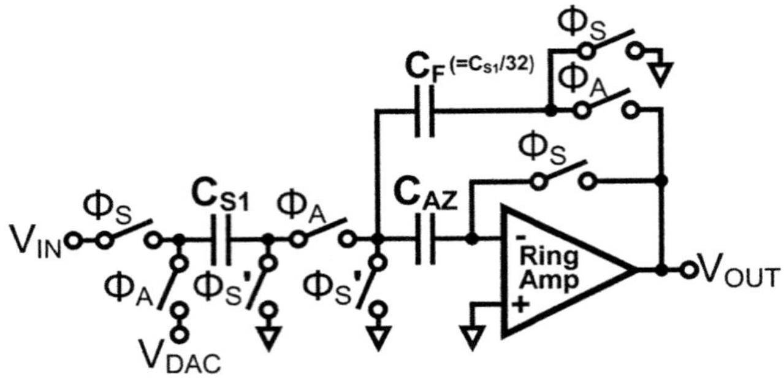

Fig. 16.10 Simplified single-ended depiction of residue gain stage structure [4]

1st stage	Tracking	Conversion	Residue transfer
Residue Amplifier	Auto-zero	Off	Residue amplification
2nd stage	Conversion	Reset	Tracking

Fig. 16.11 Simplified timing for SAR-assisted pipeline with ring amplifier [4]

Figure 16.11 shows a simplified timing diagram for the entire SAR-assisted pipeline ADC. To save power, the ring amplifier is powered down during the operation of the first-stage SAR ADC. Amplification begins after the completion of the first-stage SAR ADC conversion – this maximizes the time for residue amplification. In the prototype, an 8-bit second-stage SAR sub-ADC digitizes the amplified residue. The second stage, like the first-stage sub-ADC, uses MCS, bottom-plate input sampling, and asynchronous SAR logic. The second-stage 8-bit CDAC is reset to V_{CM} after the sub-ADC is finished so that residue amplification always starts from V_{CM}. This reset improves efficiency by halving the maximum slew rate required from the ring amplifier [18].

16.6 Conclusions

The last decade has seen a near three order-of-magnitude improvement in the energy efficiency of ADCs. Much of this can be attributed to the scaling-friendly nature of the SAR architecture. Furthermore, the SAR-assisted pipeline architecture enables SAR ADCs to dramatically improve the energy efficiency of moderately high-resolution pipeline ADCs. At the same time, the ring amplifier avoids the problems associated with OTAs in advanced CMOS nodes.

References

1. Murmann, B.: Energy limits in current A/D converter architectures, ISSCC Short Course, Feb. 2012
2. Lee, C.C., Flynn, M.P.: A SAR-assisted two-stage pipeline ADC. IEEE JSSC. **46**(4), 859–869 (2011)
3. Choo, K.D., Bell, J., Flynn, M.P.: Area-efficient 1GS/s 6b SAR ADC with charge-injection-cell-based DAC. ISSCC Digest Technical Papers. 460–461 (Feb. 2016)
4. Lim, Y., Flynn, M.P.: A 1 mW 71.5 dB SNDR 50 MS/s 13 bit fully differential ring amplifier based SAR-assisted pipeline ADC. IEEE JSSC. **50**(12), 2901–2911 (2015)
5. Le Dortz, N., et al.: A 1.62GS/s time-interleaved SAR ADC with digital background mismatch calibration achieving interleaving spurs below 70dBFS. ISSCC Digest Technical Papers. 386–388 (Feb. 2014)
6. Chan, C.-H., et al.: A 5.5mW 6b 5GS/S 4×−interleaved 3b/cycle SAR ADC in 65nm CMOS. In ISSCC Digest Technical Papers, pp. 1, 3, Feb. 2015
7. Hong, H-K., et al.: An 8.6 ENOB 900MS/s time-interleaved 2b/cycle SAR ADC with a 1b/cycle reconfiguration for resolution enhancement. In ISSCC Digest Technical Papers, pp. 470, 471, Feb. 2013
8. Liu, C.-C., et al.: A 0.92mW 10-bit 50-MS/s SAR ADC in 0.13μm CMOS process. IEEE Symposium VLSI Circuits, Digest Technical Papers, pp. 236–237, June 2009
9. Yang, W.H., Kelly, D., Mehr, I., Sayuk, M.T., Singer, L.: A 3-V 340-mW 14-b 75-Msample/s CMOS ADC with 85-dB SFDR at Nyquist input. IEEE JSSC. **36**, 1931–1936 (2001)
10. Devarajan, S., Singer, L., Kelly, D., Decker, S., Kamath, A., Wilkins, P.: A 16b 125MS/s 385mW 78.7dB SNR CMOS pipeline ADC. IEEE ISSCC Digest Technical Papers. 86–87 (Feb. 2009)
11. Lee, H.-Y., Lee, B., Moon, U.-K.: A 31.3fJ/conversion-step 70.4dB SNDR 30MS/s 1.2V two-step pipelined ADC in 0.13μm CMOS. In IEEE ISSCC Digest Technical Papers, pp. 474–475, Feb. 2012
12. Verbruggen, B., Iriguchi, M., Craninckx, J.: A 1.7 mW 11b 250 MS/s 2-times interleaved fully dynamic pipelined SAR ADC in 40 nm digital CMOS. IEEE JSSC. **47**(12), 2880–2887 (2012)
13. Verbruggen, B. et al.: A 2.1 mW 11b 410 MS/s dynamic pipelined SAR ADC with background calibration in 28nm digital CMOS. IEEE Symposium VLSI Circuits, Digest Technical Papers, pp. 268–269, June 2013
14. van der Goes, F., et al.: A 1.5 mW 68 dB SNDR 80 Ms/s 2 × interleaved pipelined SAR ADC in 28 nm CMOS. IEEE JSSC. **49**(12), 2835–2845 (2014)
15. Verbruggen, B., Deguchi, K., Malki, B., Craninckx, J.: A 70 dB SNDR 200 MS/s 2.3 mW dynamic pipelined SAR ADC in 28nm digital CMOS. In: IEEE Symp. VLSI Circuits, Digest Technical Papers, pp. 268–269, June 2014
16. Malki, B., Verbruggen, B., Wambacq, P., Deguchi, K., Iriguchi, M., Craninckx, J.: A complementary dynamic residue amplifier for a 67 dB SNDR 1.36 mW 170 MS/s pipelined SAR ADC. In: ESSCIRC Digest Technical Papers, pp. 215–218, Sep. 2014
17. Hershberg, B., Weaver, S., Sobue, K., Takeuchi, S., Hamashita, K., Moon, U.-K.: Ring amplifiers for switched capacitor circuits. IEEE JSSC. **47**(12), 2928–2942 (2012)
18. Lim, Y., Flynn, M.P.: A 100 MS/s, 10.5 bit, 2.46 mW comparator-less pipeline ADC using self-biased ring amplifiers. IEEE JSSC. **50**(10), 2331–2341 (2015)
19. Razavi, B.: Operational amplifiers. In: Design of Analog CMOS Integrated Circuits, pp. 319–324. McGraw-Hill, Boston (2001)
20. Hariprasath, V., Guerber, J., Lee, S.-H., Moon, U.-K.: Merged capacitor switching based SAR ADC with highest switching energy-efficiency. Electron. Lett. **46**(9), 620–621 (2010)
21. Chen, S.-W., Brodersen, R.W.: A 6b 600MS/s 5.3mW Asynchronous ADC in 0.13μm CMOS. IEEE ISSCC Digest Technical Papers. 2350–2351 (Feb. 2006)

22. Tai, H.-Y., Hu, Y.-S., Chen, H.-W., Chen, H.-S.: A 0.85fJ/conversion-step 10b 200kS/s subranging SAR ADC in 40nm CMOS. IEEE ISSCC Digest Technical Papers. 196–197 (Feb. 2014)
23. Enz, C.C., Temes, G.C.: Circuit techniques for reducing the effects of op-amp imperfections: autozeroing, correlated double sampling, and chopper stabilization. Proc. IEEE. **84**(11), 1584–1614 (1996)

Chapter 17
Time-Based Biomedical Readout in Ultra-Low-Voltage, Small-Scale CMOS Technology

Rachit Mohan, Samira Zaliasl, Chris Van Hoof, and Nick Van Helleputte

17.1 Introduction

Personalized healthcare applications require small and accurate sensor readout devices that consume low power and are of low cost so as to make them portable and ubiquitous. A low power consumption facilitates using smaller batteries or other energy sources which, often, are the bulkiest components of a sensor system, thereby enhancing the portability [1, 2]. Whereas, the low cost will facilitate large-scale usage and thereby help in reducing the burden on existing health infrastructure – the ultimate goal of providing personalized healthcare solutions [3].

Reliable detection of signals in biomedical applications requires high signal sensitivity and capability of handling large input signals to account for the motion. Furthermore, to obtain relevant information from the readout signal, an efficient digital signal processing is required. Integration of the sensor technology with signal processing as a system on chip (SoC) has been shown to be a great enabler for these applications [4–7]. Figure 17.1 shows a block diagram of a typical sensor SoC. An analog front end (AFE) reads the physiological signals and converts them into a digital value. The DSP processes these signals to obtain relevant information and often transmits this wirelessly for further processing.

R. Mohan (✉) • N. Van Helleputte
Imec, Leuven, Belgium
e-mail: rachit.mohan.rm@gmail.com; Nick.VanHelleputte@imec.be

S. Zaliasl
Imec-Holst Centre, Eindhoven, NL, The Netherlands

C. Van Hoof
Imec, Leuven, Belgium

Imec-Holst Centre, Eindhoven, NL, The Netherlands

P. Harpe et al. (eds.), *Hybrid ADCs, Smart Sensors for the IoT, and Sub-1V & Advanced Node Analog Circuit Design*, DOI 10.1007/978-3-319-61285-0_17

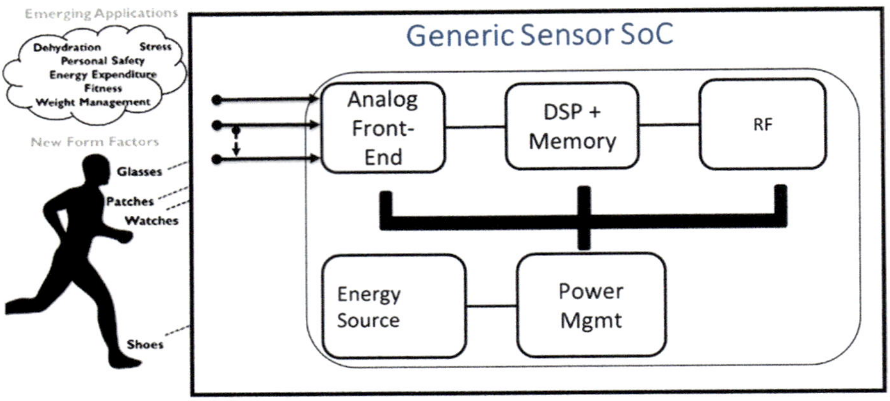

Fig. 17.1 Block diagram of a typical sensor SoC

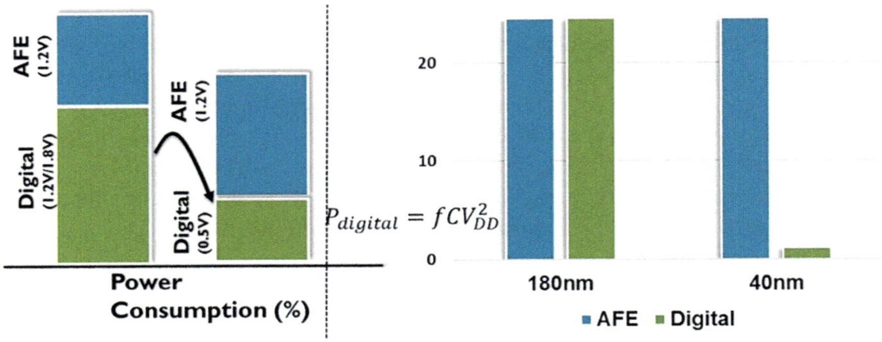

Fig. 17.2 Power and area breakdown of a sensor SoC in 180 nm CMOS technology for wearable biomedical readout application [8]

To realize power- and area-efficient SoCs in this application, both analog and digital design challenges need to be considered. To understand the challenge of obtaining a low-power, low-area sensor SoC without compromising on accuracy and large-signal handling capabilities, consider Fig. 17.2. It shows the power and area break-up of a sensor SoC [8] for a wearable ECG readout, with on the node digital signal processing including motion artefact reduction and beat detection. As can be seen in the figure, for a design in 180 nm, even fairly modest digital signal processing tasks result in a sizeable power and area consumption comparable to the AFE. The figure also plots the area of the AFE and the digital in the 180 nm technology and also an extrapolated version at 40 nm technology, just to show the magnitude of the problem. As can be seen, the digital area is hardly visible in an advanced node such as 40 nm. However, the area of the AFE does not scale so easily with technology and thereby lead to an increase in costs.

The reasons why the AFE does not scale well with technology and voltage supply are well known and have been discussed extensively in literature [9–14]. In the next section, we will briefly discuss the reasons that are relevant to biomedical sensor readout.

One of the dominant reasons that the AFE does not scale well is the reduction in available voltage headroom for signal swing due to reduction in voltage supply. To overcome this challenge, as in other design communities such as the ADCs, we look towards time-based operation. We do so because it offers the advantage, among others, of an available dynamic range that is decoupled to the VDD. This potentially allows lower voltage operation and hence a lower power consumption. Moreover, this available dynamic range even increases as we scale down in technology as opposed a voltage-based design.

However, the existing time-based implementations are not quite suitable for low-power, low-cost biomedical sensor readout. They too face challenges and often the same ones to meet the low-power and low-area constraints as a voltage-based implementation. The reasons for this are also discussed later in this paper. We propose a time-based circuit to overcome the existing challenges to meet the requirements of a sensor SoC.

This paper is structured in the following manner. In Sect. 17.2, we discuss the challenges of technology and voltage scaling of traditional AFEs. In Sect. 17.3, we will present a historical overview of existing time-based approaches. In Sect. 17.4, we will discuss the specific implementation of the proposed time-based circuit. Finally, Sect. 17.5 will list the conclusion.

17.2 Challenges of AFE Design at Low Voltages and Small-Scale CMOS Technology

In this section, we will discuss the challenges of AFE design from a biomedical perspective while also discussing some existing solutions in literature.

A summary of the main challenges is listed below:

- Voltage scaling: A lower VDD will lead to a lower available dynamic range.
- Intrinsic gain: As the channel length reduces, the intrinsic gain of the transistor reduces. This is especially problematic for design of instrumentation amplifiers (IA) which require high DC gains for power- and noise-efficient readouts.
- Flicker noise: Since biomedical signals are at low frequencies (<1 KHz), impact of flicker noise is large, and it needs to be reduced for low-noise applications. Since, flicker noise is proportional to area of the transistor, scaling down the transistor size becomes challenging.
- Gate-leakage current: As technologies scale, the gate oxide thickness reduces, leading to an increase in gate-leakage current. This is problematic for IAs which need to have extremely high input impedances (>10 MΩ).
- On-chip passives: On-chip passives are required for implementing time constants for the required signal conditioning and AC coupling. However, their area does not scale down with technology. Since biomedical signals are at low frequencies,

Table 17.1 Target AFE specifications

Parameter	Value
Technology	40 nm CMOS
VDD	<1 V
Power	<10 μW
Noise	<10 μV_{rms}
Input impedance	>50 MΩ–100 MΩ
Area	<0.1 mm^2

they require large on-chip time constants, and hence, biomedical designs are typically dominated by on-chip passive area. Therefore, this is one of the main reasons why biomedical readouts do not scale down in area with the technology.

Table 17.1 lists the rough specifications that we would like to target for an AFE design to meet the application requirements.

17.2.1 Voltage Scaling

To understand the effect of VDD scaling on a circuit's power and area consumption, consider Fig. 17.3 [15]. It shows a g_m-C circuit with an input signal with peak-to-peak voltage swing (V_{pp}) Vsig and an output voltage of V_{pp} of VDD. The g_m-C circuit is a representative conceptual model for most analog blocks.[1] The capacitance 'C' can be any capacitance – MIM cap, moscap or parasitic cap. It basically will represent the area of the circuit. For the purposes of this analysis, we assume that the gm is ideal. The only nonideality that we assume present is the *kT/C* thermal noise of the capacitor. This is because thermal noise is important for biomedical designs and often forms the fundamental limit to the circuit performance. The signal-to-noise (SNR), bandwidth (BW) and power consumption (*P*), in this case, are given by

$$SNR = \frac{\left(\frac{VDD^2}{8}\right)}{\left(\frac{kT}{C}\right)} \rightarrow C \propto \frac{SNR}{VDD^2}$$

$$BW = \frac{g_m}{2 \cdot pi \cdot C} \propto \frac{I_{DD}}{C}$$

$$P = VDD \cdot I_{DD}$$

[1]To be exact, the model and the following analysis hold true for sampled systems such as sample-and-hold circuits, ADCs. It does not hold true for some blocks such as instrumentation amplifiers (IA) and low-noise amplifiers (LNA). However, since almost any analog signal chain will contain ADCs, the overall conclusion of the analysis will still be applicable and relevant to our discussion in this paper.

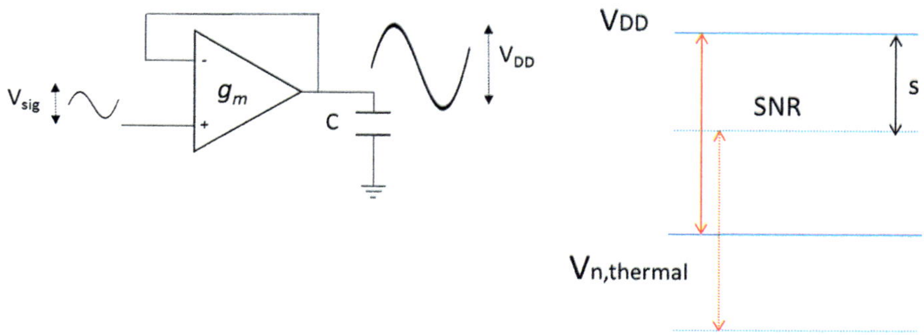

Fig. 17.3 Conceptual model to understand the effect of decrease in VDD [15]

So now, if the VDD scales down with a factor of ‘s’, then the ‘C’ (and consequently the area) will increase by ‘s^2’, and the power consumption of the circuit will increase by ‘s’ to keep the SNR and the BW constant.

A similar analysis can be performed for flicker noise and mismatch with the same result. Basically, this result means that if VDD scales down, the noise will also have to scale down to obtain the same dynamic range (or accuracy), which requires increase in power and area consumption.

One way to combat this is by representing the signal in another domain independent of VDD, that is, time. Such solutions are discussed in the next chapter.

It is important to note that in practice, the results are not so dramatic [16]. This is because the parameter that is important is not VDD but (VDD-V_{th}), where V_{th} is the threshold voltage of the transistor. A low V_{th} transistor can be used at critical places in the circuit to offset the impact of a low VDD. The increase of kT/C noise due to smaller C can be overcome by oversampling techniques to some extent. Nevertheless, the general conclusion still holds true.

17.2.2 Lower Intrinsic Gain

To understand the impact of the intrinsic gain of the transistor due to shorter channel length, consider a representative differential transistor circuit shown in Fig. 17.4 with the given design parameters. The input transistors are biased in weak-inversion region as all low VDD designs and biomedical circuits, in general, operate in.

The intrinsic gain of the input transistors is given by

$$Intrinsic\,gain \equiv \frac{g_{\mathrm{m}}}{g_{ds}} \propto L \cdot \exp\left(\frac{dV_{th}}{n \cdot U_{\mathrm{t}}}\right)$$

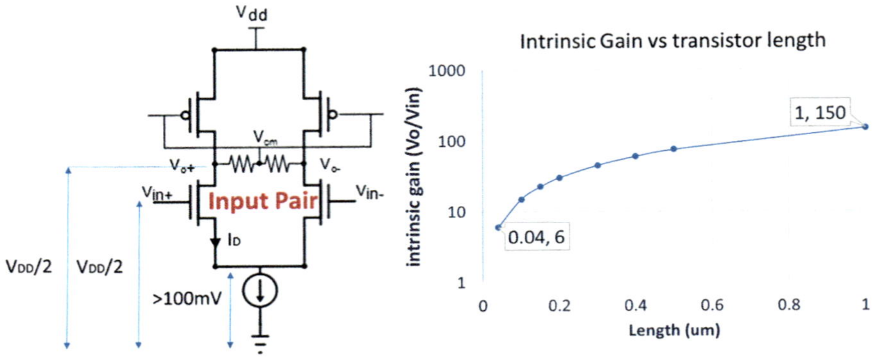

Fig. 17.4 Intrinsic gain of a representative differential opamp with different input transistor lengths

'L' is the length of the transistor, V_{th} is the threshold voltage, n is the subthreshold constant and U_t is the thermal voltage. Figure 17.4 also plots the intrinsic gain vs L, assuming all the parameters remain the same. As can be seen, the intrinsic gain reduces considerably by changing the $L = 1$ μm (which is fairly standard for biomedical designs) to $L = 40$ nm.

In practice, the exact results will vary due to secondary effects of change in V_{th} and n due to change in the drain voltage and VDD. For a more comprehensive analysis, the readers are referred to [10].

To overcome this problem, cascoding and gain-boosting techniques are typically applied to increase the intrinsic gain [17, 18]. However, for VDD < 1 V, these techniques are not so useful. Hence, existing solutions in literature apply one or multiple of the following techniques to increase gain:

- Cascading of gain stages.
- Increasing the transconductance, g_m, of the transistor by biasing the body at a voltage other than ground or source or even using the body also as input [19–21]. The possibility of latch-up in such low VDD designs is very low as VDD V_{th}.
- Using weak-positive feedback to enhance the gain [19, 22].

Figure 17.5 shows an example of a 0.5 V opamp which uses all the three techniques listed above [19]. Alternatively, it is also possible to use the phase domain [23, 24] or the time domain [25] to obtain a large gain. This will be discussed in the next chapter.

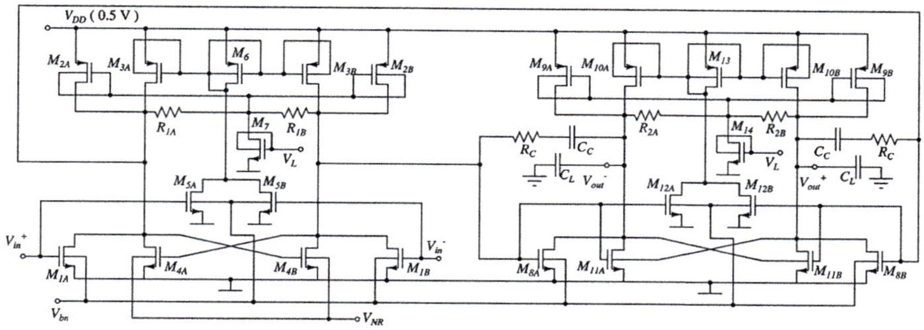

Fig. 17.5 Opamp circuit that uses cascading, body biasing and weak-positive feedback to increase the gain. It operates at 0.5 V VDD

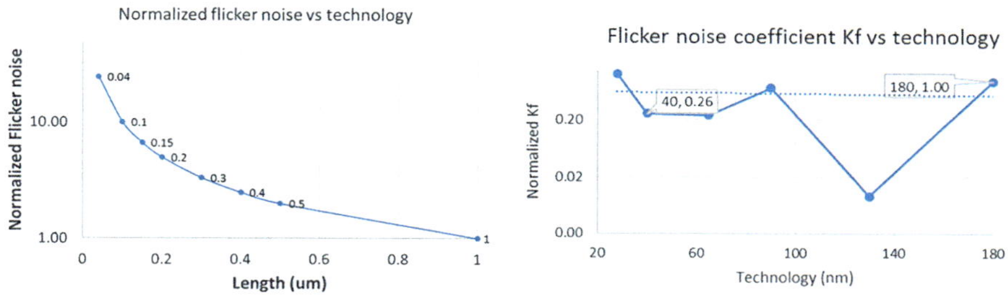

Fig. 17.6 Normalized flicker noise vs transistor channel length (a), normalized K_f vs technology

17.2.3 *Increase in Flicker Noise*

Input-referred flicker noise of the previous representative differential opamp is given by [17]

$$S_{flicker} \approx sqrt\left(\mathrm{K_f}.\frac{\mathrm{I_d^2}}{\mathrm{C_{ox}.WL.f}}\right)$$

where K_f is the flicker noise coefficient, Id is the drain current of the input-stage transistor, C_{ox} is the unit oxide capacitance of a MOSFET, f is the frequency, and W and L are the width and length of the input-stage transistors, respectively.

Figure 17.6a plots the flicker noise vs *L* to show the impact of decrease in transistor length. The values have been normalized to the value at 180 nm technology node. As can be seen, the flicker noise increases will increase >10× by reducing the transistor length down to 40 nm.

It is also of interest to know the effect of choice of technology on the flicker noise of the transistor for a given L. Figure 17.6b plots the normalized $K^2{}_f$ vs technology. The technologies chosen are sample TSMC technologies from 180 nm down to 28 nm. Interestingly, there seems to be no pattern or correlation that we can discern, making it impossible for us to predict the impact of technology on flicker noise.

Typical solutions in literature, apart from increasing the size of the transistor, to overcome the problem with flicker noise are [26]:

- Autozeroing technique
- Chopping technique

The underlying strategy for both these techniques is to have separate paths for the signal and the flicker noise and thereby be able to filter out the flicker noise after signal readout. The advantage of the chopping technique over the autozeroing is that there is no significant increase in the baseband thermal noise floor. However, the main drawback is that there can be a large output ripple in the chopping technique, which can limit its efficacy and can also be problematic for low VDD designs. This ripple voltage can be reduced by the use of a ripple feedback loop [27]. This loop can be a mixed-signal feedback loop as well [28]. Alternatively, references [29, 30] show a readout architecture that uses both autozeroing and chopping to retain the advantages of both the techniques while getting rid of flicker noise.

It is to be noted that both these techniques require the circuit to operate at a higher bandwidth than the baseband. Although at small-scale technologies this is not necessarily problematic, in older technologies with large transistor designs, this can lead to a higher power consumption. At low VDD, we also need to take care about the impedance of the switch transistors which can impact the maximum frequency of operation. However, with low V_{th} transistors and operation at frequencies <100 kHz, this is not a major problem yet [31]. However, in case it is, techniques such as clock boosting can be used to improve the speed performance [32, 33].

17.2.4 Increase in Gate-Leakage Current

One of the major effects of technology scaling is increased gate-leakage current through the oxide layer. This phenomenon originates due to carriers tunnelling through the oxide-potential barrier due to quantum mechanical effects. This carrier current increases exponentially with the reduction in oxide thickness and, hence, is prevalent in small-scale technologies. Although the phenomena and the physics have been known for decades, characterizing and modelling it is still in research phase [34].

However, we do know it is heavily dependent on the bias voltages of the transistor and the gate area. It is also highly nonlinearly dependent on temperature [10, 34].

[2] K_f has been extracted from the BSIM4 spice model parameters.

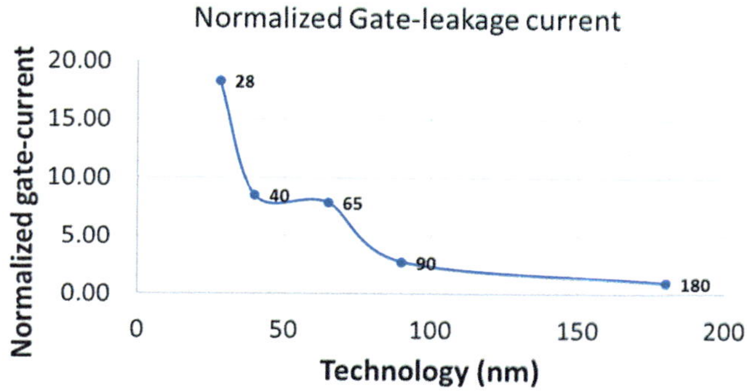

Fig. 17.7 Normalized gate-leakage current vs CMOS technology

Figure 17.7 plots the normalized (to value at 180 nm technology node) gate-leakage current of the input stage for the representative differential amplifier in Fig. 17.4, to get an idea of the increase in gate-leakage current with technology scaling. We use the following BSIM4 model to generate the plot [34]:

$$I_g \propto WL \cdot \frac{1}{t_{ox}^3 \cdot e^{t_{ox}}}$$

where I_g is the gate current, t_{ox} is the thickness of the oxide, and W and L are the width and length of the input-stage transistors, respectively. As can be seen, as we scale down to 40 nm and 28 nm technology, the gate-leakage current increases by 10 times as compared to the 180 nm technology, which is still a favourite node for biomedical designs.

The presence of gate-leakage current is problematic not just because of wasteful power consumption, but it can reduce the input impedance of the readout. It can also lead to an increase in noise, especially for sensor readouts with high source impedance.

A simple solution to overcome the challenge of gate-leakage current is to either use thick-oxide transistors or small area transistors. The drawback of using thick-oxide transistors is that they have a higher V_{th} than the typical oxide thick ones, which can be a problem at low VDD designs. On the other hand, small area transistors will be limited by their flicker noise performance. In case none of these solutions are possible, we will need to look at our bipolar counterpart for compensation circuits for gate current. Additionally, it is also possible to use overall positive DC feedback loops to compensate for it. Such feedback loops already exist in literature to compensate for input currents arising due to other effects such as chopping. Figure 17.8 shows one such architecture that incorporates such a feedback mechanism [35].

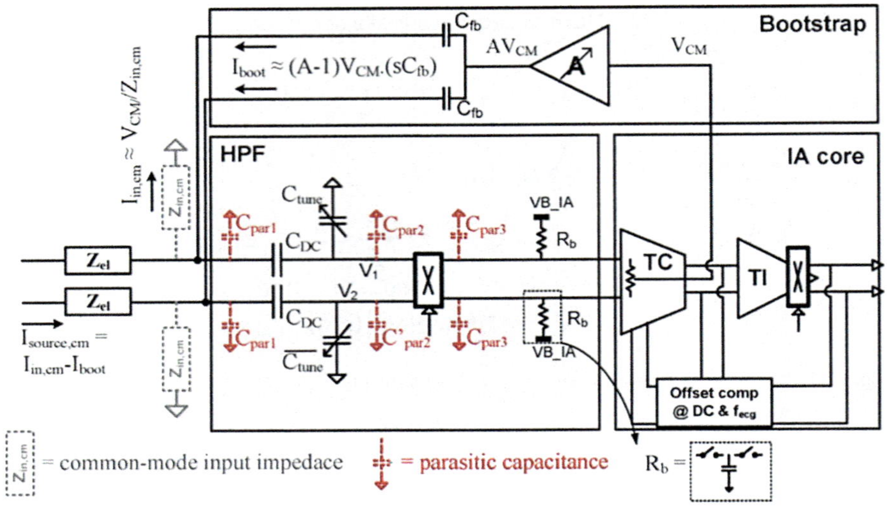

Fig. 17.8 IA architecture with positive feedback loop (bootstrap loop) to compensate for input-stage current [35]. Such feedback loops can also be utilized to compensate for gate-leakage current due to technology scaling

17.2.5 Area of On-Chip Passives

Although the size of the transistor scales down with technology, the size and density of on-chip capacitances and resistances do not. For example, the MIM (metal-insulator-metal) capacitance remains a few fF/μm [2, 12], and the sheet resistance of a hi-poly layer is around the order of a few kΩ/sq. This means that the area that is comprised of passives will remain the same as we scale down, even if we decrease the size of the transistors. Given that biomedical readouts are generally dominated by area of passives [36], we cannot scale down the area with technology.

To overcome this challenge, it is possible to implement the passives with active transistors [37]. However, such designs often face a power and dynamic range trade-off [38]. For example, reference [39] implements a large time constant by way of a switched-capacitor technique, at the expense of increase in noise. Alternatively, in certain applications, instead of an active implementation, passive elements in an unconventional configuration can be used. Reference [40] uses capacitors in a T-network configuration in the feedback to reduce the effective feedback capacitance of an IA. This allows them to reduce the input capacitance while keeping the overall gain the same. This architecture is shown in Fig. 17.9a.

It is also possible to implement the required on-chip time constant in the digital domain [36, 41]. The required time constants are implemented in the digital domain, and the information is fed back to the analog domain through DACs. Figure 17.9b shows one such architecture [36]. It implements a DC-coupled mixed-signal

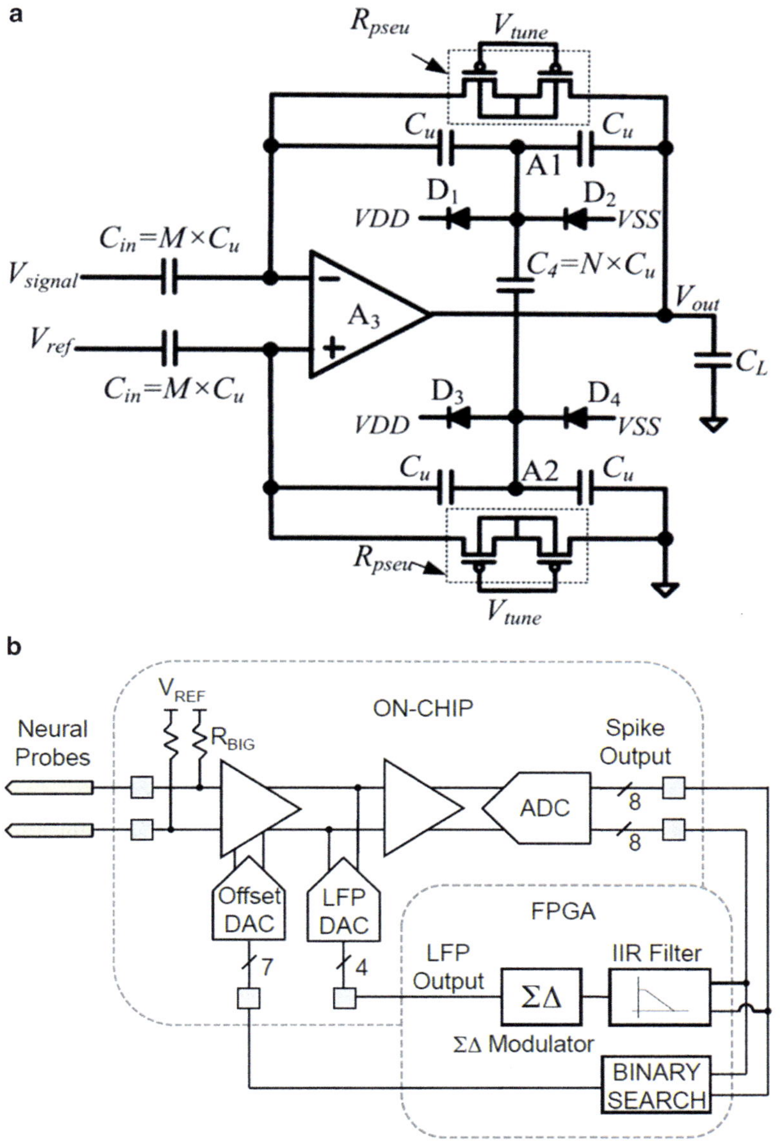

Fig. 17.9 A T-network-based capacitive feedback IA [40] (**a**) Mixed-signal feedback-based readout [36] (**b**)

feedback loop to get rid of all the capacitors required for implementing a time constant and for AC coupling. For large array readouts, it is also possible to share a common/reference terminal or even a complete block such as ADC to save area [42, 43].

17.3 Time-Based Operation: Literature Survey

In the previous section, we saw that one of the existing critical challenges is the reduction in the available signal swing due to the reducing VDD. To overcome this challenge, we need to represent the signal in a domain other than the voltage domain. We specifically look towards the time domain for the following reasons:

- A large dynamic range available: As the size of the transistor scales down, a higher speed or bandwidth is available to represent the signal as opposed to the voltage domain, wherein the dynamic range reduces.
- This dynamic range is decoupled to the voltage supply (at least to a first order).
- The signal is represented in a digital-like representation. Given the reality that the digital circuits drive the semiconductor technology, by moving to a time-based operation, we basically ensure that any changes in technology that will be beneficial to the digital circuits will be beneficial to time-based circuits as well. An important corollary of this point is that it is easy to convert a time-domain signal to a digital representation. Simple digital counters are usually sufficient as opposed to complicated voltage-based ADCs. Not only that, it will potentially allow using digital calibration and compensation techniques to correct for the errors in the analog domain.

The idea of time-domain-based operation is not new. It is explored in literature in different design communities such as ADCs [44–47], filters [23, 48], DC-DC converters [49], pixel readout [50] and resistive/capacitive sensor readouts [51, 52]. In fact, time-based operation has been existing in literature for almost as long as the semiconductor industry itself. One of the first time-interval-based ADC patents was published in the 1940s [44, 53]. By the 1960s–1980s, with the advent of digital circuits, it was clear that the time-based circuits will be more advantageous with the improvement in technology [54]. Most of the time-based circuits that will be discussed later in this section can find their origins in studies done during this period. Given this history, however, until very recently, there has almost no study done yet on time-based biomedical readout.

The reasons for this can be explained by looking at the challenges of a time-domain-based approach:

- Real-world biomedical signals are, often, voltage signals and not time signals. Thus, to implement the time-based operation, we will require a block in the signal chain to convert the information from voltage to time. This block will face the same challenges of a voltage-based design in an ultra-low-voltage, small-scale technology environment.
- In addition to the previous point, although the time-domain signals are 'digital-like' or 'pseudo-digital', they are, in essence, still analog signals. This means that they are plagued by the same concerns of mismatch, *kT/C* noise, flicker noise and technology limitations such as effect of gate-leakage currents and short-channel

effects such as low intrinsic gain of the transistor. Hence, although it is possible to operate at low VDD, scaling down in area with technology will be challenging.
- Since it is challenging to reduce the area or the capacitance, even though a larger dynamic range is available in scaled technologies, the power reduction due to low VDD operation might be minimal or worse significantly high.
- For the same reasons, although converting to the digital domain is indeed simpler, it can also be power consuming as the digital counters may require a high frequency (>100 MHz's) for sufficient resolution. This, however, may not be a problem in advanced technology nodes and low VDD.

With these pros and cons in mind, we will analyse some of the existing ideas in time-based implementations.

Although biomedical signals are in voltage domain, many physical sensors output the information in the frequency domain, or the information can be easily converted to frequency domain by integrating them with voltage-controlled oscillators (VCOs) [51, 55, 56]. Time-domain-based operation can enable large DR readout [51, 55] or alternatively ultra-low-voltage readout [56] in such applications. Figure 17.10 shows the architecture of two such capacitive readouts. The underlying principle for all the three is that the change in the capacitor value will change the frequency of the oscillator. This change is detected by a mixer and a low-pass filter. Finally, the filtered value is converted into digital via a counter.

Time-domain readout can also bring advantages of large dynamic range to sensor readout for applications which are not area sensitive [57]. Figure 17.11 shows the architecture of a resistive readout via a relaxation oscillator-based loop. The working principle is that the resistor value is sensed by the current through the resistor, which is mirrored and converted into time edges and then to a digital value via ramp and a threshold comparator.

Figure 17.12 shows a resistive/capacitive sensor readout design [58]. This circuit too, like the previous one, uses a threshold comparator to detect the rate of discharge of the RC circuit and thereby the resistance value for a known capacitor value. This design was proposed to be used in conjunction with an off-chip microcontroller. Hence, the frequency of the digital counter (200 MHz in this case) was not an issue. By implementing a time-based readout, the design could achieve a large dynamic range readout at only 1 V supply.

Figures 17.13 a and b show two VCO-based architectures in sigma-delta loop for two different applications – high PSRR-resistive readout [59] and an 800 MHz ADC [60]. However, the main idea of both the designs is the same – both use the high voltage phase gain of the VCOs to implement the integrator in the sigma-delta loop instead of the analog integrator. In both cases this leads to a low VDD operation and lower power operation.

Figure 17.14a and b show examples wherein a time-based operation is used for ease in conversion to digital [44, 61]. The digital domain is then used to calibrating the errors (nonlinearity errors in both the cases) of the analog domain.

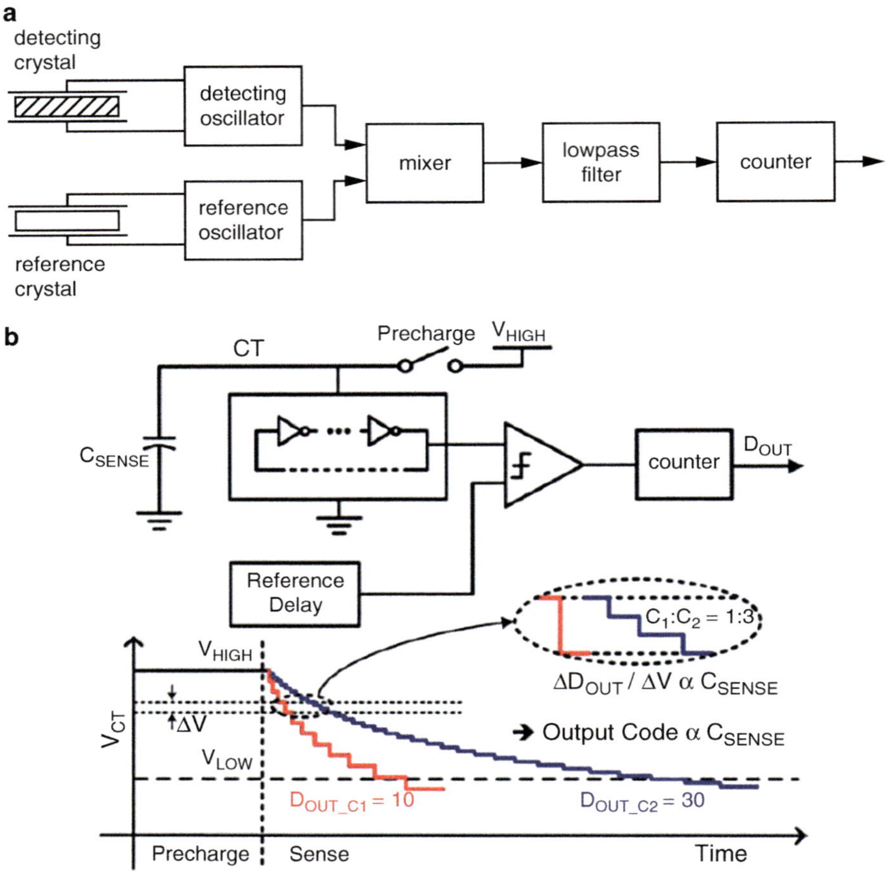

Fig. 17.10 VCO-based capacitive readout [55] (**a**), [51] (**b**)

17.4 Time-Based AFE for Ambulatory Biomedical Applications in 40 nm CMOS

In this section, we propose a time-based AFE for biomedical readout in 40 nm CMOS technology that can operate at 0.6 V. It can handle up to 40 mV_{pp} differential AC signal and 300 mV differential DC electrode offset while having an input-referred noise of 7.8$\mu$$V$rms in bandwidth of 150 Hz. It consumes 0.015 mm^2 area and 3.3 μW of power. Thus, it can handle a large dynamic range while keeping the power and area consumption low. As mentioned in the introduction, this AFE architecture will be useful in designing low-power, low-cost SoC applications such as ambulatory biomedical readout.

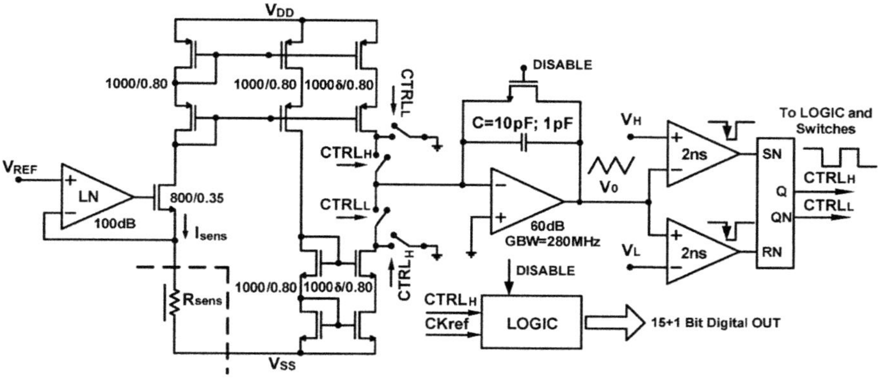

Fig. 17.11 Resistive sensor readout via a relaxation oscillator-based mechanism [57]

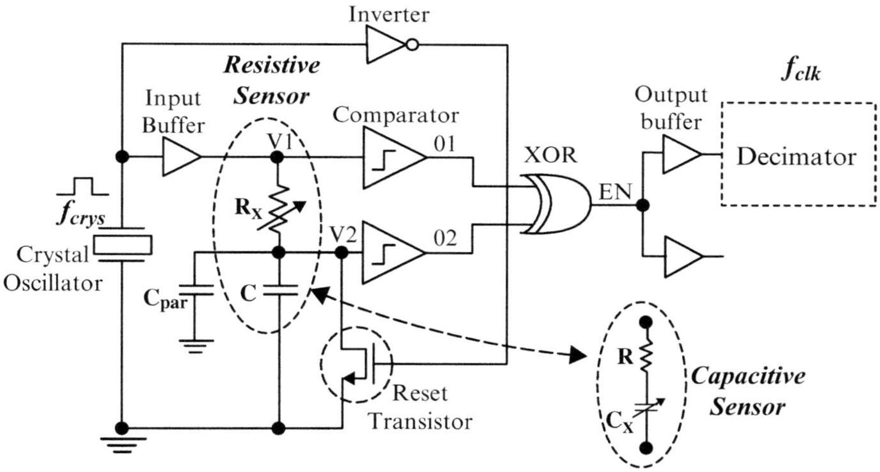

Fig. 17.12 1 V comparator-based resistive readout [58]

The main strategy of this architecture is to make use of time-domain operation to handle a larger signal range at small voltage supplies. As mentioned in the previous section, one of the main challenges of using time-based operation for biomedical readout is that the input signal is in the voltage domain. Hence, the block which needs to convert the signal in voltage to time will limit the power and area improvement. We overcome this challenge by suppressing large differential voltage swings right from the electrode input through the use of negative feedback. Furthermore, we get rid of the need for high-gain opamps and use scalable blocks such as dynamic comparators and small passives. The large time constants required are implemented in the digital domain. In this section, we present only the salient features of the design. For a more detailed discussion, we refer the readers to [31].

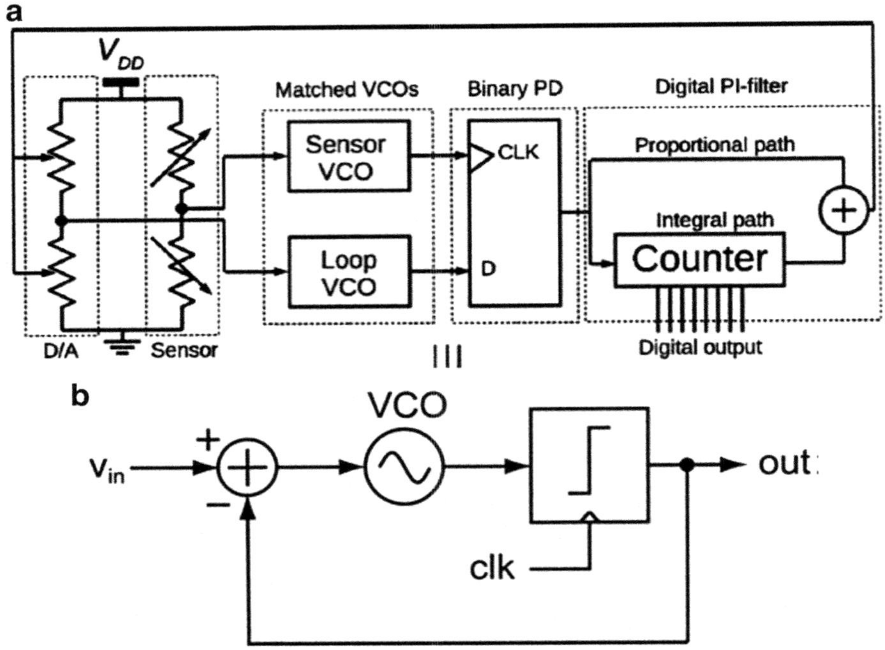

Fig. 17.13 Example architectures of VCO being used as high-gain blocks in a sigma-delta loop for (**a**) high PSRR resistive readout [52] (**b**) 800 MHz ADC [60]

Figure 17.15 shows the overview of the proposed time-based AFE architecture. The differential input signal (ECG signal in this case) is converted into 2-bit digital signal by a time-based ADC (T-ADC). This 2-bit signal is filtered off-chip via a CIC filter. This 2-bit signal is also used to filter out the DC information, which is then fed back to the input via a DAC.

Figure 17.16 shows the schematic of the T-ADC. It comprises of pseudo-differential stages, each comprising of a comparator and a charge-pump integrator in feedback. These stages are chopped to get rid of the flicker noise. The reasons for the 1MΩ resistance will be discussed later. The DC filter in the DC reduction loop is implemented using a reset counter, whose output is down sampled, quantized to 7 bit and fed back to the feedback node using a current DAC.

The comparator and the charge-pump integrator in feedback form a time-based loop. It is essentially an asynchronous delta modulator [62]. Due to the negative feedback, the input of the comparator is essentially a virtual ground, whereas the comparator output is in time domain. Hence, we eliminate any large voltage swings, allowing us to operate at low VDD and thereby consume low power. The virtual ground also allows us to implement a power- and area-efficient dynamic comparator. This is because one of the drawbacks of a dynamic comparator is noise fold-over. As will be seen later, we can suppress this by implementing a nominal gain stage

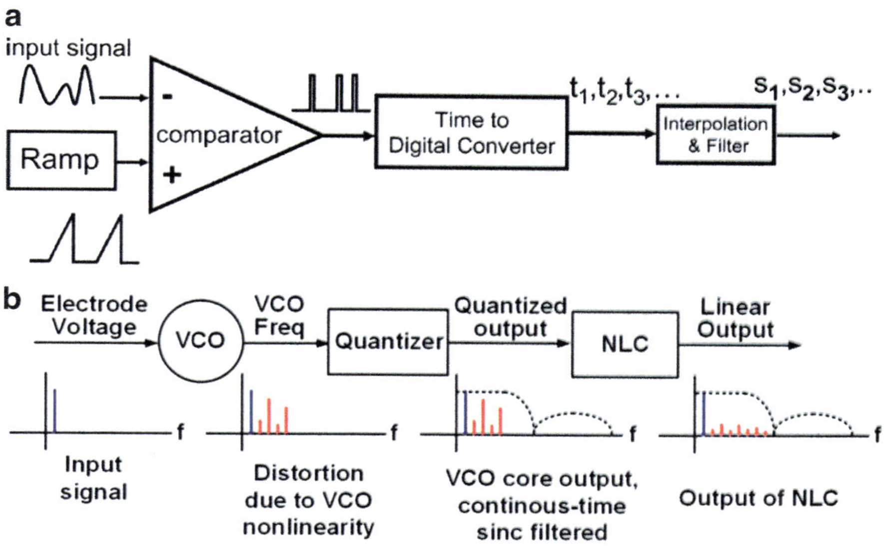

Fig. 17.14 Sample time-based architectures that also implement digital nonlinearity correction for (**a**) ADC to be used in a sensor node [44]. (**b**) Neural AFE [61]

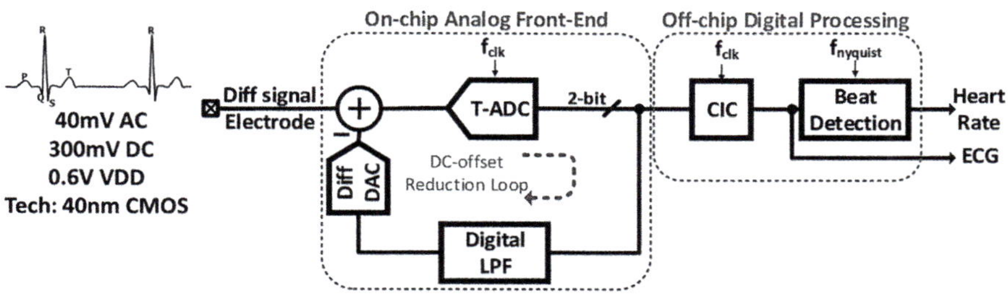

Fig. 17.15 Proposed time-based analog front end (T-AFE) architecture

and a simple anti-alias filter. This would not have been possible had the input signal swing been large. Finally, since the integrator is in feedback, and not in feedforward as seen in many time-based implementations in previous chapter, its gain need not be particularly high.

The 1MΩ resistance is implemented to reduce the noise due to the feedback integrator by shunting its current noise.

One of the drawbacks of using a delta modulator time-based loop for biomedical readout is that since it is a nonlinear loop, its analysis is very challenging. This is especially problematic if we need to estimate the amount of thermal noise and the quantization noise floor, which is important for biomedical readouts, and to optimize the loop parameters to gain the maximum power and area efficiency. Although, historically, the analysis is done by the describing function method (DF)

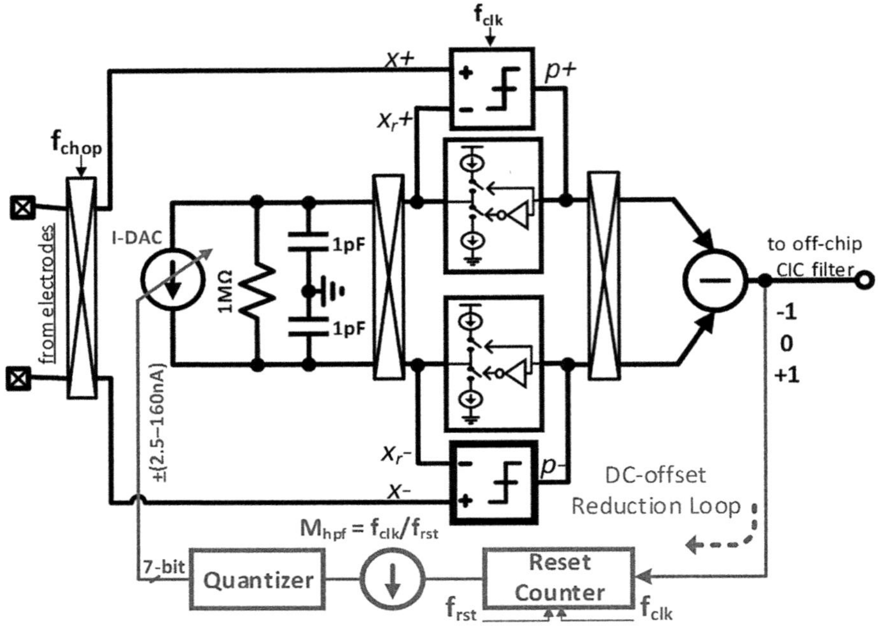

Fig. 17.16 Block-level architecture of the T-ADC

[62, 63]. However, this is quite intensive, yields very little design intuition and is quite challenging to perform a noise analysis. We, instead, propose using a pseudo-continuous analysis [64] for this time-based loop. For more details regarding this analysis and design optimization, we refer the readers to paper [31].

Figure 17.17 shows the transistor-level schematic of the implemented dynamic comparator. It is a standard architecture that comprises a preamplifier followed by a latch stage. The input-stage transistors are implemented using thick-oxide transistors to eliminate gate-leakage current. The anti-alias filter is inherently implemented by using the parasitic capacitances (C_{par}). The clock frequency of the dynamic comparator is 25 MHz. This frequency is chosen to keep the quantization noise lower than the thermal noise. The dynamic power consumption will be small due to low VDD and small parasitic capacitances.

Figures 17.18 and 17.19 show a few measurement results of the proposed T-AFE. Figure 17.18a plots the input-referred noise floor of the T-AFE. It achieves a 0.6 μVrms thermal noise floor or a 7.2 μVrms total noise in a bandwidth of 150 Hz. As expected, there is no flicker noise. Figure 17.18b plots the FFT of a 5 mV_{pp} input signal of 11 Hz frequency. We choose this frequency because most of the ECG signal energy is concentrated in this range. T-AFE achieves an SFDR of 56 dB. Figure 17.19a plots a sample ECG signal measured by the T-AFE. To validate the large-signal capabilities, Fig. 17.19b shows the readout signal in presence of motion as high as 40 mV_{pp}. As can be seen, the readout does not saturate and even maintains decent beat detection.

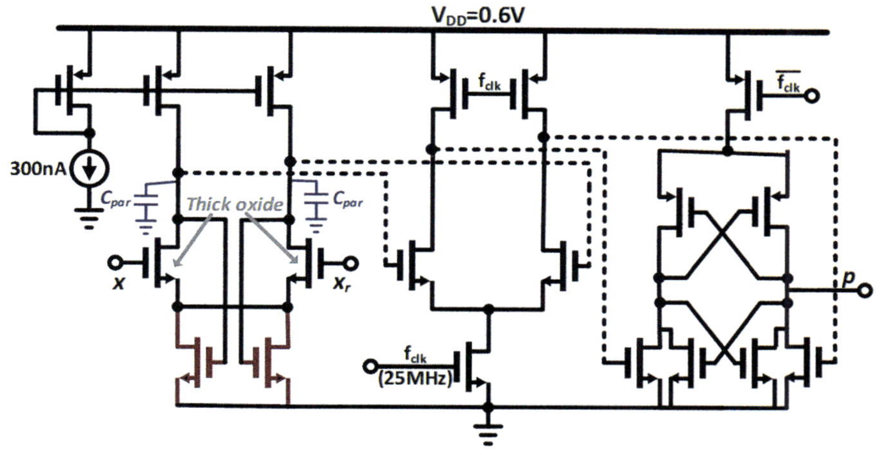

Fig. 17.17 Transistor-level schematic of the comparator

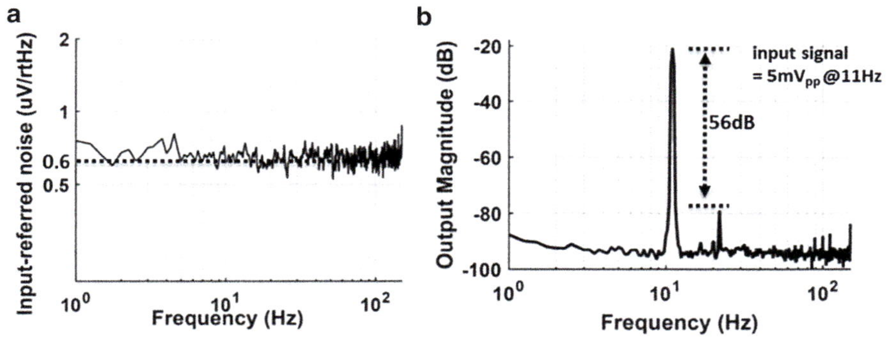

Fig. 17.18 Input-referred noise of the T-AFE (**a**) and FFT of the PWM comparator output (**b**)

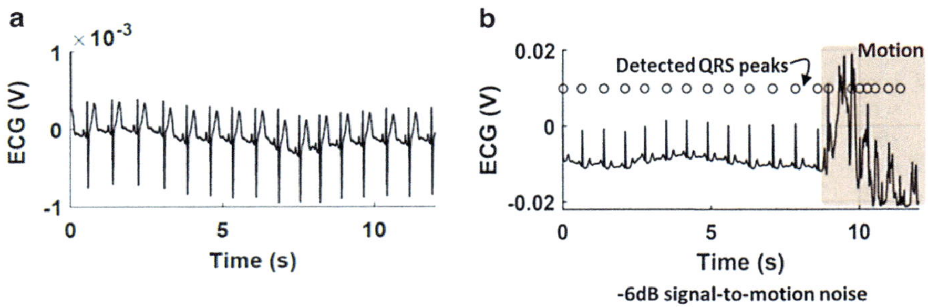

Fig. 17.19 ECG with the subject at rest (**a**), with motion noise added [65] (**b**)

17.5 Conclusions

Personalized healthcare applications are pushing the boundaries for small, low-cost, low-power and accurate devices. Existing sensor SoC designs require further improvements in terms of power consumption, accuracy and cost. One of the main reasons that this improvement is challenging is that the traditional voltage-mode AFE design techniques are not well suited for low-voltage supplies and small-scale technologies.

In this paper, we presented the challenges and existing solutions in literature for design of a low-power, low-area biomedical AFE readout. In particular, we focused on time-based operation to overcome the challenge of design with a low-voltage supply. We further discussed the reasons why the existing time-based implementations are not suitable to meet our target specifications.

To overcome these challenges, we proposed a time-based readout architecture that eliminates voltage swing from right at the electrodes itself through the use of negative feedback and focuses on scalable design techniques. By doing so, we gain significant benefits in terms of power and area consumption.

Thus, we not only come a step closer to realizing a low-power, low-cost and accurate sensor SoC, but we open further avenues for analog design in ultra-low-voltage and small-scale technology.

References

1. Patel, S., Hyung, P., Bonato, P., Chan, L., Rodgers, M.: A review of wearable sensors and systems with application in rehabilitation. J. NeuroEng. Rehabil. **9**(1), 1 (2012)
2. Blaauw, D. et al.: Iot Design Space Challenges: Circuits and Systems. VLSI-Technology, Honolulu (2014)
3. Penders, J., van Hoof, C., Gyselinckx, B.: Bio-Medical Application of WBAN: Trends and Examples. pp. 279–302. Bio-Medical CMOS ICs, Boston (2011)
4. Islam, A.B., Islam, S.K., Rahman, T.: A vertically aligned carbon nanofiber (VACNF) based amperometric glucose sensor. 2009 International Semiconductor Device Research Symposium, College Park (2009)
5. Islam, M.Z., Haider, M.R., Huque, M.A., Adeeb, M.A., Rahman, S., Islam, S.K.: A low power sensor signal processing circuit for implantable biosensor applications. Smart Mater. Struct. **16**(2), 525 (2007)
6. Baker, D.A., Gough, D.A.: A continuous, implantable lactate sensor. Anal. Chem. **67**(9), 1536–1540 (1995)
7. Jeevarajan, A.S., Vani, S., Taylor, T.D., Anderson, M.M.: Continuous pH monitoring in a perfused bioreactor system using an optical pH sensor. Biotechnol. Bioeng. **78**(4), 467–472 (2002)
8. Helleputte, N.V., Konijnenburg, M., Hyejung, K., Pettine, J., Jee, D.-W., Breeschoten, A., Morgado, A., Torfs, T., Groot, H.D., Hoof, C.V., Yazicioglu, R.F.: A multi-parameter signal-acquisition SoC for. 2014 IEEE International Solid-State Circuits Conference Digest of Technical Papers (ISSCC), San Fransisco (2014)

9. Bult, K.: The effect of technology scaling on power dissipation in analog circuits. Analog Circuit Design: RF Circuits: Wide band, Front-Ends, DAC's, Design Methodology and Verification for RF and Mixed-Signal Systems, Low Power and Low Voltage, pp. 51–294. Springer, Dordrecht (2006)
10. Annema, A.J., Nauta, B., Langevelde, R.V., Tuinhout, H.: Analog circuits in ultra-deep-submicron CMOS. IEEE J. Solid State Circuits. **40**(1), 132–143 (2005)
11. Annema, A.J.: Analog circuit performance and process scaling. IEEE Trans. Circuits Syst. II, Analog. Digit. Signal Process. **46**(6), 711–725 (1999)
12. Lee, K., et al.: The impact of semiconductor technology scaling on CMOS RF and digital circuits for wireless application. IEEE Trans. Electron Devices. **52**(7), 1415–1422 (2005)
13. Baschirotto, A., Chironi, V., Cocciolo, G., D'Amico, S., De Matteis, M., Delizia, P.: Low power analog design in scaled technologies. Proc. Topical Workshop Electron. Particle Phys.,pp. 103–110, (2009)
14. Murmann, B.: A/D converter trends: power dissipation, scaling and digitally assisted architectures. 2008 IEEE Custom Integrated Circuits Conference, San Jose (2008)
15. Enz, C.C., Vittoz, E.A.: Chapter 1.2 Cmos Low-Power Analog Circuit Design. (2001)
16. Murmann, B., Nikaeen, P., Connelly, D., Dutton, R.: Impact of scaling on analog performance and associated modeling needs. IEEE Trans. Electron Devices. **53**(9), 160–2167 (2006)
17. Razavi, B.: Design of Analog CMOS Integrated Circuits. McGraw-Hill, New York (2001)
18. Bult, K., Geelen, G.J.: The CMOS gain-boosting technique. Analog Integr. Circ. Sig. Process. **1**(2), 118–135 (1991)
19. Chatterjee, S., Tsividis, Y., Kinget, P.: 0.5-V analog circuit techniques and their application in OTA and filter design. IEEE J. Solid State Circuits. **40**(12), 2373–2387 (2005)
20. Lehmann, T., Cassia, M.: 1-V power supply CMOS cascode amplifier. IEEE J. Solid State Circuits. **36**(7), 1082–1086 (2001)
21. Stockstad, T., Yoshizawa, H.: A 0.9 V, 0.51 μA rail-to-rail CMOS operational amplifier. IEEE J. Solid State Circuits. **37**(3), 286–292 (2002)
22. Wang, R., Harjani, R.: Partial positive feedback for gain enhancement of low-power CMOS OTAs. Low-Voltage Low-Power Analog Integrated Circuits: A Special Issue of Analog Integrated Circuits and Signal Processing An International Journal **8**(1) (1995), Boston, Springer US, pp. 21–35 (1995)
23. Drost, B., Talegaonkar, M., Hanumolu, P.K.: Analog filter design using ring oscillator integrators. IEEE J. Solid State Circuits. **47**(12), 3120–3129 (2012)
24. Park, M., Perott, M.H.: IEEE J. Solid State Circuits. **44**(12), 3344–3358 (2009)
25. Mohan, R., Yan, L.L., Gielen, G., Hoof, C., Yazicioglu, R.: 0.35 V time-domain-based instrumentation amplifier. Electron. Lett. **50**(21), 1513–1514 (2014)
26. Enz, C., Temes, G.: Circuit techniques for reducing the effects of op-amp imperfections: autozeroing, correlated double sampling, and chopper stabilization. Proc. IEEE. **84**(11), 1584–1614 (1996)
27. Wu, R., Makinwa, K.K.A.A., Huijsing, J.: A chopper current-feedback instrumentation amplifier with a 1 mHz 1/f noise corner and an AC-coupled ripple reduction loop. IEEE J. Solid State Circuits. **44**(12), 3232–3243 (2009)
28. Akita, I., Ishida, M.: A chopper-stabilized instrumentation amplifier using area-efficient self-trimming technique. Analog Integr. Circ. Sig. Process. **81**(3), 571–582 (2014)
29. Tang, A.: A 3 /spl mu/V-offset operational amplifier with 20 nV//spl radic/Hz input noise PSD at DC employing both chopping and autozeroing. 2002 IEEE International Solid-State Circuits Conference, San Fransisco (2002)
30. Pertijs, M., Kindt, W.: A 140 dB-CMRR current-feedback instrumentation amplifier employing ping-pong auto-zeroing and chopping. IEEE J. Solid State Circuits. **45**(10), 2044–2056 (2010)
31. Mohan, R., Zaliasl, S.G.G.G., Hoof, C., Yazicioglu, R., Helleputte, N.: A 0.6-V, 0.015-mm2, time-based ECG readout for ambulatory applications in 40-nm CMOS. IEEE J. Solid State Circuits. **52**(1), 298–308 (2017)

32. Mesgarani, A., Alam, M., Nelson, F., Ay, S.: Supply boosting technique for designing very low-voltage mixed-signal circuits in standard CMOS. 2010 53rd IEEE International Midwest Symposium on Circuits and Systems, Seattle (2010)
33. Abo, A., Gray, P.P.R.: A 1.5-V, 10-bit, 14.3-MS/s CMOS pipeline analog-to-digital converter. IEEE J. Solid State Circuits. **34**(5), 599–606 (1999)
34. Ranuarez, J., Deen, M., Chen, C.-H.: A review of gate tunneling current in MOS devices. Microelectron. Reliab. **46**(12), 1939–1956 (2006)
35. Helleputte, N.V., Kim, S., Kim, H., Kim, J.P., Hoof, C.V., Yazicioglu, R.F.: A 160uA biopotential acquisition IC with fully Integrated IA and motion artifact suppression. IEEE Trans. Biomed. Circuits Syst. **6**(6), 552–561 (2012)
36. Muller, R., Gambini, S.S., Rabaey, J.: A 0.013 mm2, 5uW , DC-coupled neural signal acquisition IC With 0.5 V supply. IEEE J. Solid State Circuits. **47**(1), 232–243 (2012)
37. Tsividis, Y., Banu, M., Khoury, J.: Continuous-time MOSFET-C filters in VLSI. IEEE J. Solid State Circuits. **SC-1**(1), 15–30 (1986)
38. Helleputte, N.V., et al.: A 345 μW multi-sensor biomedical SoC with bio-impedance, 3-channel ECG, motion artifact reduction, and Integrated DSP. IEEE J. Solid State Circuits. **50**(1), 230–244 (2015)
39. Fan, Q., Sebastiano, F., Huijsing, J.H., Makinwa, K.A.: A 1.8 W 60 nV Hz capacitively-coupled chopper instrumentation amplifier in 65nm CMOS for wireless sensor nodes. IEEE J. Solid State Circuits. **46**(7), 1534–1543 (2011)
40. Ng, K., Xu, Y.: A compact, low input capacitance neural recording amplifier with Cin/Gain of 20fF.V/V. 2012 IEEE Biomedical Circuits and Systems Conference (BioCAS), Hsinchu (2012)
41. Bohorquez, J.L., Yip, M., Chandrakasan, A.P., Dawson, J.L.: A biomedical sensor interface with a sinc filter and interference cancellation. IEEE J. Solid State Circuits. **46**(4), 746–756 (2011)
42. Zou, X., Liew, S., Yao, L., Lian, Y.: A 1V 22μW 32-channel implantable EEG recording IC. 2010 IEEE International Solid-State Circuits Conference, San Fransisco (2010)
43. Majidzadeh, V., Schmid, A., Leblebici, Y.: Energy efficient low-noise neural recording amplifier with enhanced noise efficiency factor. IEEE Trans. Biomed. Circuits Syst. **5**(3), 262–271 (2011)
44. Naraghi, S., Courcy, M., Flynn, M.P.: A 9-bit, 14 μW and 0.06 mm pulse position modulation ADC in 90nm digital CMOS. IEEE J. Solid-State Circuits. **45**(9), 1870 (2010)
45. Hernandez, L., Prefasi, E.: Analog-to-digital conversion using noise shaping and time encoding. IEEE Trans. Circuits Syst. I, Reg. Papers. **55**(7), 2026–2037 (2008)
46. Dhanasekaran, V., et al.: A 20MHz BW 68dB DR CT ΔΣ ADC based on a multi-bit time-domain quantizer and feedback element. 2009 IEEE International Solid-State Circuits Conference, San Fransisco (2009)
47. Iwata, A., Sakimura, N., Nagata, M., Morie, T.: The architecture of delta sigma analog-to-digital converters using a voltage-controlled oscillator as a multibit quantizer. IEEE Trans. Circuits Syst. II, Analog Digit Signal Process. **46**(7), 941–945 (1999)
48. Kuppambatti, J., Vigraham, B., Kinget, P.: 17.9 A 0.6V 70MHz 4th-order continuous-time butterworth filter with 55.8dB SNR, 60dB THD at +2.8dBm output signal power. 2014 IEEE International Solid-State Circuits Conference Digest of Technical Papers, San Fransisco (2014)
49. Cuk, S.M.: Modeling, Analysis and Design. Thesis – California Institute of Technology, Pasadena (1977)
50. Hanson, S., Foo, Z., Blaauw, D., Sylvester, D.: 0.5 V sub-microwatt CMOS image sensor with pulse-width modulation read-out. IEEE J. Solid State Circuits. **45**(4), 759–767 (2010)
51. Jung, W., Jeong, S., Oh, S., Sylvester, D., Blaauw, D.: 27.6 A 0.7pF-to-10nF fully digital capacitance-to-digital converter using iterative delay-chain discharge. 2015 IEEE International Solid-State Circuits Conference, San Fransisco (2015)

52. Rethy, J., Smedt, V., Dehaene, W., Gielen, G.: Supply-noise-resilient design of a BBPLL-based force-balanced Wheatstone bridge Interface in 130-nm CMOS. IEEE J. Solid State Circuits. **48**(11), 618–627 (2013)
53. Yurish, S.: Sensors and transducers: frequency output vs voltage output. Sens. Transducers Mag. **49**(11), 302–305 (2004)
54. Middlehoek, S., French, P., Huijsing, J., Lian, W.: Sensors with digital or frequency output. Sensors Actuators. **15**(2), 119–133 (1988)
55. Kindlund, A., Sundgren, H., Lundstrom, I.: Quartz crystal gas monitor with a gas concentrating stage. Sensors Actuators. **6**(1), 1–17 (1984)
56. Danneels, H., Coddens, K., Gielen, G.: A fully-digital, 0.3V, 270 nW capacitive sensor interface without external references. 2011 Proceedings of the ESSCIRC (ESSCIRC), Helsinki (2011)
57. Grassi, M., Malcovati, P., Baschirotto, A.: A 141-dB dynamic range CMOS gas-sensor interface circuit without calibration with 16-bit digital output word. IEEE J. Solid State Circuits. **42**(7), 1543–1554 (2007)
58. Lu, J.H.-L., Inerowicz, M., Joo, S., Kwon, J.-K., Jung, B.: A low-power, wide-dynamic-range semi-digital universal sensor readout circuit using pulsewidth modulation. IEEE Sensors J. **11**(5), 1134–1144 (2011)
59. Rethy, J.V., Danneels, H., Smedt, V.D., Dehaene, W., Gielen, G.E.: Supply-noise-resilient design of a BBPLL-based force-balanced wheatstone bridge interface in 130nm CMOS. IEEE J. Solid Stage Circuits. **48**(11), 2618–2627 (2013)
60. Park, M., Perrott, M.: A VCO-based analog-to-digital converter with second-order Sigma-Delta noise shaping. 2009 IEEE International Symposium on Circuits and Systems, Taipei (2009)
61. Jiang, W., Hokhikyan, V., Chandrakumar, H., Karkare, V., Markovic, D.: 28.6 A $+-50$mV linear input-range VCO-based neural-recording front-end with digital non-linearity correction. 2016 IEEE International Solid-State Circuits Conference, San Fransisco (2016)
62. Booton, R.: Nonlinear control systems with random inputs. IRE Trans. Circ. Theory. **1**(1), 9–18 (1954)
63. Ouzounov, S., Hegt, H., Roermund, A.V.: Sigma–delta modulators operating at a limit cycle. IEEE Trans. Circuits Syst. II. **53**(5), 399–403 (2006)
64. Perrott, M., Trott, M., Sodini, C.: A modeling approach for Σ-Δ fractional-N frequency synthesizers allowing straightforward noise analysis. IEEE J. Solid State Circuits. **37**(8), 1028–1038 (2002)
65. Goldberger, A., et al.: PhysioBank, PhysioToolkit, and PhysioNet: components of a new research resource for complex physiologic signals. Circulation. **101**, e215–e220 (2000)

Chapter 18
A 4.4 mW-TX, 3.6 mW-RX Fully Integrated Bluetooth Low Energy Transceiver for IoT Applications

Masoud Babaie, Sandro Binsfeld Ferreira, Feng-Wei Kuo, and Robert Bogdan Staszewski

18.1 Introduction

Ultra-low-power (ULP) wireless transceivers for Internet of things (IoT) are a subject of intensive research in both industry [1–7] and academia [8–10]. Bluetooth Low Energy (BLE) [11] is currently the most popular standard for short-range IoT communications. BLE is an extension of the conventional Bluetooth (BT) that specifies an increased channel spacing of 2 MHz and a relaxed interference tolerance to allow for low-power implementations. This work focuses on implementing such a transceiver (TRX) in the most advanced low-cost bulk CMOS technology node: low-power (LP) polysilicon gate version of 28-nm CMOS with nine metal layers. The key objective is to maximally reduce the system cost by fully integrating all the RF transceiver building blocks *while* maximally reducing the power consumption.

Figure 18.1 shows a block diagram of the BLE transceiver [8, 9]. It successfully integrates all the required RF and IF building blocks and further adds a transmit/receive (T/R) antenna switch with adjustable digital PA (DPA) and low-noise transconductance amplifier (LNTA) matching networks such that the RF input-output (RFIO) pin can be directly connected to the antenna. Several new

M. Babaie (✉)
Delft University of Technology, Delft, The Netherlands
e-mail: m.masoud.babaie@ieee.org

S.B. Ferreira
Federal University of Rio Grande Do Sul (UFRGS), Porto Alegre, Brazil

F.-W. Kuo
Taiwan Semiconductor Manufacturing Company (TSMC), Hsinchu, Taiwan

R.B. Staszewski
Delft University of Technology, Delft, The Netherlands

University College Dublin (UCD), Dublin, Ireland

P. Harpe et al. (eds.), *Hybrid ADCs, Smart Sensors for the IoT, and Sub-1V & Advanced Node Analog Circuit Design*, DOI 10.1007/978-3-319-61285-0_18

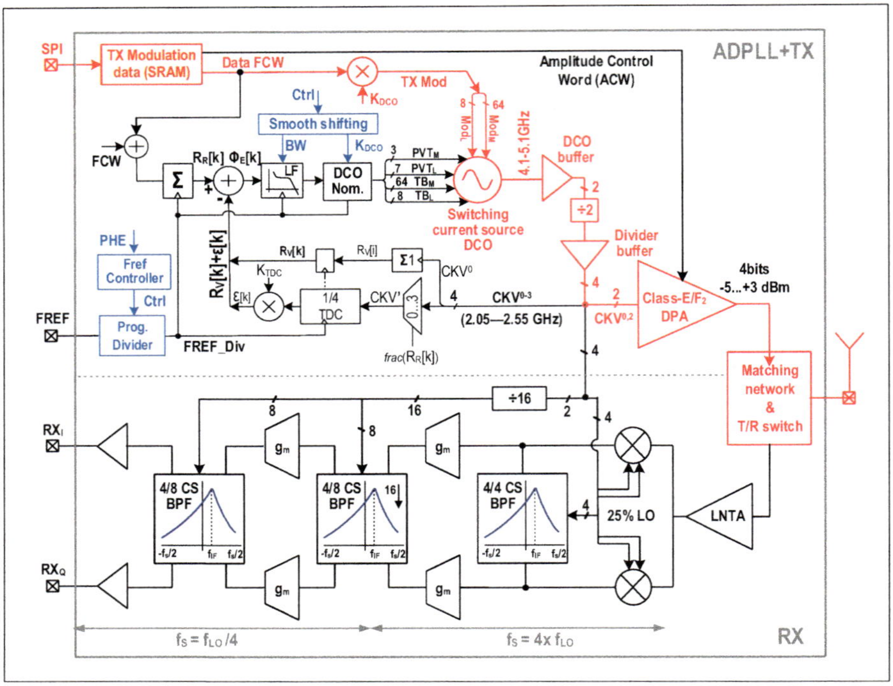

Fig. 18.1 Block diagram of the proposed Bluetooth Low Energy transceiver

system and circuit techniques are exploited here to enhance the power and area efficiency of an ULP transceiver: First, the most energy-hungry circuitry, such as a digitally controlled oscillator (DCO) and an output stage of a power amplifier (PA), can operate directly at the low voltage of harvesters [12, 13]. Second, a new switching current source oscillator reduces power and supply voltage without compromising the robustness of its start-up [14]. Third, thanks to the low wander of the DCO, digital power consumption of the rest of all-digital PLL (ADPLL) is saved by scaling the rate of a sampling clock to the point of its complete stillness [12, 13]. Fourth, a fully integrated differential class-E/F_2 switching PA is utilized to optimize high power-added efficiency (PAE) at a low output power of 0–3 dBm [13, 15]. Fifth, the RX architecture was derived from realizing that the best devices and basic building blocks in low-voltage deep-nanoscale CMOS are logic gates, transistor switches, inverter-like g_m transconductors, and metal-oxide-metal (MOM) capacitors. Hence, the most logical topology would be a charge-domain switched-capacitor network operating in discrete time. However, to maximally reduce power consumption, MOS devices would need to be remarkably small, which would invariably increase their flicker noise corner. To mitigate that, we propose to increase the RX intermediate frequency (IF) to just beyond the flicker corner frequency [9, 16]. Sixth, new multistage multi-rate charge-sharing band-pass filters are adapted to achieve high out-of-band linearity, low noise, and low power consumption [9, 17]. Last, an integrated on-chip matching network serves both PA and low-noise transconductance amplifier (LNTA), thus allowing a one-pin direct antenna connection with no external band-selection filters [9].

The paper is organized as follows: Sect. 18.2 introduces a new RF oscillator topology that is suitable for ultra-low-voltage/power applications. The ADPLL-based TX architecture is discussed in Sect. 18.3. The trade-offs between the output power, matching network insertion loss, drain, and power-added efficiency (PAE) of the class-E/F_2 PA are investigated in Sect. 18.4. Section 18.5 reveals the RF input/output matching and switching. Section 18.6 details the discrete-time receiver implementation. In Sect. 18.7, the experimental results are discussed.

18.2 Digitally Controlled Oscillator

The phase noise (PN) of the traditional oscillator (i.e., class B) with an ideal current source at an offset frequency $\Delta\omega$ from its resonating frequency ω_0 can be expressed as

$$\mathcal{L}(\Delta\omega) = 10\log_{10}\left(\frac{KT}{2\,Q_t^2\,\alpha_I\,\alpha_V\,P_{DC}}(1+\gamma)\left(\frac{\omega_0}{\Delta\omega}\right)^2\right) \tag{18.1}$$

where K is the Boltzmann's constant; T is the absolute temperature; Q_t is the LC tank quality factor; α_I is the current efficiency, defined as a ratio of the fundamental current harmonic I_{ω_0} over the oscillator DC current I_{DC}; and α_V is the voltage efficiency, defined as a ratio of the single-ended oscillation amplitude, $V_{Osc}/2$, over the supply voltage V_{DD} [18]. The oscillator I_{DC} may be estimated by one of the following equations:

$$I_{DC} = \frac{I_{\omega_0}}{\alpha_I} \xrightarrow{I_{\omega_0}=\frac{V_{Osc}}{R_{in}}} I_{DC} = \frac{V_{Osc}}{R_{in}}\cdot\frac{1}{\alpha_I} \xrightarrow{V_{Osc}=2\alpha_V V_{DD}} I_{DC} = \frac{2V_{DD}}{R_{in}}\cdot\frac{\alpha_V}{\alpha_I}. \tag{18.2}$$

where R_{in} is an equivalent differential input parallel resistance of the tank's losses. The RF oscillator's P_{DC} is derived by

$$P_{DC} = \frac{2V_{DD}^2}{R_{in}}\cdot\frac{\alpha_V}{\alpha_I}. \tag{18.3}$$

Equation (18.3) indicates that the minimum achievable P_{DC} can be expressed in terms of a set of optimization parameters, such as R_{in}, and a set of topology-dependent parameters, such as minimum supply voltage ($V_{DD\min}$) and current and voltage efficiencies.

Lower P_{DC} is typically achieved by scaling up $R_{in} = L_p\omega_0 Q_t$ simply via a large multi-turn inductor [5]. For example, by continuously increasing the inductance by 2× at constant Q_t, R_{in} could theoretically enhance by 2×, which would reduce P_{DC}

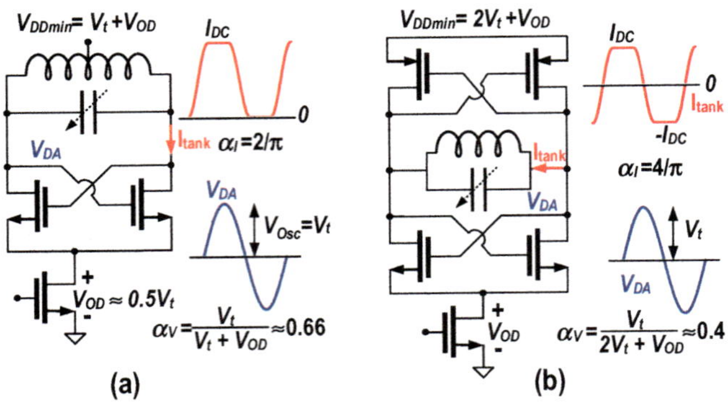

Fig. 18.2 $V_{DD\min}$, α_I, and α_V parameters for (**a**) cross-coupled NMOS and (**b**) complementary push-pull oscillators

Table 18.1 Minimum P_{DC} for different RF oscillator topologies

Topology	$V_{DD\min}$	α_V@$V_{DD\min}$	α_I	$P_{DC\min}$
Osc_N	$V_t + V_{OD} \approx 1.5V_t$	~0.66	$2/\pi$	$4.66\ V_t^2/R_{in}$
Osc_{NP}	$2V_t + V_{OD} \approx 2.5V_t$	~0.4	$4/\pi$	$3.92\ V_t^2/R_{in}$
This work	$V_t + V_{OD} \approx 1.5V_t$	~0.33	$4/\pi$	$1.2\ V_t^2/R_{in}$

by half with a 3 dB PN degradation. However, at some point, that trade-off stops due to a dramatic drop in the inductor's self-resonant frequency and Q factor. This constraint sets an upper limit on maximum R_{in}, which is a function of technology.

The topology parameters also play an important role in the minimum achievable P_{DC}. Figure 18.2 shows such effects for the traditional cross-coupled NMOS-only (Osc_N) and complementary push-pull (Osc_{NP}) structures. The $V_{DD\min}$ of Osc_N can go lower than in Osc_{NP}. However, the current efficiency of Osc_{NP} is doubled due to the switching of tank current direction every half period. Its voltage efficiency is also smaller. Hence, Osc_{NP} offers ~3× lower α_V/α_I. Consequently, each of these structures has its own set of advantages and drawbacks such that the minimum achievable P_{DC} according to (18.3) is almost identical, as shown in Table 18.1.

In this paper, we propose to convert the fixed current source of the traditional NMOS topology into a structure with alternating current sources such that the tank current direction can change every half period. Consequently, the benefits of low supply of the Osc_N topology and higher α_I of Osc_{NP} structure are combined to reduce power consumption further than practically possible in the traditional oscillators.

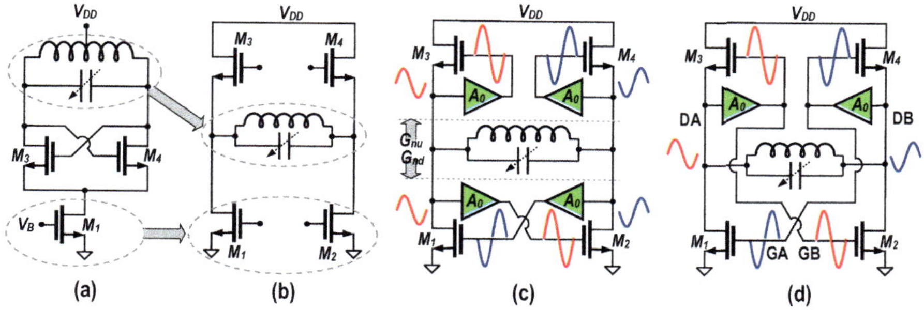

Fig. 18.3 Evolution toward the switching current source oscillator

18.2.1 Switching Current Source Oscillator

Figure 18.3 shows an evolution toward the switching current source oscillator. The $\mathrm{Osc_N}$ topology is chosen as a starting point due to its low V_{DD} capability. To reduce P_{DC} further, it is desired to switch the direction of the LC tank current in each half period, which will double α_I. Consequently, we propose to split the fixed current source M_1 in Fig. 18.3a into two switchable "current sources," M_1 and M_2, as suggested in Fig. 18.3b. This allows for the tank to be disconnected from the V_{DD} feed and be moved in between the upper and lower NMOS transistor pairs to give rise to an H-bridge configuration. In the next step, the passive voltage gain blocks, A_0, are added to the NMOS gates, as shown in Fig. 18.3c. Both upper and lower NMOS pairs should each individually demonstrate *synchronized* positive feedback to realize the switching of the tank current direction. The "master" positive feedback enforces the differential-mode operation and is realized by the lower-pair transistors configured in a conventional cross-coupled manner. Since the lower pair is *voltage biased*, its negative conductance seen by the tank may be estimated as $G_{nd} = -0.25 \cdot A_0[g_{m1}(\phi) + g_{m2}(\phi)]$.

On the upper side, the differential-mode oscillation of the tank is *reinforced* by the $M_{3,4}$ devices which realize the second positive feedback.[1] The negative conductance seen by the tank into the upper pair can be calculated as $G_{nu} = -0.25 \cdot (A_0 - 1)[g_{m3}(\phi) + g_{m4}(\phi)]$, which clearly indicates that the voltage gain block is necessary and A_0 must be safely larger than 1 to be able to present a negative conductance to the tank, thus enabling the H-bridge switching. By merging the redundant voltage gain blocks, the proposed switching current source oscillator is arrived at in Fig. 18.3d.

Figure 18.4 illustrates the proposed oscillator schematic and simulated waveforms indicating various operational regions of M_{1-4} transistors. The two-port

[1]It should be noted that the "master/slave" view is mainly valid from a small-signal standpoint. Both are equally important when considering the large-signal switching operation.

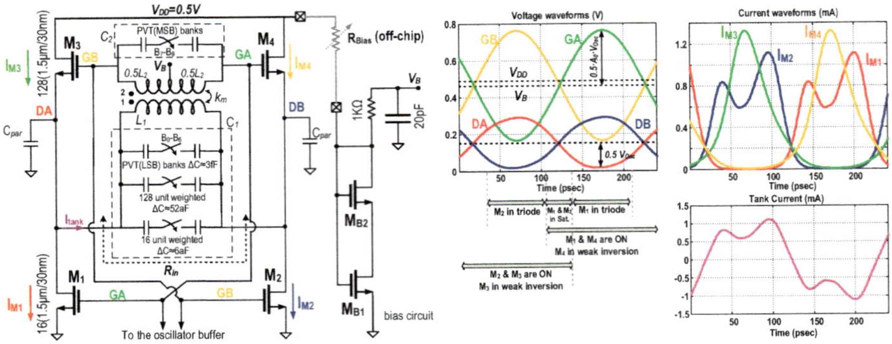

Fig. 18.4 Schematic and waveforms of the proposed oscillator

resonator consists of a step-up 1:2 transformer and tuning capacitors, $C_{1,2}$, at its primary and secondary windings. The current-source transistors $M_{1,2}$ set the oscillator's DC current. Along with M_{3-4}, they play a vital role of switching the tank current direction. As can be gathered from Fig. 18.4, G_B oscillation voltage is high within the first half period. Hence, only M_2 and M_3 are on and the current flows from the left to right side of the tank. However, M_1 and M_4 are turned on for the second half period, and the tank's current direction is reversed, thus doubling α_I to $4/\pi$.

V_{DD} of the proposed oscillator can be as low as $V_{OD1} + V_{OD3} \approx V_t$, which is extremely small given the capability of switching the tank current direction. Note that the oscillation swing cannot exceed $V_{OD1,2}$ at DA/DB nodes and is chosen 150 mV to satisfy the PN requirements with a margin. However, it is the bias voltage $V_B \approx V_{OD1} + V_{gs3}$ that limits the minimum supply. Hence, $M_{3,4}$ should work in weak inversion keeping $V_{gs3} < V_t$ to achieve lower $V_{DD,\min}$. Consequently, even by considering the tougher V_B requirement, the proposed structure can operate at V_{DD} as low as 0.5 V, on par with $\mathrm{Osc_N}$.

As evident from Fig. 18.4, $M_{3,4}$ transistors operate in a class C manner as in a Colpitts oscillator, meaning that they deliver somewhat narrow-and-tall current pulses. However, their conduction angle is quite wide, $\sim\pi$, due to the low overdrive voltage in the subthreshold operation. On the other hand, $M_{1,2}$ operate in a class B manner like cross-coupled oscillators, meaning that they deliver square-shaped current pulses. Hence, the shapes of drain currents are quite different for the lower and upper pairs. However, their fundamental components demonstrate the same amplitude and phase to realize the constructive oscillation voltage across the tank. The higher drain harmonics obviously show different characteristics. However, they are filtered out by the tank's selectivity characteristic. Note that the current through a transistor of the upper pair will have two paths to ground: through the corresponding transistor of the lower pair and through the single-ended capacitors. Consequently, the single-ended capacitors sink the higher current harmonics of $M_{3,4}$ transistors.

18.3 All-Digital Phase-Locked Loop

Figure 18.1 shows a block diagram of the proposed ultra-low-power (ULP) all-digital PLL (ADPLL), whose architecture is adapted from a high-performance cellular 4G ADPLL disclosed in [19]. Due to the relaxed PN requirements of BLE, the DCO $\Sigma\Delta$ dithering [20] was removed thanks to the fine switchable capacitance of the tracking bank varactors producing a fine step size of 4 kHz. The DCO features two separate tracking banks (TB): (1) phase-error correction and (2) direct FM modulation. Each bank is segmented with LSB (i.e., $1\times \equiv 4$ kHz) and MSB (i.e., $8\times$) unit weights. Each TB range is $4\,\text{kHz} \times (8 + 8 \times 64) = 2.08$ MHz.

The DCO clock is divided by two to generate four phases of a variable carrier clock, CKV^{0-3}, in the Bluetooth frequency range of $f_V = 2402$–2478 MHz. Two of its phases, $\text{CKV}^{0,2}$, are fed as differential clock signals to the digital PA (DPA). The four CKV^{0-3} phases are routed to the phase detection circuitry, which selects the phase whose rising clock edge is *expected* to be the closest to the rising clock edge of a frequency reference (FREF) clock. This prediction is based on two MSB bits of a fractional part of reference phase, $R_R[k]$, which is an accumulated frequency command word (FCW). By means of this prediction, the selected TDC input clock CKV spans a quarter of the original required TDC range, i.e., $T_V/4$, where T_V is the CKV clock period. This way, the long string of 417 ps/12 ps >35 TDC inverters is shortened by $4\times$, improving INL linearity and power consumption by the same amount.

The TDC output, after decoding, is normalized to T_V by the Δ_{TDC}/T_V multiplier, and the quadrant estimation, normalized to $T_V/4$, is added to produce the phase error ϕ_E. The DCO tuning word is updated based on ϕ_E. The $\phi_E[k]$ is fed to the type II loop filter (LF) with fourth-order IIR. The LF is dynamically switched during frequency acquisition to minimize the settling time while keeping phase noise (PN) at optimum. The built-in DCO gain, K_{DCO}, and TDC gain, K_{TDC}, calibrations are autonomously performed to ensure the wideband FM response.

The following architectural innovations allow the ADPLL to support ULP operation (highlighted in blue): The effective sampling rate of the phase detector and its related DCO update is dynamically controlled by scaling down the FREF clock and simultaneously adjusting the LF coefficients in order to keep the same bandwidth and LF transfer function characteristics. During the ADPLL settling, the full FREF rate is used, but afterward, its rate could get substantially reduced (e.g., $8\times$), or completely shut down, thus saving power consumption of the digital circuitry. The resulting in-band PN degradation is tolerable due to low PN of the DCO. In fact, freezing FREF would incur sufficiently low-frequency drift during the BLE 376 μs packets while keeping in operation the bare minimum of circuitry highlighted in red.

18.4 Digitally Controlled Switched Mode Power Amplifier

Designing a fully integrated PA optimized for low output power (P_{out} < 3 dBm) with PAE > 40% is very challenging, especially when differential structure is needed to satisfy the stringent second harmonic emissions. In this work, a fully integrated differential class-E/F$_2$ PA is exploited to address the aforementioned challenges.

Figure 18.5a illustrates a general schematic of a transformer-based matching network of a switched mode PA, which performs simultaneously m-series (i.e., voltage) and p-parallel (i.e., current) combining [21, 22]. As proven in [13], the equivalent resistance r_L seen by PA switching transistors may be estimated by

$$r_L \approx R_L \cdot \frac{p}{m}\left(\frac{k_m}{n}\right)^2 \cdot \frac{Q_s + Q_L/\xi}{2\xi Q_L + Q_s + Q_L^2 Q_s(\xi - 1)^2} \tag{18.4}$$

where Q_s is the Q factors of the transformer's secondary windings and ξ is defined as $p/(mL_s C_L \omega_0^2)$. We also define $Q_L = R_L C_L \omega_0$ as the loaded Q factor of the secondary side of the matching network. To deliver the relatively low $P_{out} \leq 3$ dBm to the antenna, realizing a larger r_L is desired. Unfortunately, as can be gathered from (18.4), the voltage summation ($m > 1$) and imperfect magnetic coupling k_m exhibit a reverse effect of reducing r_L. The p-way current combining enhances r_L but at the price of ($p-1$) extra transformers and thus a dramatic increase in the PA die area [15, 22]. Hence, the parallel combining is not considered in this work. Equation (18.4) further indicates that a step-down transformer (1:n) with a small turns ratio ($n < 1$) could be used to enhance r_L. However, the Q factor of transformer windings, and thus its efficiency, η_p, drops dramatically as n reduces. Consequently, the turns ratio of 1:$\frac{1}{2}$ was chosen in consideration of both the r_L enhancement and η_p optimization scenarios. P_{out} is further reduced by using V_{DD} = 0.5 V (i.e., roughly half the nominal supply) for the drains of switching transistors with the side effect of ~6 dB lower-power gain for PA's transistors. However, the power gain of 28 nm NMOS devices is high enough at a relatively low frequency of 2.4 GHz such that the 6 dB power gain penalty has a negligible effect on the total system efficiency. Furthermore, the drain voltage peak of the switching transistors is ≤1.5 V, thus alleviating reliability issues due to a gate-oxide breakdown [18, 23].

18.4.1 Class-E/F$_2$ Operation

Figure 18.5b, c respectively illustrate a schematic of the proposed switched mode PA and an equivalent circuit of its matching network in the differential mode at the fundamental frequency ω_0. At all higher odd harmonics, L_{ser} presents a high impedance, and thus the only load seen by the switch is its parallel capacitance, C_s, just the way it is in the traditional class-E PAs.

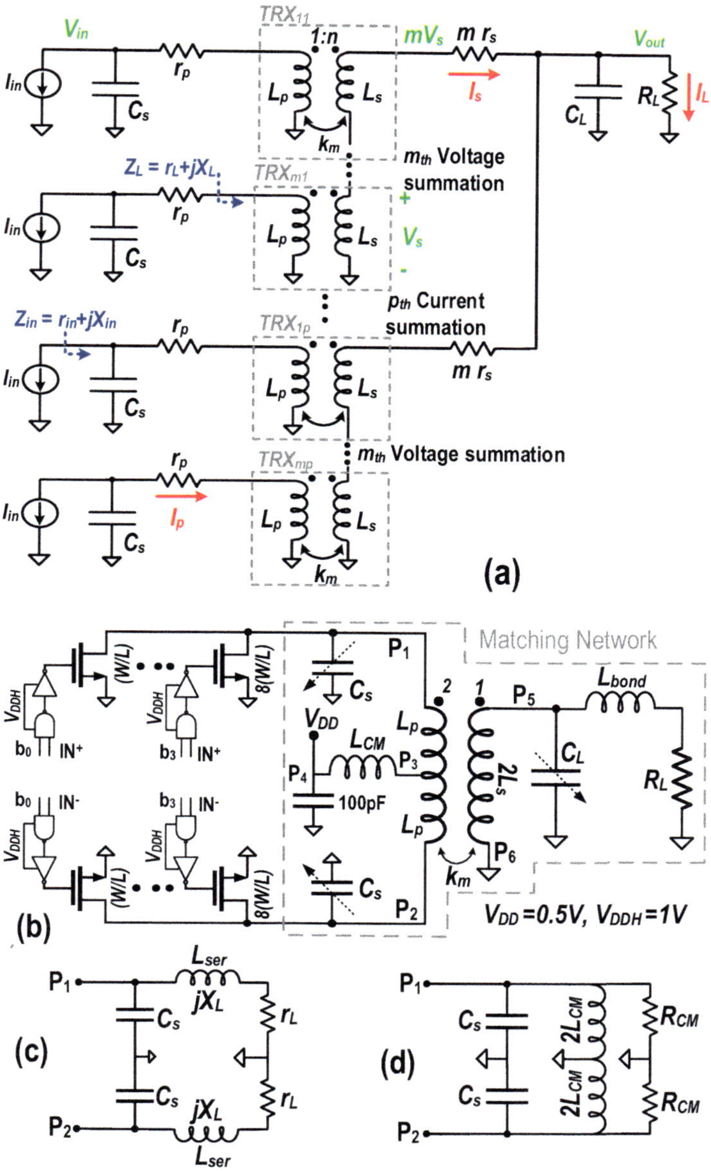

Fig. 18.5 (**a**) Transformer-based matching network with m-way voltage and p-way current summation; (**b**) schematic of the proposed class-E/F$_2$ PA. Equivalent circuit PA's matching network for (**c**) differential and (**d**) common-mode excitations

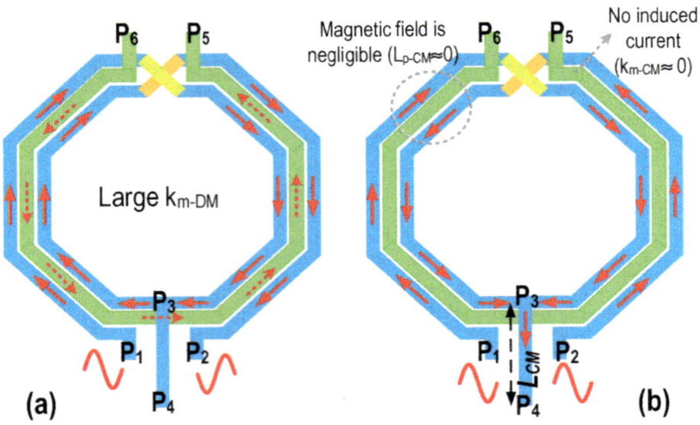

Fig. 18.6 Behavior of a 2:1 step-down transformer in (**a**) differential-mode and (**b**) common-mode excitations

Table 18.2 Design sets for different flavors of class-E/F PA

Tuning	K_C	K_L	K_P
E	0.184	1.152	0.577
E/F_2	0.337	1.609	0.381
E/F_3	0.209	0.961	0.657
E/$F_{2,3}$	0.323	0.832	0.747
E/F_4	0.218	1.173	0.533
E/$F_{2,4}$	0.361	1.667	0.350

As illustrated in Fig. 18.6, the step-down 2:1 transformer acts differently to the common-mode (CM) and differential-mode (DM) input signals. When the transformer's primary is excited by a CM signal (Fig. 18.6b), the magnetic flux within the primary's two turns cancels itself out [24]. Consequently, the transformer's L_p is negligible, and no current is induced at the transformer's secondary ($k_{m-CM} \approx 0$). Hence, R_L, L_s, and C_L cannot be seen by even harmonics of drain current.

Furthermore, the CM inductance, $2L_{CM}$, seen by the switching transistors, is mainly determined by the dimension of the trace between the transformer's center-tap and decoupling capacitors at the V_{DD} node. Together with C_s, $2L_{CM}$ realizes a CM resonance, ω_{CM}. Note that P_{out} of the class-E PA can be reduced by $\sim$2 dB at the same r_L and V_{DD} by means of an additional open circuit acting as the switches' effective load at $\sim 2\omega_0$ (i.e., class-E/F_2 operation [25]), as supported in the power factor, K_p, column in Table 18.2. Consequently, this PA needs smaller impedance transformation ratio for P_{out} <3 dBm, which results in a lower insertion loss for its matching network and thus higher system efficiency. However, in practice, limited value of an equivalent parallel resistance of the CM resonance, R_{CM}, leads to a power loss at the second harmonic and thus a penalty on the PA's efficiency if ω_{CM} is set at precisely $2\omega_0$. Consequently, in this design, we adjust the CM resonance slightly lower (i.e., at $\sim 1.8\omega_0$) to benefit from the lower K_p of semi-class-E/F_2 operation while avoiding the additional power loss at even harmonics.

18.5 Radio-Frequency Input-Output Switching and Matching

Figure 18.7a illustrates the proposed implementation of the on-chip matching networks with a '*soft*' T/R switch (i.e., without any explicit switches carrying RF signals) between the transmitter (TX) and receiver (RX) paths. In the RX mode (see Fig. 18.7b), the PA's transistors are off, and consequently, the TX is simplified to the PA's transformer-based TX matching network (TXMN) acting as a second-order resonator. In this mode, the ultimate goal is to alleviate the side effects of TXMN on the receiver noise figure (NF) and input return loss. The receiver noise factor (F) can be calculated by

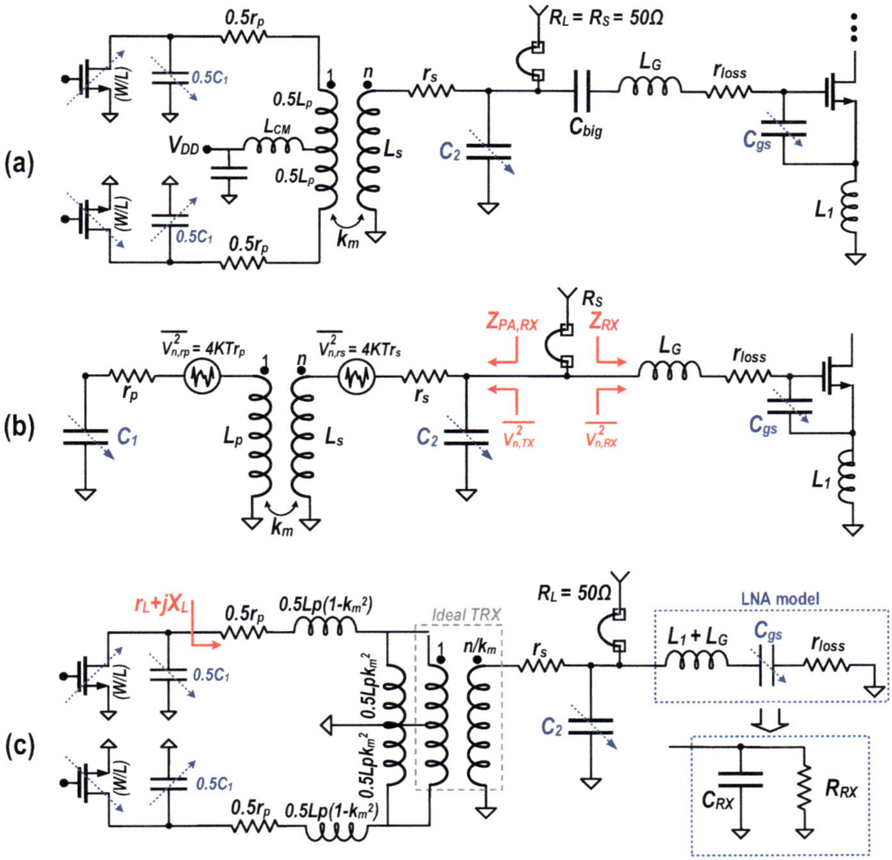

Fig. 18.7 (**a**) RF input/output block (RFIO) including the first stage of LNTA and the last stage of class-E/F$_2$ PA; (**b**) RFIO in the RX mode; (**c**) RFIO in the TX mode

$$F = 1 + \frac{R_S}{4KT} \cdot \frac{\overline{V_{n,\mathrm{RX}}^2}}{|Z_{\mathrm{RX}}|^2} + \frac{R_S}{4KT} \cdot \frac{\overline{V_{n,\mathrm{TX}}^2}}{|Z_{\mathrm{PA,RX}}|^2} \tag{18.5}$$

where $\overline{V_{n,\mathrm{RX}}^2}$ and Z_{RX} are, respectively, the equivalent input noise and input impedance of LNTA at the operating frequency ω_0. Furthermore, $Z_{\mathrm{PA,RX}}$ and $\overline{V_{n,\mathrm{TX}}^2}$ are, respectively, the output impedance and equivalent output noise of the PA's matching network. As shown in [9], the contribution of TXMN to the system noise figure may be estimated by

$$\frac{\overline{V_{n,\mathrm{TX}}^2}}{|Z_{\mathrm{PA,RX}}|^2} = \frac{4KT}{L_s\omega Q_s} \cdot \frac{\frac{Q_s}{Q_p}k_m^2 L_p^2 C_1^2 \omega^4 + \left(1 - L_p C_1 \omega^2\right)^2}{\frac{1}{Q_s^2} \cdot \left(1 - L_p C_1 \omega^2 \left(1 + \frac{Q_s}{Q_p}\right)\right)^2 + \left(1 - L_p C_1 \omega^2 \left(1 - k_m^2\right)\right)^2} \tag{18.6}$$

It can be shown that (18.6) reaches its minimum at

$$L_p C_1 \approx \frac{1}{\omega_0^2} \cdot \frac{Q_p}{Q_p + Q_s} \tag{18.7}$$

To achieve the minimum noise figure penalty, one should tune C_1 switchable capacitor to roughly satisfy (18.7). The optimum $\overline{V_{n,\mathrm{TX}}^2}/\,|Z_{\mathrm{PA,RX}}|^2$ is then obtained by inserting (18.7) into (18.6)

$$\left(\frac{\overline{V_{n,\mathrm{TX}}^2}}{|Z_{\mathrm{PA,RX}}|^2}\right)_{\min} = \frac{4KT}{L_s\omega Q_s} \cdot \frac{k_m^2 Q_p Q_s + Q_s^2}{\left(Q_s + k_m^2 Q_p\right)^2} \xrightarrow{Q_p = Q_s} \left(\frac{\overline{V_{n,\mathrm{TX}}^2}}{|Z_{\mathrm{PA,RX}}|^2}\right)_{\min} = \frac{4KT}{L_s\omega Q_s\left(1 + k_m^2\right)} \tag{18.8}$$

As a result, the noise factor penalty reduces with increasing Q_s and k_m, which fortunately coincides with efforts to optimize the efficiency of the PA's matching network [13]. However, a step-down transformer must be employed for the PA's matching network to scale up the load resistance seen by PA's transistor in order to achieve the highest possible efficiency at relatively low output power of 3 dBm. It is against the noise factor optimization, as evident from (18.8), and clearly demonstrates a trade-off between TX efficiency and RX noise factor. The total noise factor may be estimated by

$$F = 1 + \frac{r_{\mathrm{loss}}}{R_s} + \gamma g_m R_s \left(\frac{L_1 \omega_0}{R_s}\right)^2 + \frac{R_s}{L_s \omega Q_s \left(1 + k_m^2\right)} \tag{18.9}$$

By considering $L_s = 880\,\mathrm{pH}$, $Q_s = 11$, and $k_m = 0.75$, the noise factor penalty in (18.9) can be as low as 0.22.

Now, moving attention to the TX mode, the LNTA's transistor is off, and consequently, the RX path can be simplified to a series RLC network (RXMN) as shown in Fig. 18.7d. In this mode, the ultimate goal is to alleviate the side effects of RXMN on the efficiency of the PA. To analyze this efficiency drop, it is more convenient to replace the RLC series network with its equivalent parallel capacitance (C_{RX}) and resistance (R_{RX}), as illustrated in Fig. 18.7d. It can be shown that

$$R_{\text{RX}} = r_{\text{loss}} \cdot \left(1 + Q_{\text{RX}}^2 \cdot \left(\frac{\omega_{RX}}{\omega_0} - \frac{\omega_0}{\omega_{\text{RX}}}\right)^2\right) \tag{18.10}$$

and

$$C_{\text{RX}} = C_{gs} \cdot \frac{Q_{\text{RX}}^2 \cdot \left(\left(\frac{\omega_{\text{RX}}}{\omega_0}\right)^2 - 1\right)}{1 + Q_{\text{RX}}^2 \cdot \left(\frac{\omega_{RX}}{\omega_0} - \frac{\omega_0}{\omega_{\text{RX}}}\right)^2} \tag{18.11}$$

where Q_{RX} and ω_{RX} are, respectively, the RXMN's quality factor and its resonant frequency $1/\sqrt{(L_1 + L_G)C_{gs}}$. Due to R_{RX} power dissipation, the PA's efficiency scales down with

$$\eta_{\text{RX}} = \frac{R_{\text{RX}}}{R_L + R_{\text{RX}}}. \tag{18.12}$$

As a result, the side effect of RXMN on the TX efficiency can be minimized by having a larger R_{RX}. As can be gathered from Eq. (18.10), this can be achieved by pushing the resonant frequency of LNTA's matching network to a much lower frequency than ω_0 via the C_{gs} switchable capacitor bank. Consequently, at the operating frequency ω_0, PA sees the RX path as a small negative capacitor in parallel with $R_{\text{RX}} \approx 1\,\text{k}\Omega$ modeling LNTA matching network losses. This negative capacitance is absorbed by the PA's matching network, while R_{RX} creates a large resistance path for the TX signal (compared to the $50\,\Omega$ load), which leads to a negligible penalty (<5%) in the efficiency of the transmitter.

18.6 Discrete-Time Receiver

Recent ULP receivers for BLE achieve significant power reduction [2, 26] and higher level of integration [4, 7, 10] primarily using sliding intermediate frequency (IF) and low-IF continuous-time (CT) architectures. To reduce the RX power consumption beyond state of the art, we propose a discrete-time (DT) high-IF or superheterodyne RX architecture with complex-signaling band-pass filters (BPF) and a progressively reduced sampling rate.

The front-end section of DT-RX is presented in Fig. 18.8a. It consists of the narrow-band LNTA, a single-ended-to-differential quadrature sampling mixer, and a DT 4/4 CS-BPF [16]. The LNTA is composed of two stages (see Fig. 18.8c): a single-input/single-output common-source cascode LNA and a common-source transconductance (g_m) amplifier. Both stages operate in moderate inversion as opposed to a strong inversion operation in prior reports, in order to reduce power consumption with I_d = 400 and 100 μA, respectively. Capacitors C_g and C_d are

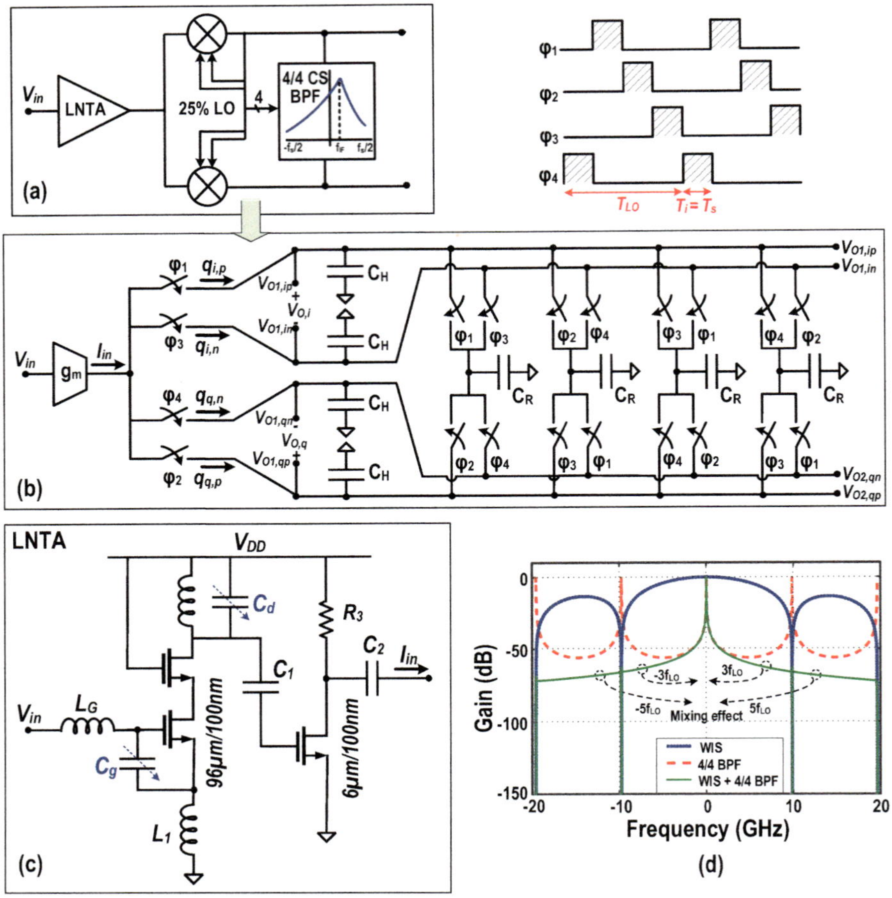

Fig. 18.8 (**a, b**) Full-rate receiver strip; (**c**) LNTA schematic; (**d**) transfer function of 4/4 CS-BPF also showing windowed integration sampling (WIS)

4-bit programmable to tune the LNTA input matching network and its tank load over process, voltage, and temperature (PVT), as well as over package parasitics.

The LNTA is connected to a 25% quadrature passive mixer, which implicitly acts as a balun, converting its single-ended input to differential output signals (see Fig. 18.8b). The passive mixer works in the current mode, which results in a low quadrature imbalance, low noise, and high linearity [27]. The sampling mixer is then cascaded with the first complex BPF as shown in Fig. 18.8b. Transfer function (TF) of the 4/4 CS-BPF in z-domain, from the charge input $Q_{\text{in}}(z)$ to the voltage output $V_{\text{out}}(z)$ is given by

$$H_{4/4\ \text{CS-BPF}}(z) = \frac{V_{\text{out}}(z)}{Q_{\text{in}}(z)} = \frac{1/(C_H + C_R)}{1 - [\alpha + j(1-\alpha)]\,z^{-1}} \tag{18.13}$$

where C_R is the rotating capacitor, C_H is the history capacitor, and $\alpha = \frac{C_H}{C_H+C_R}$. As shown in [17], an equivalent input resistance, R_{eq}, and center frequency (f_c) of the filter can be estimated by

$$R_{\text{eq}} = \frac{1}{C_R f_s}, \qquad f_c = \frac{f_s}{2\pi}\arctan\left(\frac{C_R}{C_H}\right). \tag{18.14}$$

In this design, the center frequency of the BPFs is adjusted just beyond the flicker noise corner of active devices (i.e., IF = 5 MHz). Equation (18.14) clearly demonstrates a trade-off between capacitor size (area), sampling frequency (power), and linearity. The gain of the first stage is given simply by a product of effective transconductance of LNTA and R_{eq}. In this low-power application, the strategy is to reduce C_R as much as possible in order to increase R_{eq} and, consequently, the gain of the first stage. This strategy enables high gain with lower GMs, hence with a lower current at the transconductors. The increase of the input impedances also allows for the use of smaller switches (with higher resistances) both in the mixer and in the filters, with a consequent reduction in the power consumption of the clock generation.

Figure 18.8d plots a transfer function (TF) of the infinite impulse response (IIR) filter of 4/4 CS-BPF (18.13). As expected of any DT filter, TF reveals repetition peaks (replicas) at multiples of f_s ($\approx$ 9.8 GHz). Repetition peaks are folded to DC, but not before being attenuated by a windowed integration sampling (WIS) effect of the current-mode sampling, which creates an inbuilt sinc filter response, also shown on the plot. Combination of these two effects determines the filtering shape of the 4/4 CS-BPF [28, 29].

The next stage of the receiver is a 4/8 CS-BPF. Its schematic is shown in Fig. 18.9. It is based on eight rotating capacitors sampled at eight phases with D = 12.5%, which results in a quadrature filter with higher quality factor (Q = 1.14) [17]. Its z-domain TF of charge input to voltage output is

$$H_{4/8\ \text{CS-BPF}}(z) = \frac{V_{\text{out}}(z)}{Q_{\text{in}}(z)} = \frac{1/(C_H + C_R)}{(1-\alpha z^{-1})^2 - j\left[(1-\alpha)\,z^{-1}\right]^2} \tag{18.15}$$

where $\alpha = \frac{C_H}{C_H+C_R}$. Note that the first BPF stage runs at the full rate since it uses the same $4\times f_{LO}$ sampling clock rate as that of the mixer. This is done to increase blocker protection and to reduce noise folding. However, the second BPF stage implements a decimation by 16 to drastically reduce power consumption of the clock generator [29, 30]. This decimation process is done by allowing multiple input samples to be integrated onto the rotating capacitors creating a finite impulse response (FIR) filter described by

$$H(z) = \frac{1 - z^{-16}}{1 - z^{-1}} \Rightarrow |H(f)| = \left|\frac{\sin(\pi f\, 16/fs)}{\sin(\pi f/fs)}\right|. \tag{18.16}$$

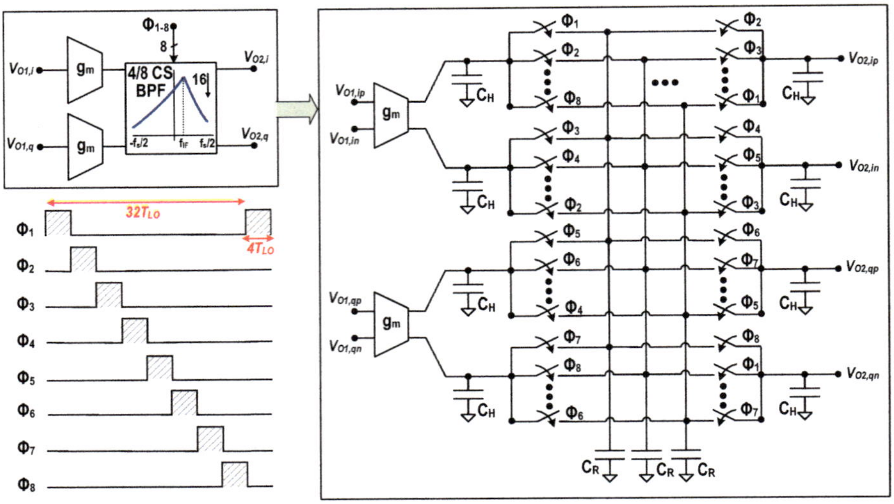

Fig. 18.9 4/8 CS-BPF filter schematics and its clock waveforms

The third stage of the receiver is also a 4/8 CS-BPF. However, there is no further decimation between the second and third filter stages. It is done to avoid any additional clock generation circuitry since the power consumption of these blocks is already very low (around 160 μA in simulations for both 4/8 CS-BPF clock generation, including buffers). The sufficient front-end filtering provided by the three-stage CS-BPF allows to directly digitize the IF signal using a low-power ADC and move the second mixer and baseband filtering into the digital domain [17]. Based on BLE requirements, two 9-bit 20 MS/s ADCs would be sufficient to digitalize the IF output signals of the proposed receiver [31].

18.7 Experimental Results

Figure 18.10 shows the die photo of the proposed transceiver implemented in TSMC 1P9M 28 nm digital CMOS. The total core area, including empty space between the subblocks, is merely 1.9 mm^2.

Figure 18.11 displays the phase noise of the proposed oscillator at the lowest and highest tuning frequencies for $V_{DD} = 0.5$ and 0.8 V. The measured PN is -111 dBc/Hz at 1 MHz offset from 5.1 GHz carrier while consuming ~0.35 mW at 0.5 V. As justified in [13, 32, 33], the 1/f^3 PN corner of the oscillator is extremely low (i.e., ≤100 kHz) across the tuning range (TR) of 22% (i.e., from 4.1 to 5.1 GHz). Its average FoM is 189 dBc and varies ±1 dB across the TR.

Figure 18.12 plots the measured phase noise at different configurations for both integer-N and fractional-N BLE channels. When used as an LO at *undi-*

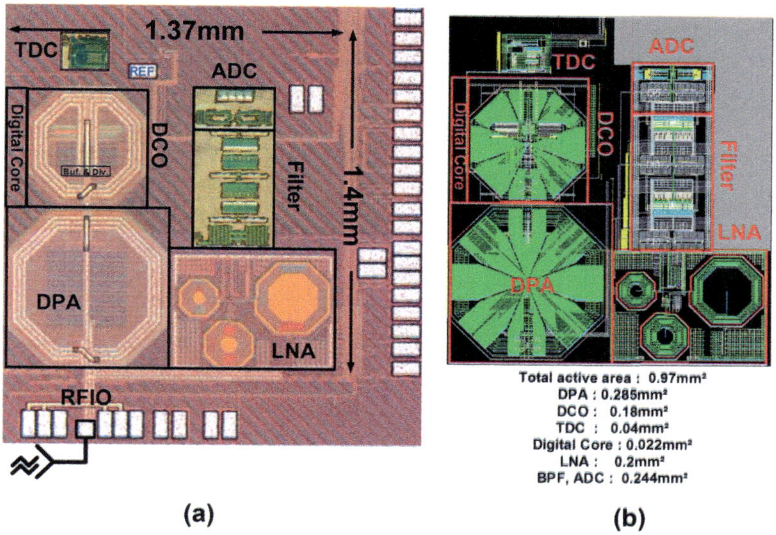

Fig. 18.10 (**a**) Die micrograph of the proposed BLE transceiver; (**b**) its layout with breakdown of subblock areas

vided 40 MHz FREF, the ADPLL consumes 1.4 mW with an integrated PN of 0.87° (yellow line in Fig. 18.12). It exhibits in-band PN of −101 dBc/Hz, which corresponds to an average TDC resolution of ∼12 ps. Thanks to the low wander of the DCO, digital power consumption of the rest of ADPLL can be saved by scaling the rate of sampling clock to 5 MHz. However, the in-band PN increases by $10\log_{10}(40/5) = 9$ dB to −92 dBc/Hz with an integrated PN of 1.08° (blue line in Fig. 18.12).

Figure 18.13 shows the TX spectra for 1 Mb/s GFSK modulation at different modulation indexes and its burst modulation quality. All spectral mask requirements are fulfilled, while the FSK error is 2.7%.

The measured lock-in time is less than 15 μs for f_{REF} of 40 MHz as shown in Fig. 18.14a. Thanks to the low flicker noise, frequency pushing, and pulling of the DCO, its frequency drift is extremely small, as demonstrated in Fig. 18.14b. The maximum difference between 0/1 symbol frequency at the start of the BLE packet and 0/1 frequencies within the packet payload should be less than ±50 kHz. This specification is properly satisfied with over an order-of-magnitude margin even while in the open-loop operation, as shown in Fig. 18.14b, c.

The RX average performance figures are 46 dB of gain, >42 dB of image rejection, 6.5 dB of noise figure, and −19 dBm of IIP3. The measured RX noise figure and filter characteristics are respectively plotted in Fig. 18.15a, b. They demonstrate excellent correlation with simulations using a linear PSS model. The NF variation is less than 1 dB for various IF offsets. Figure 18.15b shows a gain of 46 dB at the −5 MHz IF with an image attenuation of 26 dB at $f_{LO} + 5$ MHz.

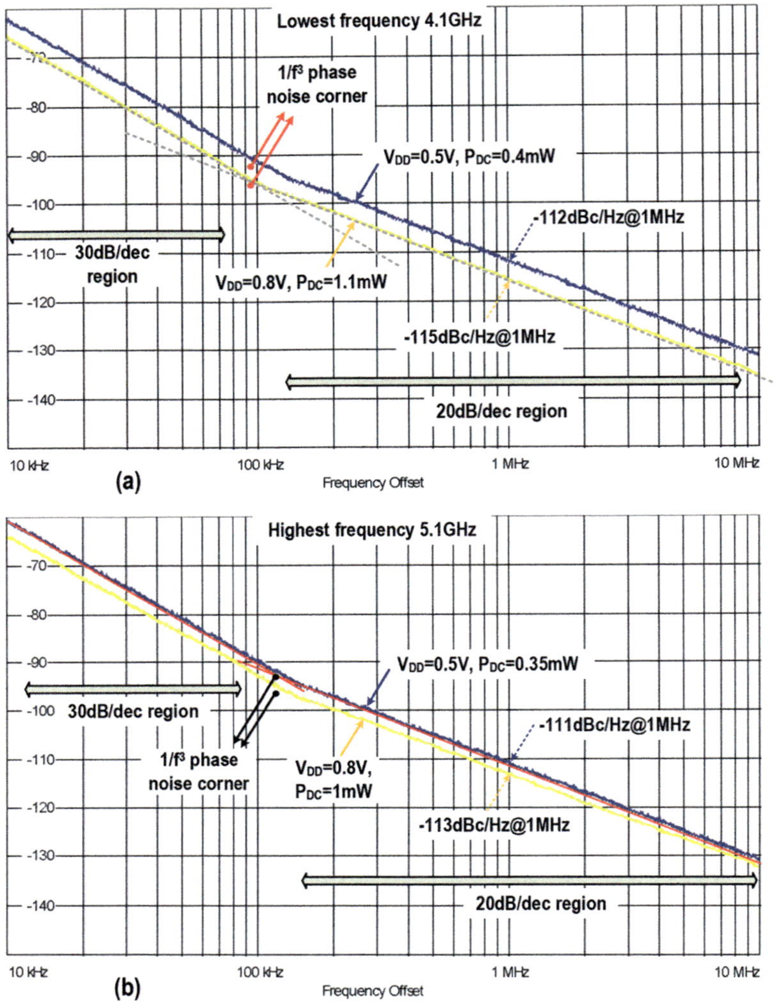

Fig. 18.11 Measured phase noise of the proposed oscillator at (**a**) the lowest and (**b**) the highest frequency

Figure 18.15c shows the BLE receiver packet error rate (PER) versus the input signal power. The sensitivity is −95 dBm at 30.8% PER. For the OOB blocking measurement shown in Fig. 18.15d, the desired BLE signal is fixed at channel 12 with an input power of −67 dBm. Both the desired signal and out-of-band CW blocker are injected into the receiver. The OOB blocker power is recorded when the PER reaches 30.8%. Results corroborate with the proposed full-rate DT-RX strategy and show that the receiver is able to tolerate the OOB BLE blocker mask.

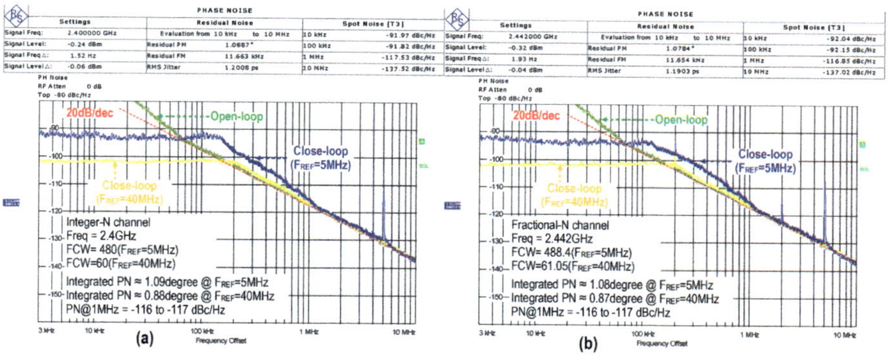

Fig. 18.12 Measured transmitter PN in open-loop and different close-loop configurations for (**a**) integer-N and (**b**) fractional-N channels

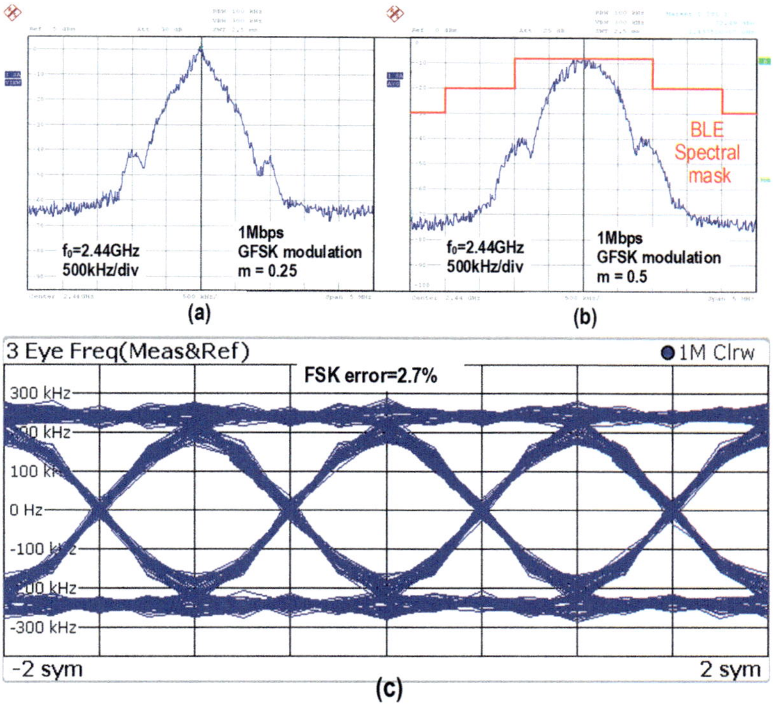

Fig. 18.13 Bluetooth GFSK modulation spectrum for modulation index of (**a**) $m = 0.25$, (**b**) $m = 0.5$, and (**c**) burst-mode modulation accuracy

Table 18.3 summarizes the proposed transceiver and compares it with recent state-of-the-art BLE designs. It is the first implemented in the 28-nm CMOS node. It reaches similar RX performance (NF, linearity, and sensitivity) and better TX

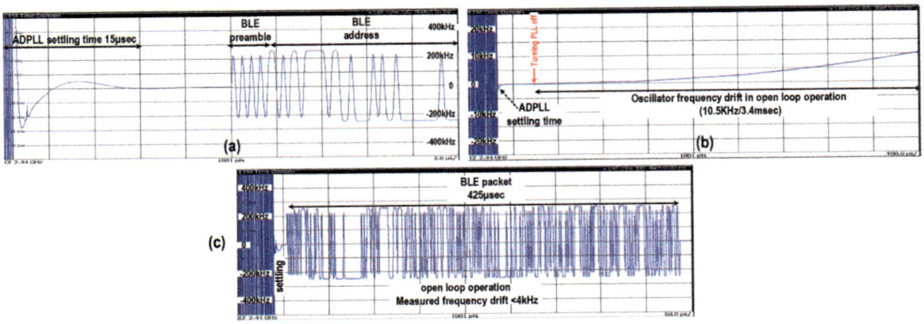

Fig. 18.14 (**a**) ADPLL settling, (**b**) oscillator frequency drift, and (**c**) demodulated TX frequency for 425 μs BLE packet in the open-loop operation

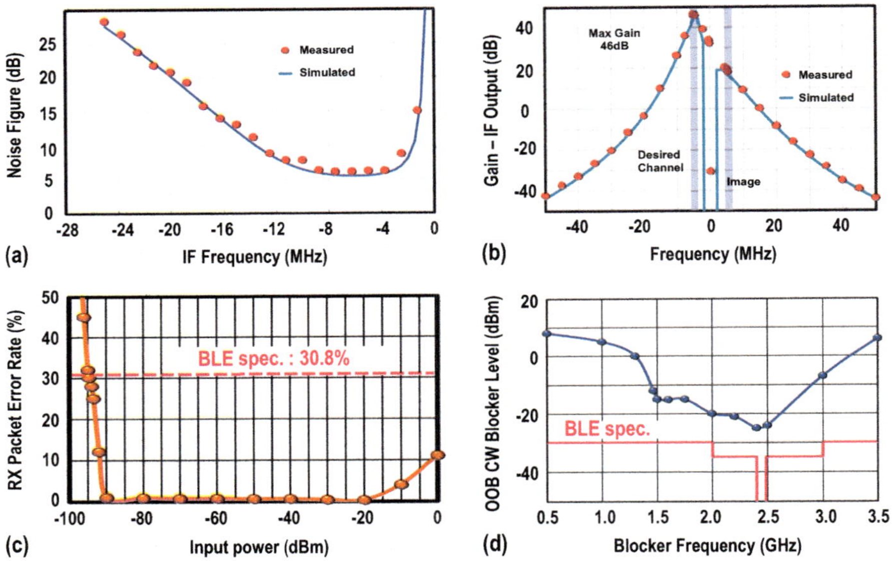

Fig. 18.15 Summary of the RX measurements: (**a**) Noise figure of the RX at various IF frequencies; (**b**) RX filtering characteristics; (**c**) RX packet error rate across input power (**d**) OOB blocking performance

performance (max P_{out}, PLL PN) but at a much lower power consumption, even better than [5], which uses off-chip matching network and T/R switch. When compared with the other two designs with fully integrated on-chip T/R switch [4, 7], the power efficiency is over 2× better for both the TX and RX.

Table 18.3 Performance summary and comparison with state-of-the-art transceivers

	This work	ISSCC'15 IMEC [5]	JSSC'15 Dialog [7]	ISSCC'15 Renesas [4]	TMTT2013 [10]	CC2640 TI
CMOS technology	28 nm	40 nm	55 nm	40 nm	130 nm	N/A
Osc PN @1MHz (dBc/Hz)	−116 to −117	−110	−111.5	N/A	−110	−109
Osc FoM (dB)	188–189	183	179	N/A	185	N/A
Osc tuning range	2.05–2.55 GHz (22%)	25%	20%	N/A	N/A	N/A
PLL in-band PN (dBc/Hz)	−92 @ FREF = 5 MHz	−90	N/A	N/A	−100	N/A
	−101(a) FREF = 40 MHz					
Integrated PN (degree)	1.08 @ FREF = 5 MHz	1.5	N/A	N/A	N/A	N/A
	0.87 @ FREF = 40 MHz					
PLL FoM[a] (dB)	−238.65	−236	N/A	N/A	N/A	N/A
PLL settling time (μs)	15	15	15	N/A	N/A	N/A
Reference/Fractional spurs (dBc)	−80/−60	−70/−38	N/A	N/A	−75/−37	N/A
TX modulation error	2.70%	5%	N/A	N/A	N/A	N/A
Output power (dBm)	−5 to +3	−2/1	−20 to 0	0	1.6	−21 to +5
Total PA efficiency	41%	25%	30%	<30%	26.80%	N/A
Strongest harmonic emission	HD3/−47 dBc	HD2/−49 dBc	HD2/−50 dBc	HD3/−48 dBc	HD3/−34 dBc	−46 dBc
RX noise figure (dB)	6.5	6	5.5	6.5	16	N/A
RX sensitivity (dBm)	−95	−98	−94.5	−94.5	−81.4	−97

(continued)

Table 18.3 (continued)

	This work			ISSCC'15 IMEC [5]	JSSC'15 Dialog [7]	ISSCC'15 Renesas [4]	TMTT2013 [10]	CC2640 TI	
RX IIP3 (dBm)	−19			−19	N/A	N/A	2.9	N/A	
Supply voltage (V)	0.5/1			1	0.9–3.3	1.1	1	1.8–3.8	
RX power consumption (mW)	Open-loop	2.75		3.7	11.2	6.3	1.1	17.7	
	Close-loop	3.55							
TX power consumption (mW)	@P_{OUT}	0 dBm	3 dBm	−2 dBm	0 dBm	0 dBm	1.6 dBm	0 dBm	5 dBm
	Open-loop	3.7	5.6	4.2	10.1[b]	7.7[b]	5.9	18.3	27.3
	Close-loop	4.5	6.4						
TX efficiency (P_{OUT}/P_{DC})	Open-loop	27%	35%	15%	10%	13%	25%	5.5%	11.5%
	Close-loop	22%	31%						
Direct antenna connection	Yes			No	Yes	Yes	Yes	Yes	
TRX active area (mm^2)	1.9			1.2	1.5[c]	1.1[c]	1.5[c]	N/A	

[a] $FoM = 10\log_{10}[\sigma^2_{jittler} \cdot (P_{DC,PLL}/1mW)]$
[b] Including DC-DC converters
[c] Graphically estimated

18.8 Conclusions

A single-chip ultra-low-power transceiver for IoT applications, fully compliant with the Bluetooth Low Energy standard, is demonstrated in a digital 28-nm CMOS technology. The transceiver demonstrates the best-ever-reported system efficiency and phase purity by exploiting new system and circuit techniques. First, a new switching current source oscillator is introduced to reduce the oscillator power consumption further than practically possible in the traditional oscillators. Second, due to the low wander of DCO, digital power consumption of ADPLL can be significantly saved by scaling down the rate of sampling clock after settling or even shutting it down entirely during direct DCO data modulation. Third, a fully integrated differential class-E/F_2 switching PA is utilized to improve system efficiency at low output power of 0–3 dBm while fulfilling all in-band and out-of-band emission masks. Fifth, the transmitter and receiver share a single pin for a direct connection to an antenna. Sixth, the receiver is a discrete-time superheterodyne architecture performing amplification and filtering using charge-sharing complex-signaling band-pass filters.

References

1. Wong, A., et al.: A 1V 5mA multimode IEEE 802.15.6/Bluetooth Low-Energy WBAN transceiver for biotelemetry applications. In: IEEE International Solid-State Circuits Conference Digest of Technical Papers, pp. 300–301, Feb 2012
2. Liu, Y.-H., et al.: A 1.9nj/b 2.4GHz multistandard (Bluetooth Low Energy/Zigbee/IEEE802.15.6) transceiver for personal/body-area networks. In: IEEE International Solid-State Circuits Conference Digest of Technical Papers, pp. 446–447, Feb 2013
3. Devita, G., et al.: A 5mW multi-standard Bluetooth LE/IEEE 802.15.6 SoC for WBAN applications. In: IEEE European Solid State Circuits Conference, pp. 283–286, Sept 2014
4. Sano, T., et al.: A 6.3mW BLE transceiver embedded RX image-rejection filter and TX harmonic-suppression filter reusing on-chip matching network. In: IEEE International Solid-State Circuits Conference Digest of Technical Papers, pp. 240–241, Feb 2015
5. Liu, Y.-H., et al.: A 3.7mW-RX 4.4mW-TX fully integrated Bluetooth Low-Energy/IEEE802.15.4/proprietary SoC with an ADPLL-based fast frequency offset compensation in 40 nm CMOS. In: IEEE International Solid-State Circuits Conference Digest of Technical Papers, pp. 236–237, Feb 2015
6. Bachmann, C., et al.: A 3.5mW 315/400MHz IEEE802.15.6/proprietary mode digitally-tunable radio SoC with integrated digital baseband and MAC processor in 40 nm CMOS. In: IEEE International Solid-State Circuits Conference Digest of Technical Papers, pp. 94–95, June 2015
7. Prummel, J., et al.: A 10 mW Bluetooth Low-Energy transceiver with on-chip matching. IEEE J. Solid-State Circuits **50**(12), 3077–3088 (2015)
8. Kuo, F.-W., et al.: A Bluetooth Low-Energy (BLE) transceiver with TX/RX switchable on-chip matching network, 2.75 mW high-IF discrete-time receiver, and 3.6 mW all-digital transmitter. In: Proceedings of IEEE Symposium on VLSI Circuits, pp. 64–65, June 2016

9. Kuo, F.-W., et al.: A Bluetooth Low-Energy (BLE) transceiver with 3.7mW all-digital transmitter, 2.75mW high-IF discrete-time receiver, and TX/RX switchable on-chip matching network. IEEE J. Solid-State Circuits **52**(4), 1144–1162 (2017)
10. Masuch, J., et al.: A 1.1-mW-RX 81.4 dBm sensitivity CMOS transceiver for Bluetooth Low Energy. IEEE Trans. Microw. Theory Tech. **61**(4), 1660–1673 (2013)
11. Bluetooth specification version 4.2, in Available: http://www.bluetooth.com (2014)
12. Kuo, F.-W., et al.: A fully integrated 28 nm Bluetooth Low-Energy transmitter with 36% system efficiency at 3 dBm. In: IEEE European Solid State Circuits Conference, pp. 356–359, Sept 2015
13. Babaie, M., et al.: A fully integrated bluetooth low-energy transmitter in 28-nm CMOS with 36% system efficiency at 3 dBm. IEEE J. Solid-State Circuits **51**(7), 1547–1565 (2016)
14. Babaie, M., Shahmohammadi, M., Staszewski, R.B.: A 0.5 V 0.5 mW switching current source oscillator. In: IEEE RFIC Symposium, pp. 183–186, May 2015
15. Babaie, M., Staszewski, R.B., Galatro, L., Spirito, M.: A wideband 60 GHz class-E/F_2 power amplifier in 40 nm CMOS. In: IEEE RFIC Symposium, pp. 215–218, May 2015
16. Tohidian, M., Madadi, I., Staszewski, R.B.: A fully integrated discrete-time superheterodyne receiver. IEEE Trans Very Large Scale Integr. (VLSI) Syst. **25**(2), 635–647 (2017)
17. Madadi, I., Tohidian, M., Staszewski, R.B.: A high IIP2 saw-less superheterodyne receiver with multistage harmonic rejection. IEEE J. Solid-State Circuits **51**(2), 332–347 (2016)
18. Babaie, M., Staszewski, R.B.: An ultra-low phase noise class-F_2 CMOS oscillator with 191 dBc/Hz FoM and long-term reliability. IEEE J. Solid-State Circuits **50**(3), 679–692 (2015)
19. F.-W. Kuo, et al.: A 12mW all-digital PLL based on class-F DCO for 4G phones in 28 nm CMOS. In: Proceedings of IEEE Symposium on VLSI Circuits, pp. 1–2, June 2014
20. Staszewski, R.B., et al.: All-digital PLL and transmitter for mobile phones. IEEE J. Solid-State Circuits **40**(12), 2469–2482 (2005)
21. Aoki, I., Kee, S.D., Rutledge, D.B., Hajimiri, A.: Distributed active transformer–a new power-combining and impedance-transformation technique. IEEE Trans. Microw. Theory Tech. **50**(1), 316–331 (2002)
22. Kim, J., et al.: A fully-integrated high-power linear CMOS power amplifier with a parallel-series combining transformer. IEEE J. Solid-State Circuits **47**(3), 599–614 (2012)
23. Babaie, M., Staszewski, R.B.: A study of RF oscillator reliability in nanoscale CMOS. In: Proceedings of IEEE 21st European Conference on Circuit Theory and Design, pp. 243–246, Sept 2013
24. Chen, J., et al.: A digitally modulated mm-Wave Cartesian beamforming transmitter with quadrature spatial combining. In: IEEE International Solid-State Circuits Conference Digest of Technical Papers, pp. 232–233, Feb 2013
25. Kee, S., Aoki, I., Hajimiri, A., Rutledge, D.: The class-E/F family of switching amplifiers. IEEE Trans. Microw. Theory Tech. **51**(6), 1677–1690 (2003)
26. Selvakumar, A., Zargham, M., Liscidini, A.: Sub-mW current re-use receiver front-end for wireless sensor network applications. IEEE J. Solid-State Circuits **50**(12), 2965–2974 (2015)
27. Mirzaei, A., Darabi, H., Leete, J.C., Chang, Y.: Analysis and optimization of direct-conversion receivers with 25% duty-cycle current-driven passive mixers. IEEE Trans. Circuits Syst. I: Regul **57**(9), 2353–2366 (2010)
28. Staszewski, R.B., et al.: All-digital TX frequency synthesizer and discrete-time receiver for Bluetooth radio in 130-nm CMOS. IEEE J. Solid-State Circuits **39**(12), 2278–2291 (2004)
29. Karvonen, S., Riley, T.A.D., Kurtti, S., Kostamovaara, J.: A quadrature charge-domain sampler with embedded FIR and IIR filtering functions. IEEE J. Solid-State Circuits **41**(2), 507–515 (2006)
30. Bagheri, R., et al.: An 800 MHz to 6 GHz software-defined wireless receiver in 90 nm CMOS. IEEE J. Solid-State Circuits **41**(12), 2860–2876 (2006)

31. Ferreira, S.B., Kuo, F.-W., Babaie, M., Bampi, S., Staszewski, R.B.: System design of a 2.75 mW discrete-time superheterodyne receiver for Bluetooth Low Energy. Submitted for review to a special issue in IEEE Trans. Microw. Theory Tech. **65**(5), 1904–1913 (2017)
32. Shahmohammadi, M., Babaie, M., Staszewski, R.B.: A 1/f noise upconversion reduction technique applied to class-D and class-F oscillators. In: IEEE International Solid-State Circuits Conference Digest of Technical Papers, pp. 444–445, Feb 2015
33. Shahmohammadi, M., Babaie, M., Staszewski, R.B.: A 1/f noise upconversion reduction technique for voltage-biased RF CMOS oscillators. IEEE J. Solid-State Circuits **51**(11), 2610–2624 (2016)

Printed by Printforce, the Netherlands